Reproductive Behaviour of Insects

Pachylomera femoralis is a very large (45 mm long) dung beetle found in southern Africa. Females construct nests deep in the sand, where they bury a single brood ball containing one egg. Up to 50 adults may congregate at a single mass of dung, feeding or cutting up balls of dung and removing them. All reproductive activity occurs away from the dung pad. Competition for the dung resource occurs at the pad, while that for mates occurs at sites away from the pad. (P. B. Edwards, unpublished data).

Reproductive Behaviour of Insects

Individuals and populations

Edited by

W. J. BAILEY

Senior Lecturer in Zoology,
University of Western Australia

and

J. RIDSDILL-SMITH

Principal Research Scientist,
CSIRO Division of Entomology,
Perth, Western Australia

CHAPMAN & HALL
London · New York · Tokyo · Melbourne · Madras

UK   Chapman and Hall, 2–6 Boundary Row, London SE1 8HN

USA   Chapman and Hall, 29 West 35th Street, New York NY10001

JAPAN   Chapman and Hall Japan, Thomson Publishing Japan, Hirakawacho
            Nemoto Building, 7F, 1-7-11 Hirakawa-cho, Chiyoda-ku, Tokyo 102

AUSTRALIA   Chapman and Hall Australia, Thomas Nelson Australia, 102 Dodds
            Street, South Melbourne, Victoria 3205

INDIA   Chapman and Hall India, R. Seshadri, 32 Second Main Road, CIT East,
            Madras 600 035

First edition 1991

© 1991 W. J. Bailey and J. Ridsdill-Smith

Typeset in 10/12 pt Palatino by
Rowland Phototypesetting Ltd, Bury St Edmunds, Suffolk
Printed in Great Britain by St Edmundsbury Press Ltd, Bury St Edmunds,
Suffolk

ISBN 0 412 31280 8

British Library Cataloguing in Publication Data

Reproductive behaviour of insects.
  1. Insects. Reproduction
  I. Bailey, Winston J.   II. Ridsdill-Smith, J. (James)
  595.7016

  ISBN 0 412 31280 8

Library of Congress Cataloging-in-Publication Data

Reproductive behaviour of insects : individuals and populations /
  edited by W. J. Bailey and T. J. Ridsdill-Smith.
      p.   cm.
    Includes bibliographical references and index.
    ISBN 0-412-31280-8
    1. Insects—Behavior.   2. Insects—Reproduction.   3. Sexual
  behavior in animals.   I. Bailey, Winston J.   II. Ridsdill-Smith,
  T. J., 1942–
  QL496.R38   1991   67648
  595.70556–dc20   90-47563
                                                   CIP

Contents

List of contributors

John Alcock, Department of Zoology, Arizona State University, TEMPE AZ 85287, USA.

Winston Bailey, Department of Zoology, University of Western Australia, NEDLANDS WA 6009, AUSTRALIA.

Gary Fitt, CSIRO Division of Entomology, Cotton Research Unit, PO Box 59, NARRABRI NSW 2390, AUSTRALIA.

Brian Fletcher, CSIRO Division of Entomology, PO Box 1700, CANBERRA ACT 2601, AUSTRALIA.

Darryl Gwynne, Department of Biology, Erindale Campus, University of Toronto, Mississauga, ONTARIO, CANADA L5L 1C6.

Rhondda Jones, Department of Zoology, James Cook University, TOWNSVILLE QLD 4811, AUSTRALIA.

John Lawton, Centre for Population Biology, Imperial College, Silwood Park, ASCOT, BERKS, SL5 7PY, UNITED KINGDOM.

Ron Prokopy, Department of Entomology, University of Massachusetts, Amhurst, MASSACHUSETTS 01002, USA.

Duncan Reavey, Department of Biology, University of York, York YO1 5DD, UNITED KINGDOM.

James Ridsdill-Smith, CSIRO Division of Entomology, Private Mail Bag, PO WEMBLEY, WA 6014, AUSTRALIA.

Philip Spradbery, CSIRO Division of Entomology, PO Box 1700, CANBERRA, ACT 2601, AUSTRALIA.

Seamus Ward, Department of Zoology, La Trobe University, BUNDOORA, VIC 3083, AUSTRALIA.

Paul Wellings, CSIRO Division of Entomology, PO Box 1700, CANBERRA, ACT 2601, AUSTRALIA.

Acknowledgements

It is difficult in a book of this nature to give adequate thanks to all those colleagues who have helped with encouragement, ideas and criticism, and not least the contribution of the many anonymous reviewers. In producing the book we would thank Lisa Masini, Tanya Nicholson and Adrian Vlok, while special thanks go to Bob Carling, Biological Editor of Chapman and Hall (London) for encouraging us in the production of this book.

Preface

Interest in editing a book of this nature started with a symposium on Insect Behaviour at the annual meeting of the Australian Entomological Society held in Perth, Western Australia. The meeting linked papers on sexual selection and mate choice with those on host finding and oviposition behaviour in species relevant to pest management. Although the symposium was somewhat limited in both the range of taxa treated and the reproductive behaviour covered, it stimulated interest on the importance of an evolutionary approach to applied problems of insect management.

Authors in this book, however, cover a much wider range of topics under the general title of 'Reproductive Behaviour of Insects' than was possible at the Perth symposium. And here subjects range from mate finding and host selection to oviposition behaviour and competition for resources, to the behaviour of feeding larvae. The spectrum of papers reflects both a diversity of approach as well as differing degrees of conviction as to the relevance of evolutionary theory to the study of insect populations. We have not produced a stamp collection of everything that is known, but we hope that the examples in this book will encourage the reader to think more critically about the evolution of behaviour, and in particular about the importance of small differences in individual behaviour within natural populations.

1

Individual perspectives on insect reproductive behaviour

Winston J. Bailey and James Ridsdill-Smith

There are more known species of insects than of any other life form, and this, with their numerical abundance and short life cycles, makes them excellent subjects for studying the mechanisms by which populations are regulated.

Historically, studies concerning abundance have focused on extrinsic factors that cause change—climate and resources—and those within a population controlling these numerical changes—competition for space and food. Thus, recognized key factors influencing life history patterns, and hence insect abundance, are fecundity, survival of young, the age to reproduction, the interaction between reproductive effort and adult mortality and finally, any variation in these traits between individual progeny (Stearns, 1976; Denno and Dingle, 1981). When we consider these factors, an insect population has an obvious homogeneity, and any measure of a particular variable results in a mean value with which we may describe the population at any one moment. Variations in behaviour may be more difficult to interpret at a population level, and here one may take the line that 'behavioural idiosyncrasies at the individual level (are) . . . ironed out statistically at the population level and can therefore be ignored' (Dunbar, 1983). Thus it is hard to contemplate variation in behaviour between individuals, or even the possibility of multiple behaviour patterns within an individual, as having any significant bearing on the behaviour of the entire population, when faced with a swarm of locusts, or an outbreak of armyworm.

The measurement of a mean value implies that it represents a normal behaviour for the population; again variation or flexibility is largely ignored (Dingle, 1981). Hence, concentration on the behaviour of the 'population' rather than the 'individual' may promote the uncritical view that selection acts at the 'population' level: there is an autoregulatory mechanism that ensures its (the population or species) survival. For example, dispersive behaviour following intraspecific competition between aphids massing on a plant stem suggests that the development of winged individuals moving to a new plant acts to maintain the population's continued use of the plant resource (van Emden, 1978; see Ward, Chapter 8 in this volume, for critical comment). A similar function could be attributed to the chorusing behaviour of acoustic insects, or the massed luminescent flashing of fireflies. Individuals, by calling or flashing, inform their neighbours of the population's density and either reduce their reproduction or disperse (Wynne-Edwards, 1962). The genes that code for such behaviour would be selected out of the population (Alexander, 1975; Lloyd, 1979). For a more complete comment on the differences between group and individual selection, see Williams (1966), Lomnicki (1988) and Alcock and Gwynne (Chapter 2, this volume).

If the focus is placed on the individual genome, then inevitably it is the success of this replicating unit that will determine the population's eventual structure. The place to start such a study of individual behaviour is with mating itself, for ultimately the success of the male's genotype is usually dependent on the number of females it is able to fertilize, whereas for the female it is on the number and size of the eggs she is able to lay. In this context selection should act differently on the sexes, producing male traits important in competing for the opportunity to mate, and female traits important in choosing between partners that are best able to produce offspring (Darwin, 1874; Trivers, 1972). Thus on the one hand, selection may be on elaborate structures and behaviours associated with male aggression, and on the other, similarly elaborate structures may be associated with female choice. Structures associated with reproductive fitness will become increasingly apparent in the population. For example, the use by beetles of horns and snouts to remove competing males improves the individual's chance of securing a mate (Johnson, 1982). And male nuptial offerings of bittacid hangingflies (Thornhill, 1980), or the spermatophore gift associated with sperm transfer in katydids (Gwynne, 1988), increases the opportunity for female choice.

However, selection does not rest here, and the search for suitable egg-laying sites, ensuring that the progeny at least have the chance of survival, will be subject to a different set of structural and social factors. Usually it is the female alone who searches for the best place to lay her eggs. However, the male may contribute by defending a resource needed by the female for oviposition, thereby not only demonstrating to a choosy female his ability to ward off competitors, but in doing so enhancing the chance of offspring survival by providing the female with a suitable site in which to lay the eggs he has fertilized (Thornhill and Alcock, 1983).

In parasitoid wasps, ovipositional strategies influence both the sex of the progeny through the control of fertilization (haplodiploidy), and the number of eggs laid in the host. These factors in turn influence the offspring's size and competitive status as an adult (Wagge and Ming, 1984). In gregarious species, larger males emerging from the host will out-compete smaller males and gain access to more, and perhaps larger, females. Larger females are more fecund and produce larger eggs that in turn have a better chance of survival (see Wellings, Chapter 4, this volume), which will have its inevitable effect on the abundance and behaviour of the ensuing parasite populations.

Along the path between oviposition and eventual adult mating, strategies should evolve that increase the efficiency of larval feeding. These may be from two sources: the ovipositional behaviour of the female, and the feeding behaviour of individual larvae. Those larvae best able to detect a suitable source of food will have the better chance of maturing to become reproducing adults.

An evolutionary view of insect populations includes selection on adaptations to changing environments for larvae and adults; the development of life history 'traits' (e.g. Denno and Dingle, 1981). There is, however, a gap in our thinking between this position and a consideration of the consequences of variation between individuals on their reproductive success. Just how successful are the larvae of parasitoids and Lepidoptera whose parents have made the right decision in placing their eggs?

The reproductive behaviour of insects is taken to include the finding of mates, choice of mates, selection of oviposition sites and the factors affecting the fitness of larvae. By combining in one volume contributions from those having the broader, more theoretical, approach of evolutionary biology, with contributions from those who mainly study the population ecology of economically important pests, the aim is to introduce selectionist thinking to a wider audience of entomologists.

Alcock and Gwynne, in Chapter 2, present the argument for selection acting at the level of the individual. They illustrate their point by focusing on the 're-discovery' of sexual selection, where males may evolve elaborate weapons to combat rivals, and those who lose the battle may adopt alternative strategies to achieve mating. One of the central elements of sexual selection is female choice, and here they point out that empirical studies are a relatively recent phenomenon. A reason for this is that, apart from the reluctance of behavioural scientists to accept selection through female choice, it is difficult to separate those behavioural traits that are distinctly involved in 'choice' from those of the other, more easily recognized, components of sexual selection, notably male–male competition.

The functional elements of mate choice are the cues on which female preference is based, and as with the myriad of cues available to the searching female for suitable oviposition sites for her eggs, there are an equally large number of cues available to her for distinguishing between males. Where searching is based on 'distance' cues, such as smell, vision or hearing, we would expect selection to act on both the signaller and receiver to highlight differences between males. In Chapter 3 Bailey examines possible routes selection may have taken in fashioning one of these cues—sound. He uses this as a model to show the ways in which social signals are influenced, not only by selection on the signaller to increase effective transmission, but also by selection on the receiver to recognize and locate the call. Since any advertising is open to predators as well as to competing conspecifics, this too may have an effect on both calling and searching behaviour.

Wellings, in considering the role of individual behaviour in Chapter 4, suggests that host location and exploitation by parasitoids will affect the population dynamics of both host and parasitoid (cf. Hassell and May, 1985). Thus, where hosts are patchily distributed, the ways in which parasitoids respond to this distribution will affect the mortality of host populations. As with mate searching, distance cues produced by the host are used by searching females, but here they are often kairomones. Again, the parasitoid shows a preference, but may alter her searching behaviour with differing host densities in any patch. Despite the strenuous efforts of workers to isolate cues and establish patterns of searching behaviour in the laboratory, and the development of models describing host–parasitoid relationships, Wellings suggests that there are too few studies that critically test these models, and even fewer that are based on field observations.

Phytophagous insects are in essence plant predators, and as with parasitoids, the female needs to search for and locate a suitable host for her future progeny. Thus, cues for oviposition are provided by the range of plant forms, colours and chemicals. The problem in host location is similar to that of a female searching for its mate; the insect 'must move, encounter cues from the target, and then respond to the cues appropriately' (Jones, Chapter 5, this volume). Although providing a review of plant-searching behaviour for a range of insect species, Jones concentrates on Lepidoptera. The butterfly is presented with a mosaic of plants, and despite the assumption by early workers that plant preference is simply a matter of sifting through chemical cues, there appears to be a far more sophisticated array of signals available to the female, including colour and shape. However, not all plants are obvious, even to a small insect, and Jones discusses the question of 'apparancy' in plant host location. The female is provided with a cascade of cues varying in the senses with which they can be perceived, and the distance from the plant at which they can be detected. Hence, Jones deflects attention from a critical 'cue', to the more subtle relationship between both positive and negative signals that eventually encourage the female to persist with her quest for a place to lay her eggs.

Chapter 6, by Fletcher and Prokopy, examines those factors influencing selection of oviposition site by tephritid fruit flies, an economically important group of fruit pests. They emphasize the hierarchy of foraging cues females utilize in nature during the location and selection of oviposition sites: tree shape, fruit, size, colour and chemistry. The chapter describes the behavioural strategies used by females to locate their hosts; external factors, such as fruit marking compounds, and internal factors, including motivation level and learning, are included. The phenomena of host-shifts, particularly in response to the introduction of new hosts, may shed some light on the current debate on host race formation and sympatric speciation.

The noctuid moths, belonging to the sub-family Heliothinae, are severe agricultural pests. It is therefore appropriate, in a book on insect reproductive behaviour, that this family receives special attention. In Chapter 7 Fitt describes host selection in these polyphagous and highly mobile pests. They are able to utilize a range of crop species, and also move with ease from preferred native hosts to introduced cultivated plants. There has been little work on host selection at the individual level within this group of insects, and although it is easy to assume polyphagy by all individuals within a population, Fitt suggests an alternative explanation. Individual members of a population may in fact exhibit more specialized host preferences, but these appear at the population level as polyphagy. Although primarily night-flying insects, they use visual cues to locate their hosts. However, during long distance movements, often at altitudes between 200 and 1000 m, host cues are unlikely to affect where the moth lands. Fitt suggests that large stands of a crop may produce a sufficiently large attractant plume to arrest low-flying moths in the vicinity of the crop. Certainly, during flights 15 m or so over stands of crops such as maize, there are repeated turns up wind, ensuring that the flying insects remain within the host patch.

Unlike the strong flying moths, when aphids take off they have so little control on the direction of their flight that they appear at a severe disadvantage in terms of host plant location. In Chapter 8 Ward suggests that the chance of a migrant aphid reaching a suitable host plant is about 1 in 1000. Even given the dangers involved in moving between hosts, and the frequency with which migrant aphids land in an almost random manner on suitable and unsuitable hosts, they still change hosts. Finding suitable hosts will depend on visual cues that mainly involve colour, and once on a plant they appear to work through a series of suitability checks before attempting to feed: they become very good 'botanists'. Many aphids are highly host specific, and as one host species becomes unavailable they can move to a second, quite different, host. Sexual reproduction occurs on the host plant, and for sexes to meet both must arrive on this plant. Ward points out that in species that migrate before mating, the risks of not mating at all now increase even further, for not only must a new plant be found, but also one on which there is a receptive mate. These he calls 'rendezvous' hosts. If there is any change in plant preference by individual aphids, then Ward argues that this may well be a mechanism leading to sympatric speciation; the formation of discrete host-specific 'biotypes'. Once a biotype has been established, host specificity through assortative mating will maintain species isolation.

There is a significant change in direction with Spradbery's description in Chapter 9 of oviposition in social wasps. Here the onus is on the female to provide a protein food source for her egg production, as well as for developing young. The rate of egg-laying can vary between genera, with some *Vespa* species producing up to 300 eggs per day. Oviposition and protection of the brood appear to be essential elements of female behaviour in social insects, and Spradbery provides a review of the mechanisms and behaviour used by social wasps. More subtle than the dangers of external interference is the struggle within the colony for egg-laying sites by lower queens competing with the foundress queen. Although there are some reported cases of egg-laying by a group of queens within the colony, most species exhibit a social hierarchy with only one female laying the eggs. In some species, however, females have evolved a more distant control over the egg-laying by their rivals, suppressing the rival's ovarian development through pheromones. This behaviour reaches its pinnacle in vespid colonies, where the queen may have control over many thousands of female workers. In social wasps, lower females appear to adopt alternative strategies, secreting eggs in peripheral parts of the colony and can be seen guarding their brood. Even when these lower order queens are able to lay eggs, competition is high, and the dominant queen may eat the eggs of subdominants.

The final two chapters are largely about larval fitness; in Chapter 10 Ridsdill-Smith examines the complex interactions of insects in dung, and in Chapter 11 Reavey and Lawton examine the broader relationships between larvae and plants. Dung is a highly ephemeral resource, which at high temperatures may become unusable by insects within hours of dropping. In addition to rapid changes occurring in the physical state of the dung, its quality is influenced by the herbage consumed by the animal producing the dung. This is influenced by season. For some insects, dung is the 'rendezvous' (*vide* Ward) for the sexes, and mating occurs quickly allowing the female to make maximum use of the dung as a food source for its larvae. Male dung flies arrive at the pad soon after it is dropped, competing with other males for the best sites to intercept the arriving females. Ridsdill-Smith places the strategies of mating in dung flies within the context of the entire dung habitat.

Dung beetles are being used in a biological control programme in Australia, where their introduction has been an effective control over fly abundance, and the interaction between flies and beetles forms the central theme of Chapter 10. The integration of life history studies with a recent study of dung beetle mating strategies provide an example of how population studies may form the framework for more focused work on single-species mating patterns.

The final chapter belongs to larval fitness in the plant feeders, and it is appropriate to note the challenge Reavey and Lawton give to those whose interest is in populations—'There is no substitute for time spent watching what larvae actually do in natural conditions'—the essential element of a study of the individual. As with the chapter by Jones (Chapter 5), these authors concentrate on the Lepidoptera. Their discussion ranges from life history patterns to how much time should be spent in one stage rather than another—trade-off strategies for survival between the adult, egg and larva and between larva and pupa. They also emphasize how the detail of an insect's feeding behaviour can contribute to overall fitness. Rather than seeing a group of caterpillars feeding on a host plant, they view this behaviour from the perspective of the single larva feeding within a highly heterogeneous habitat, sampling leaves of different ages, varying nitrogen concentrations and with changing levels of plant defence compounds. As with adults searching for suitable oviposition sites, larvae have the ability to discriminate between leaves of the same food plant, avoiding damaged leaves. Even this simple observation can be used to highlight the paucity of good observational data on larval feeding—what is missing are data on individual caterpillars. Indeed, many of the questions raised in this book are most likely to be answered by detailed studies of insects in nature.

References

Alexander, R. D. (1975) Natural selection and specialized chorusing behavior in acoustic insects. In *Insects, Science and Society* (ed. D. Pimental), Academic Press, New York, pp. 35–77.

Darwin, C. (1874) *The Descent of Man, and Sexual Selection in Relation to Sex*, 2nd edn, John Murray, London.

Denno, R. F. and Dingle, H. (1981) Considerations for the development of a more general life history theory. In *Insect Life History Patterns: Habitat and Geographic Variation* (eds R. F. Denno and H. Dingle), Springer-Verlag, New York, 1–6.

Dingle, H. (1981) Geographic variation and behavioural flexibility in milkweed bug life histories. In *Insect Life History Patterns: Habitat and Geographic Variation* (eds R. F. Denno and H. Dingle), Springer-Verlag, New York, 57–74.

Dunbar, R. I. M. (1983) Life history tactics and alternative strategies of reproduction. In *Mate Choice* (ed. P. Bateson), Cambridge University Press, Cambridge, 423–434.

van Emden, H. C. (1978) Insects and secondary plant substances—an alternative viewpoint with special reference to aphids. In *Biochemical Aspects of Plant and Animal Coevolution* (ed. J. B. Harborne), Academic Press, New York, 309–322.

Gwynne, D. T. (1988) Courtship feeding and the fitness of female katydids (Orthoptera: Tettigoniidae). *Evolution*, **42**, 545–555.

Hassell, M. P. and May, R. M. (1985) From individual behaviour to population dynamics. In *Behavioural Ecology: Ecological Consequences of Adaptive Behaviour* (eds R. M. Sibly and R. H. Smith), Blackwell, Oxford, 3–32.

Johnson, L. K. (1982) Sexual selection in the tropical brentid weevil. *Evolution*, **36**, 251–262.

Lloyd, J. E. (1979) Sexual selection in luminescent beetles. In *Sexual Selection and Reproductive Competition in Insects* (eds M. S. Blum and N. A. Blum), Academic Press, New York, pp. 293–342.

Lomnicki, A. (1988) *Population Ecology and Individuals*, Princeton University Press, Princeton, NJ.

Stearns, S. C. (1976) Life history tactics: A review of the ideas. *Q. Rev. Biol.*, **51**, 3–47.

Thornhill, R. (1980) Mate choice in *Hylobittacus apicalis* (Insecta: Mecoptera) and its relation to some models of female choice. *Evolution*, **34**, 519–538.

Thornhill, R. and Alcock, J. (1983) *The Evolution of Insect Mating Systems*, Harvard University Press, Cambridge, MA.

Trivers, R. L. (1972) Parental investment and sexual selection. In *Sexual Selection and the Descent of Man, 1871–1971* (ed. B. Campbell), Aldine, Chicago, pp. 136–79

Wagge, J. K. and Ming, N. S. (1984) The reproductive strategy of a parasitic wasp: 1. Optimal progeny and sex allocation in *Trichogramma evanescens. J. Anim. Ecol.*, **53**, 401–405.

Williams, G. C. (1966) *Adaptation and Natural Selection*, Princeton University Press, Princeton, NJ.

Wynne-Edwards, V. C. (1962) *Animal Dispersion in Relation to Social Behaviour*, Hafner, New York.

2

Evolution of insect mating systems: the impact of individual selectionist thinking

John Alcock and Darryl T. Gwynne

2.1 INTRODUCTION

The Darwinian revolution in biology occurred twice. Darwin (1859) initiated the first revolution himself by writing *On the Origin of Species*, which inspired T. H. Huxley (1910) and others to promote the theory of evolution by natural selection to a large audience. Although there was resistance to Darwinism, this faded quickly in biological circles and soon most biologists counted themselves as Darwinists.

The second revolution occurred nearly a century later in response to the gradual and largely 'unconscious' replacement of Darwinian theory, even among biologists, with a belief that traits advantageous to the species as a whole would spread through populations. This argument finally received its formal presentation in V. C. Wynne-Edwards's (1962) *Animal Dispersion in Relation to Social Behaviour*. This book offered an analysis of the evolution of social behaviour via what can be labelled species-benefit group selection. It inspired G. C. Williams (1966) and David Lack (1966) to write rebuttals in which they pointed out that: (1) group selectionist thinking was not based on the logic of Darwinian natural selection; and (2) the species-benefit version of group selection was unlikely to provide a mechanism for evolutionary change sufficiently strong to override the effects of natural selection operating at the level of individuals.

In formulating his version of group selection Wynne-Edwards envisioned competition amongst groups, primarily species, as the driving force of evolution. According to this view, those species endowed with properties that enabled them to avoid extinction would survive, leading to the evolution of attributes 'designed' by group selection to promote species survival. Wynne-Edwards explicitly argued that individuals will sacrifice some current reproductive opportunities if this sacrifice promotes the long-term survival of the species as a unit.

In contrast, Darwin proposed that lifetime differences in *individual* reproductive success *within* a species would lead to evolutionary change, if the differences in fitness were correlated with hereditary differences in the attributes of individuals. Darwinian selection operates on the short-term at the level of individuals, and the logic of the process is that individuals that leave more descendants than others, whatever the long-term consequences for the species as a whole, will control the course of evolution of their species.

Williams and others argued that even if group selection favoured reproductive altruism among the members of a species as a means to prevent future extinction, individual selection would act to eliminate the altruists. Or stated in another way, a species or group composed of individuals with self-sacrificing, species-benefiting characteritics would be vulnerable to invasion by a 'selfish' mutant individual whose individual-benefiting characteristics resulted in a higher rate of reproduction. Thus there is strong reason to believe that species-benefiting altruism would disappear in the face of competition from superior reproducers, which would inevitably arise by chance within species.

2.1.1 The impact of individual selectionist thinking

The simple argument that selection at the level of individuals is likely to be more potent than selection at the level of groups, which refreshed the biological community's memory of Darwinian theory, has had remarkable impact on the development of the fields of ecology and behaviour in the last two decades. After 1966 it was no longer possible for researchers to ascribe a group-benefiting function to a trait in a casual manner, as had been common practice in the decades previously. Instead, the force of Williams's argument led almost every evolutionary biologist to agree that, as a matter of logic, hypotheses that proposed a species benefit were improbable. This led to a sweeping reappraisal of a huge battery of traits, especially those related to sexual and social behaviour, that had until then been loosely interpreted as adaptive at the group level.

Biologists quickly discovered that alternative hypotheses on the function of these traits could be easily generated using Darwinian individual selection as the theoretical foundation. If European swifts generally laid three eggs, not four, the group selectionist pointed to reproductive restraint that helped keep the population from outgrowing its food supply. The individual selectionist, however, was forced to assume that three eggs was actually the best number for promoting individual reproductive success. This might be true if, in this species, parents that tried to rear four offspring would often fail to find enough food for their large brood, with consequent high mortality rates for their young. Or parents that were forced to work exceptionally hard to get enough food for four nestlings in one year might suffer high mortality themselves, and so fail more often than three-egg adults to return to breed again another year. The existence of two (or more) individual selectionist hypotheses for a given trait helped make researchers aware of the need for testing competing ideas. In the swift case, experiments revealed that adding a youngster to a brood of three actually reduced the number of fledglings produced by the parents on average (Perrins, 1964), a result that supports the Darwinian hypothesis that having three eggs maximizes a swift's annual output of young.

2.1.2 The rediscovery of sexual selection

Nowhere was the effect of discriminating between group selectionist and individual selectionist hypotheses more dramatic than in the area of sexual behaviour. Whereas group selectionists took it for granted that males and females reproduced cooperatively for the greater good of the species, the renewed interest in individual selection led biologists to anticipate conflict and competition at all levels of the reproductive enterprise. It is surely not coincidental that by the early 1970s there was a rebirth of enthusiasm for sexual selection theory (Campbell, 1972), which emphasizes precisely these things.

In developing the concept of sexual selection, Darwin (1871) employed essentially the same logic that he had used in his earlier analysis of natural selection. He proposed that the number of descendants an individual left might not be affected only by the individual's interaction with predators, climatic factors, and competitors of other species, but might also be affected by the results of social engagements within a species. The social effects on reproductive success that occur in the context of acquiring mates creates sexual selection. Variation in the number of mates acquired (and thus, the number of offspring produced) can occur either because of differences among males in their ability to compete with each other for access to females, or because of differences among males in their ability to attract or to 'charm' females.

Sexual selection theory had relatively little immediate impact on biologists compared with natural selection theory, in part because it was not as revolutionary nor as wide-ranging a concept. In addition, critics claimed that Darwinian female choice implied a level of decision-making likely to exceed the capacities of non-human animals (Thornhill, 1980). Thus, the theory was never applied with the enthusiasm that it deserved, until the new climate of the 1960s and 70s, when researchers began to look in earnest for the effects of male–male competition and female choice, and to find widespread evidence for their occurrence.

2.2 SEXUAL SELECTION THEORY AND INSECT REPRODUCTIVE BEHAVIOUR

Whether one employs Darwinian natural selection or sexual selection as the theoretical basis for analysing behaviour, the fundamental underlying assumption is the same—namely, that individuals will do that which yields the greatest number of surviving progeny. This expectation has been responsible for a major reinterpretation of several elements of insect reproductive behaviour, including the significance of communal signalling by males. The approach to the topic of social signals, which is analysed fully in Chapter 3, beautifully illustrates the change wrought by using sexual selection theory to develop working hypotheses about reproductive behaviour. Whereas earlier workers attempted to explain such things as synchronized acoustical or visual signalling in terms of various species benefits, such as improved communication with females, modern analysis suggests that synchrony can be the outcome of competition among males to avoid masking of their individual signal, or alternatively, may arise as some males attempt to prevent others from monopolizing the most effective channel of communication to females (Lloyd, 1971, 1979; Otte, 1980; review in Thornhill and Alcock, 1983).

Fig. 2.1 Two male Japanese scarab beetles fighting for possession of a female beneath them. (Photography by J. E. Lloyd.)

2.2.1 Reproductive aggression and the evolution of weapons

Although competition for access to mates may be subtle, as in the strategic delivery of attractant messages when other males are also calling for females, it can also be obvious, as when males fight directly for a potential mate or attempt to interfere with another male as he copulates with a female (Fig. 2.1). The point we wish to make here is that it is only in the post-group selection years that many diverse structures on male insects have been shown to play a role in fighting for mates. Despite the fact that the elaborate horns and 'antlers' on certain male beetles were known to naturalists for centuries, not until the past two decades has their function as sexually selected weapons become unequivocally clear. Darwin (1871) suggested that scarab beetle horns and other outgrowths of the male beetle exoskeleton evolved, not in the context of competition among males, but because of female preference for elaborately ornamented males. We know of only one study that supports Darwin's viewpoint. Working with the CSIRO dung beetle group in Western Australia, Cook (1988) has shown that female *Onthophagus binodis* mated to horned males produce more progeny than those mated to males with no horns in this highly variable species. Horned males may assist females in parental duties, and therefore the female preference for such individuals yields reproductive benefits for the choosy female (Cook, 1990).

Most studies, however, of the function of horns (e.g. Eberhard, 1979, 1982; Otte and Stayman, 1979) have demonstrated that these devices

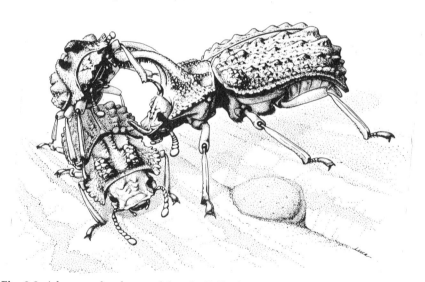

Fig. 2.2 A large male of a scarab beetle (*Bolitotherus cornutus*) using its thoracic horn to pry a smaller rival from the back of a female. (Drawing courtesy of D. Luce.)

almost invariably come into play when males are competing for access to mates in the manner characteristic of their species. Thus, there are beetles with stout horns for prying rival mounted males from the backs of their mates (Fig. 2.2); elongate snouts that are used to flip opponents into the air and away from females the males are guarding; elaborate but seemingly delicate jaws that are used to pick up an opponent and drop him from a tree trunk to the ground; thoracic antlers that interlock with those of rivals when the combatants are within a burrow in wood; and so on.

Enough evidence now exists to make the claim that whenever there is a striking sexual dimorphism in body shape, the hypothesis that male morphology is used in combat with other males over females or access to them must be considered. For example, Windsor (1987) found that the odd elytral configuration of males of a Panamanian chrysomelid was related to reproductive competition. Males use an elytron and pronotum as a vice to lock onto the elytron of a male interfering with them during courtship or copulation. When used successfully, their distinctive elytral morphology enables them to lift the intruder into the air where he is helpless for as long as the 'winner' chooses to hold him up (Fig. 2.3).

2.2.2 Alternative male mating tactics

Just as adoption of Darwinian sexual selection has proved to be the key for a full understanding of many cases of distinctive male morphology, so too it has led to the productive exploration of male behavioural variation, a topic that was almost completely ignored prior to 1965. Persons viewing

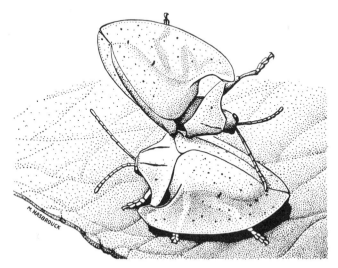

Fig. 2.3 A male chrysomelid using its elytral pincers to hold an opponent male high in the air. (Drawing courtesy of Donald Windsor.)

animal reproductive behaviour from a group-selectionist perspective had no reason to consider what losers in aggressive male–male competition might do to salvage their reproductive chances. To the extent that group selectionists acknowledged variation in male behaviour, it was to claim that the interactions among males permitted the selection of the genetically superior individuals to propagate the genes of the species. By this reasoning, we would expect subordinates and defeated individuals to refrain from copulation for the good of the species, and thus there was little cause to examine the behaviour of these males closely. In contrast, Darwinian thinking suggests that males excluded from reproduction as a result of being defeated in combat or other interactions with the dominant males in their population should still attempt to acquire mates. In the past, losers that succeeded to even a small degree would clearly leave more descendants than males that completely ceded reproduction to their 'superiors'. There are many situations in which genotypes that endowed their owners with the capacity to adjust their behaviour to variable environmental conditions ought to do better than genotypes that contribute to the development of inflexible behaviour. It is now clear that males of many insect species that face variable social conditions have evolved conditional strategies—inherited rules of behaviour that permit individuals to adopt more than one tactic, with the choice of tactic influenced by a host of variables depending on the species (Alcock, 1979; Cade, 1980; Dawkins, 1980; Dominey, 1984).

Consider the behaviour of small males in the curculionid *Rhinostomus barbirostris*, a species in which males use their elongate rostrum to flip male rivals from the log on which receptive females are mating and ovipositing (Eberhard, 1983). As is common among insects, there is substantial environmentally-induced size variation among males. Small males with their small snouts are very inferior fighters relative to the larger males in the population. Small males can somehow assess their fighting ability relative to others, and they adjust their behaviour to salvage some reproductive opportunities. They avoid combat for the most part, and instead stealthily approach males in attendance with a female that is about to oviposit. If the larger attendant becomes involved in a struggle with yet another male, the small male seizes the opportunity to copulate with the female during the confusion (Fig. 2.4).

The number of cases of species with known or suspected alternative male mating tactics has grown exponentially in the last two decades, and the variety of these alternatives is increasingly impressive. The 'sneaky' non-combative tactic of smaller males (or older weaker ones) is very widespread (e.g. Alcock and Houston, 1987; Forsyth and Montgomerie, 1987; see Fig. 2.5), but it is far from the only evolved attribute that enables some males to reproduce in a social environment in which they are faced with more powerful or energetic opponents. In the apid bee *Panurgus*

Fig. 2.4 A small male bottle brush weevil sneakily copulates with a female while two large rivals fight for her possession. (Photograph by W. G. Eberhard.)

banksianus (Meyer-Holzapfel, 1984), some males, perhaps the older ones, do not patrol the emergence site in flight, but instead sit in a flower likely to be visited by a foraging female, with whom they attempt to mate. Similarly in a species of *Philanthus* wasp, larger males fly on patrol over the region from which receptive females are emerging, while smaller (weaker?) males wait in small territories on the periphery of the site for any females their rivals have missed (O'Neill and Evans, 1983).

A still more distinctive alternative is exhibited by young, immature males of the rove beetle *Aleochara curtula*, which avoid combat with older, dominant males at carrion where the beetles feed on fly maggots and mate. The subordinate males release cuticular pheromones that mimic those produced by females, and so are tolerated, even courted, by rival males (Peschke, 1987). This enables them to remain in a superior feeding site where they can acquire the food needed for their maturation and the production of a spermatophore, a necessary adjunct for successful copulation. Female mimicry that promotes a male's mating opportunities is known for a number of other insects as well (Wendelken and Barth, 1985).

Our goal here is not to analyse the various proximate and evolutionary mechanisms that can generate diversity in the reproductive tactics of a species, but simply to note that these phenomena were not considered

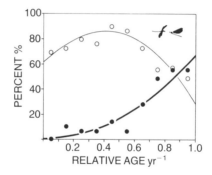

Fig. 2.5 In the damselfly *Calopteryx maculata* older males are more likely to practise the sneaker alternative tactic than are younger individuals, which instead guard mating territories —○— territorial, —●— sneaker. (Graph courtesy of A. Forsyth and R. L. Montgomerie.)

interesting until it was recognized that reproduction is not a cooperative enterprise designed for the welfare of the species. Now alternative mating tactics are the focus of intensive study, thanks to the revival in sexual selection theory with its focus on the competition among males for mates.

2.2.3 Post-copulatory competition among males

In addition to sexual dimorphism in morphology and variation in male reproductive tactics, another fascinating evolutionary topic whose study has blossomed with the rediscovery of Darwinian approaches to behaviour is post-copulatory competition for egg fertilizations. This now active field of inquiry, which has spread into many taxa other than insects (reviewed in Smith, 1984), had its origins in the paper 'Sperm competition and its evolutionary consequences in the insects' (Parker, 1970).

As a rule, sexual selection favours males that copulate with as many females as possible because such males fertilize more eggs than males that mate less often. Parker pointed out, however, that the fundamental currency of reproductive success for males is eggs fertilized, not females inseminated, although the two may often be correlated. Most insect females store sperm received from male partners, thus creating the possibility of competition among rival ejaculates to fertilize the eggs when females oviposit. Parker hypothesized that many attributes of males had evolved by sexual selection acting in the context of sperm competition.

For example, in many odonates, males sacrifice time that might be spent searching for additional mates to remain with a female to whom the male has just transferred sperm (e.g. Waage, 1986, see Fig. 2.6). Post-copulatory association between males and females takes many forms in odonates (and other insects), in itself an interesting problem, but the general phenomenon was largely overlooked until Parker developed a sexual selectionist hypothesis for it. He proposed that males guarded ovipositing mates to prevent them from receiving still more sperm from other males, sperm that might take precedence in fertilizing the female's eggs and thus reduce the prior male's fitness.

The interest in sexual selection's effects on post-copulatory competition has been responsible for many remarkable discoveries since 1970 (Smith, 1984). One of the more dramatic of these was Waage's (1979) demonstration that the penis of the damselfly *Calopteryx maculata* is used to remove stored sperm from the bursa and spermatheca of females before a male transfers his sperm to a mate (Fig. 2.7). Certainly this case exemplifies the extreme effects of sexual selection for characters that spread after they originate because they give individuals an advantage in the competition to fertilize eggs, even though they do nothing to enhance the population's productivity as a whole.

The evolutionary consequences of sperm competition include effects ranging from penis morphology to guarding behaviour, from copulation duration to the structure and shape of those parts of the male that grip the female during mating. Again, our goal is not to review these traits

Fig. 2.6 A male damselfly grasps his ovipositing female in the tandem position after copulating with her earlier, presumably to prevent her from mating with another male.

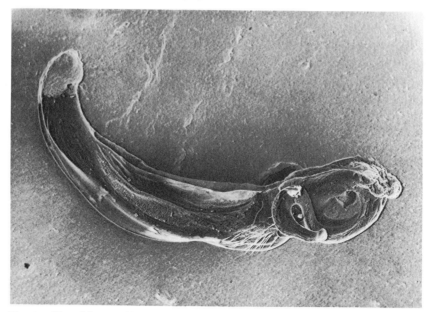

Fig. 2.7 The elaborate horned penis of the damselfly *Calopteryx maculata* is used to remove sperm stored within the female prior to the transfer of the male's own sperm to his mate. (Photograph by J. K. Waage.)

extensively, but to note that their function remained hidden until behavioural researchers began to use working hypotheses derived from sexual selection theory.

2.3 FEMALE CHOICE OF MATES

The empirical study of mate choice is also a relatively recent phenomenon. As we have already mentioned, the early critics of Darwinian sexual selection theory claimed that any type of mate preference required an improbable aesthetic sense in lower animals similar to that of humans. Because female choice among conspecific mates was rejected as an evolutionary force by many of Darwin's contemporaries, little effort was made to study this behaviour. Ironically, the post-Darwinian lapse into 'species-benefit thinking' did include discussion of mate preferences, but only in the sense of females selecting males of the 'correct' species. The revival of selectionist thinking, particularly with respect to sexual selection, placed female choice as a central part of the theory. If individuals do that which yields the greatest number of surviving progeny, then it follows that, if given a choice, females should choose mates in a way that enhances their reproductive output.

Despite the upsurge of interest in female choice, it has remained a controversial issue for a number of reasons (Thornhill and Alcock, 1983). One difficulty lies in demonstrating that differential mating by males is due to female discrimination rather than because of male–male competition for mates. For instance, Partridge (1980) showed that female *Drosophila* allowed a choice of several males produced larvae with increased competitive abilities relative to females not given a choice. The design of Partridge's experiment did not, however, exclude the possibility that the males that successfully mated in the 'choice' group may have not been actively chosen by females but instead had simply outcompeted the other males. Indeed, there is evidence from Partridge *et al.* (1987) that large males of *D. melanogaster* enjoy higher mating success than small ones because they court females more often and more vigorously. These authors were unable to find any indication that females discriminated between large and small males, suggesting that male–male interactions alone generated the differences in fitness of the two kinds of males.

One of the first demonstrations of adaptive mate choice by females was Thornhill's (1976) classic study of scorpionflies (*Hylobittacus (Bittacus) apicalis*). Females of this species go some distance from one male to another one and so the effects of female mate choice can be separated from those of male–male competition. Male *H. apicalis* provide an arthropod prey item for their mates (Fig. 2.8). Detailed laboratory and field

Fig. 2.8 A male *Harpobittacus* scorpionfly holding a nuptial gift of a small insect prey in its mouthparts. The male has also everted abdominal glands that release sex pheromones attractive to females.

observations by Thornhill showed that females favoured males that provided large nuptial offerings. First, females rarely copulated with males that offered very small prey items. Second, even when mating occurred, the duration of copulation was determined by the size of prey offered; females terminated copulation sooner when males provided smaller prey. Thus males presenting small gifts might not mate at all, or if they did, they would often lose in sperm competition with other males able to provide larger prey.

Female preference for mates that supply important resources at mating has subsequently been observed in other insects. For example, males of some species of katydid produce a large proteinaceous spermatophore which, when eaten by the female after mating, results in an increase in fecundity (Gwynne, 1984a). In at least one species in this group, females prefer to mate with the larger of two singing males and these larger males produce a larger spermatophore gift (Gwynne, 1982).

Courtship feeding of females can be more subtle than that observed in scorpionflies and katydids. Males of a number of insect groups pass nutrients with the ejaculate into the female genital tract (e.g. Boggs and Gilbert, 1979). Although *Drosophila* are unusually well-studied insects, only recently has it been discovered that males of some fruit fly species feed females by exuding nutritive droplets from the mouth or anus (Steele, 1986a). In *D. subobscura*, females are observed to discriminate against males that produce small or 'watery' droplets (Steele, 1986b).

A study of tobacco moths (*Ephestia elutella*) (Phelan and Baker, 1986) provides another dimension to the mate choice story. Research has shown that size-correlated mating success by males occurs because females reject small males. Mate choice is adaptive because females that mate with larger males produce more offspring and also progeny that are significantly larger in size. These results suggest that female tobacco moths select genetically superior males, whose sons will be more fit than those of smaller males.

Besides a female preference for useful resources, some theories predict that females should select as mates those males with above-average genotypes. In contrast to food resources that can be converted into immediate reproductive gains for females, such as increases in fecundity, the benefits of acquiring 'good genes' during mating will only be expressed in the next generation, when the male's genetic donation might increase the fitness (reproductive success) of his mate's offspring. This delay in the beneficial effects of female choice is the basis for a great deal of controversy over whether females in natural populations can evolve to prefer certain male genotypes over others (reviewed in Bradbury and Andersson, 1987; Kirkpatrick, 1987). The problem lies in the inheritance of a male's superior attributes by his offspring. Some theorists argue that constant sexual selection by females for traits that indicate male genetic quality would reduce genetic variation underlying these traits so that eventually there would be no differences among males, and thus, no superior genes for offspring to inherit. Other theorists have postulated mechanisms for maintaining genetic variance underlying sexually-selected traits (e.g. Cade, 1984) and have pointed to the evidence showing that many of these traits, such as body size, are indeed heritable (Mousseau and Roff, 1987).

The controversy over whether females of some species choose males on the basis of heritable indicator traits that signal the adaptive value of male genes is less likely to be settled by theory than by careful study of animals in nature. A study of *Colias* butterflies by Watt *et al.* (1986) exemplifies the latter approach. Intensive research on the ecological genetics of these insects has shown that variation at specific genetic loci leads to differences in the ability of individuals to fly at various temperatures. Watt *et al.* (1986) established that older female *Colias* are more discriminating of mates than virgins, which cannot fly away from unwanted suitors for some time after eclosion. Older females mating for a second time are much more likely to mate with males carrying the genotypes that are associated with the most effective flight under natural thermal conditions. Because the sperm of a second mate supplant those of the first mate, females produce more fit offspring as a result of their discriminating copulation: progeny raised from older females showed a significant preponderance of the favoured flight genotype compared to the progeny of younger (once-mated) females.

2.4 SEXUAL SELECTION AND THE EVOLUTION OF MATING SYSTEMS

The problem of why animal species exhibit so many different mating systems has been dealt with effectively by using sexual selection theory to generate testable hypotheses on the relation between ecological variables and male mating success (Emlen and Oring, 1977). Rather than viewing the mate-locating tactics of males as designed to promote efficient ferti-lization of the female population with consequent species advantages in population growth or regulation, the new approach employs the Darwinian expectation that individual males are in sexual competition with each other for mates.

Emlen and Oring (1977) recognized that the key to determining what would be most productive for a mate-seeking male depended in substan-tial measure on the ecological factors that affected the distribution of receptive females. When females are clustered in space, the likelihood increases that some males can economically defend the clusters, reaping great reproductive rewards by excluding other males from the females under their control. Clusters arise, for example, when females live together in groups, often for protection against certain predators. In this case, males can productively exhibit female defence territoriality. Alternatively, the resources females use may draw females from a wide area to a small one; if the resources are patchily distributed, we expect males to practise resource defence territoriality.

There are many species, however, whose females do not live in potential harems, nor do they use resources that are distributed in monopolizable patches. Under these circumstances, males that attempted to defend a territory embracing the home range of several females would be forced to cover a large area with all the energetic disadvantages this would entail and with the practical impossibility of monitoring the activities of females living in different places at the same time. Emlen and Oring argued that these conditions favoured the evolution of either non-aggressive attempts by wandering males to locate as many females as possible (via scramble competition) or the evolution of systems of symbolic competition among males in 'leks' or swarms of flying males. Although lekking and swarming males appear to defend the area immediately around them, they do so solely to advertise their social status to females that arrive at these locations to compare the males present, and perhaps to select one as a mate.

2.4.1 Female ecology and insect mating systems

The argument that differences in female ecology and distribution play a dominant role in the evolution of diversity in male mating systems has been examined by analysing insect reproduction. Although Emlen and Oring developed their ideas by comparing vertebrate mating systems, especially those of birds and mammals, their approach should be applicable to insects as well, and it is. The fundamental point that female defence and resource defence by males reliably occur when females or resources used by females are clustered in space appears to hold well for insects; when females and resources are scattered male insects often resort to scramble competition and lekking mating systems (for review, see Thornhill and Alcock, 1983).

Rather than broadly sketch the variety of insect mating systems to show that they do indeed correspond to those identified by Emlen and Oring, we have chosen to focus on just one important category—resource defence polygyny. Our goal shall be to examine how the renewed interest in sexual selection theory has led to a re-evaluation and new understanding of the structures and tactics that males use in species in which male competition revolves around access to the resources females need.

The first point is that resource defence polygyny has evolved repeatedly within insect groups when critical resources, typically oviposition sites or food, occur in small and defendable patches. The discrete nature of these patches enables males to identify productive locations at which to wait for females; the relatively small size of the sites enables some males to monopolize them in competition against rivals. This is perhaps clearest in species whose males collect useful prey items or seeds for transfer to their mates, such as empidid flies (Downes, 1970; Svensson and Petersson, 1987), various hangingflies (Thornhill, 1976, 1981) and lygaeid bugs (Carayon, 1964), or in species whose males construct burrows as oviposition sites for their mates, a group that includes many crickets (West and Alexander, 1963) and some bark beetles (Birch, 1978). In all these insects, males are able to defend a valuable resource that comes in a small enough packet to be grasped physically or that can be fully occupied by the burrow builder.

But naturally distributed resources used by females may come in units of suitable dimensions for effective defence by one male, and it is under these circumstances that we expect male territoriality to evolve. For example, in the dragonflies, there are species whose females oviposit in small clumps of floating vegetation or in very shallow water above limited substrates, others that utilize vast algal mats, and still others that broadcast their eggs in water of variable depths. The cost-benefit approach to territoriality suggests that males in the first category should be far more likely to exhibit strong resource defence than males in the other groups. Only in the first group is defence of a small area likely to yield disproportionate access to females attracted to oviposition substrate, thus repaying the time and energy investment needed for successful territoriality. The classic examples of territorial odonates, such as the damselfly *Calopteryx maculata* (Waage, 1973; Fig. 2.9) and the dragonfly *Plathemis lydia* (Jacobs, 1955; Koenig and Albano, 1985), are species whose females lay their eggs in discrete small patches of aquatic vegetation.

Comparative studies of cerambycid beetles reinforce the conclusion that female requirements and resource distribution determine the nature of resource-based mating systems. Many cerambycids mate on their foodplant, but far from all are territorial. Males of the flower-feeding *Perarthrus linsleyi* are non-territorial, with individuals employing scramble competition to locate mates, which occur on the flowers of creosote bush, a resource that is abundant and evenly distributed in most Sonoran desert habitats (Goldsmith, 1987a). In contrast, some males of the sap-feeding cerambycid, *Dendrobias mandibularis*, defend oozes in the foodplant *Baccharis*. The rare oozes offer a major and highly localized source of food for females, permitting some males to gain access to numbers of mates by guarding rich sources of plant sap (Goldsmith, 1987b).

Fig. 2.9 Resource defence polygyny in the damselfly *Calopteryx maculata*. A territorial male (second from left) guards a patch of aquatic vegetation that has attracted three receptive females to him.

2.4.2 Resource quantity and male reproductive success

In mating systems in which males control access to resources useful to females, female mating preferences should revolve primarily around the material benefits of accepting a certain male, rather than the male's genetic contribution. A fundamental prediction is that the quantity (and quality) of material resources within a male's domain will affect the numbers of females that he attracts (e.g. Emlen and Oring, 1977; Borgia, 1979). For example, the 'polygyny threshold model' developed for birds (Verner and Willson, 1966; Orians, 1969) argues that some males achieve polygyny because females gain more by joining an already mated male on his resource-rich territory than by entering into monogamy with a male that controls a resource-poor site. Some studies of birds, mammals, and fish have uncovered a correlation between the quantity or quality of a resource controlled by males and the number of mates they secure (see reviews in Wittenberger, 1981; Searcy, 1982; Davies and Houston, 1984).

Corresponding studies of insects are scarce, but as noted already, Thornhill (1976, 1981) working with scorpionflies and Steele (1986a,b) with *Drosophila subobscura* have both shown that males able to offer their mates larger quantities of collected food, in the form of nuptial gifts, attract more females and transfer more sperm than males that have less to offer. Likewise experimental tests of the effect of the quantity of food controlled by *Panorpa* scorpionflies on male reproductive success revealed that males controlling large items of carrion (a dead cricket) copulated with more females than males offering much smaller self-manufactured salivary masses, and these individuals in turn mated more frequently than males that neither defended a cricket nor made a salivary present (Thornhill, 1981).

The fact that oviposition substrates appear to be the key resources in the territories of odonates suggests the possibility of adding or subtracting these materials from territories in order to examine the effect this has on female behaviour and male reproductive success. Recently Waage (1987) and Alcock (1987) have completed parallel studies of the damselfly *Calopteryx maculata* in which they were able to induce males to defend territories containing different quantities of oviposition resources. In *C. maculata* suitable oviposition sites must contain floating emergent aquatic vegetation (Waage, 1973), and generally this resource is distributed in a highly patchy manner. Both workers found that adding plant material in stream sections that previously lacked floating vegetation attracted territory-defending males to areas where they had previously been absent. Two somewhat different experimental procedures showed that more females came to sites with more egg-laying materials. This in turn influenced male copulatory success and, almost certainly, the egg-fertilization success of males as well.

2.4.3 Male responses to variation in territory quality

If it is generally true that locations with large amounts of resources attract disproportionate numbers of females, then we can predict that males will compete more intensely for resource-rich locations. This simple argument receives support from a variety of insect studies. Thus, Shelly *et al.* (1987) have found that the females of the grasshopper *Ligurotettix coquilletti* feed preferentially on the leaves of certain creosote bushes. Although there is still some uncertainty about the basis for feeding preferences in this grasshopper, the evidence points to differences in the level of a toxic secondary chemical on the surface of *Larrea* foliage that presumably affects the nutritional quality of the leaves for *L. coquilletti*, which are extreme specialist consumers of creosote. In any event, certain bushes consistently attract larger numbers of males than others, with individuals competing to control the favoured shrubs. Some males try to defend their shrubs and signal loudly for females, while other males may wait silently in the most preferred creosote bushes, also seeking to mate with females that come to the bush.

By following male residence patterns over several years, Shelly *et al.*

(1987) were able to select two sets of creosote bushes that were equivalent in size and all other parameters except that one set had been regularly held by males in the past, and the other had not. During one summer, they were able to keep one and only one male singing per bush by moving males about during the course of the experiment. This equalized the strength with which the two groups of bushes were advertised by males. Six of 12 bushes that had been popular with males in previous years attracted three or more resident females, whereas ten of 12 unpopular bushes attracted fewer than two resident females. Females of this grasshopper evidently prefer certain bushes as feeding sites, independent of male activity, and males distribute themselves and compete territorially in ways that reflect the probability of encountering potential mates.

Similarly, in the cerambycid beetle *Monochamus scutellatus*, males attempt to control sections of fallen white pine that have the greatest circumference (Hughes and Hughes, 1982). It is these sections where larvae are believed to have the greatest chance of survival, and gravid females concentrate in these places. Males fight most strongly for these sections, with the result that large males (which are the superior fighters) come to occupy these parts of the tree disproportionately, excluding smaller males to some degree. The same kind of result has been obtained in a study of a lygaeid bug *Neacoryphus bicrucis* (McLain, 1984). Here there were also differences in resource attractiveness with females preferring patches of ragwort where flower heads (and the seeds of the plant) were most numerous. Again, large males won fights for possession of feeding territories, and they came to dominate the locations with rich clusters of flower heads.

These results illustrate how an analysis of resource-based mating systems can proceed based on sexual selection theory and its assumption that males are in competition with each other, with individuals expected to adjust their tactics to the ecological factors that determine the distribution of receptive conspecific females.

2.5 PARENTAL INVESTMENT AND INSECT MATING SYSTEMS

Our discussion so far has examined the variety of male mating strategies and mating systems as consequences of different sexual selection pressures on males and females. We now turn to the question of what controls sexual selection and thus the differences in reproductive strategies used by the sexes. Although the work of Emlen and Oring (1977) showed how ecological factors can determine the degree of sexual competition among males, it did not explain why males are competitive in the first place. Moreover, the competitive male–choosy female pattern does not always apply in nature; the 'typical' reproductive roles are reversed in a number of animals. Wynne-Edwards (1962) noted reversed sex roles in his group-selectionist tome. He could not, however, explain such mating biologies and concluded that 'there must be some kind of natural disadvantage or handicap' in these species. Once again it was an analysis of strategies of individual selection that dealt with this problem adequately. Williams (1966) was the first to consider the question from a Darwinian perspective asking (in a chapter title), 'Why are males masculine, females feminine and occasionally vice-versa?' The examples discussed by Williams were vertebrates, but in recent years there have been several well-documented cases of sex role reversals in mating behaviour recorded among insects. Before discussing the factors thought to control sexual differences in reproductive behaviour we will describe three cases of sex-role reversal that occur in three very different groups of insects (Table 2.1).

R. L. Smith (1979a) studied the mating behaviour of giant waterbugs *Abedus herberti* (Belostomatidae). He found that the female, not the male, first and repeatedly approached the initially passive male. Males usually displayed during courtship but occasionally showed signs of rejecting females by opting not to display when approached, thus preventing mating from proceeding. The male also controlled the mating sequence by guiding the female into the copulatory position and signalling the end of copulation. After uncoupling, he directed and assisted the female during oviposition. Females laid at most three eggs before they copulated again prior to another bout of egg laying. In these waterbugs, females lay their eggs on the backs of males, a most unusual oviposition site (Fig. 2.10). After mating and egg laying, males make an important contribution to the survival and development of offspring by doing 'push-ups' near the surface of the water which appear to function in maintaining an oxygen-rich environment around the eggs (Smith, 1976).

Table 2.1 Insect species showing evidence of a sex-reversal in courtship roles

Species	Form of male investment	Male choice of mates	Female–female competition
Abedus herberti (Hemiptera)	Male care of eggs	Some male coyness	
Several katydids (Orthoptera)	Large, edible spermatophore (courtship feeding)	Males prefer large females	Females fight for access to males
Empis borealis (Diptera)	Prey item (courtship feeding)	Males interact with several females before mating	Females assemble in lek-like swarms

Fig. 2.10 Sex-role reversal in the belostomatid waterbug. A male broods a cluster of eggs glued to his back by his mates. A freshly hatched nymph appears on the right. (Photograph by R. L. Smith.)

A second group of insects with 'atypical' mating behaviour are the katydids. Two species have been studied in detail, the North American *Anabrus simplex* (the Mormon cricket, Fig. 2.11) (Gwynne, 1981) and a Western Australian species, *Metaballus litus* (Gwynne, 1985). Males of both species are similar to other species of singing insects in producing a calling song that attracts receptive females. In many singing insects males call for long periods of time before attracting a female and the male call is the basis for a territorial male competitive system (e.g. Alexander, 1961). This is not the case, however, in certain populations of the two katydid species. Males appear to have very little time to compete with each other since receptive females are attracted quickly, often less than a minute after the start of a calling bout. Frequently, several females are attracted to the caller and if the females make contact, they grapple and fight in much the same way as described for 'border disputes' among territorial male crickets (Alexander, 1961). The female katydid winning the interaction courts the male but a large percentage of females are rejected as mates by males. Males prefer to mate with large females, possibly because these females have more eggs on average than smaller females, additional eggs that could be fertilized by the male's sperm (Gwynne, 1981, 1985). If the female is accepted as a mate, the pair couple and sperm is transferred in a spermatophore, part of which is inserted into the female gonopore. From the male's perspective the transfer of a spermatophore is a very costly part of the mating process because attached to the sperm-containing ampulla is a large, metabolically expensive mass, the spermatophylax. After mating, the female grasps the spermatophylax with her mandibles and eats it. This 'nuptial gift' from the male is an important source of nutrition for the female. Laboratory experiments with another Western Australian katydid, *Requena verticalis*, have shown that consumption of the spermatophylax meal increases the size of eggs later produced by the female. These larger eggs have a higher probability of surviving the winter, and therefore the meal donated by the male is important to the fitness of the offspring (Gwynne, 1984a, 1988a).

The most recent example of sex-role reversal described in insects concerns a dance fly, *Empis borealis* (Empididae). The observations of Svensson and Petersson (1987) show that this species forms competitive mating swarms above landmarks (bushes or trees) similar to those described for other flies. In contrast to other species, however, swarm members are females not males, thus making these flies the first example of a female 'lekking swarm'! Sexually active males approached the swarm carrying prey items. Mating followed contact between the sexes after the male handed the prey item to the female as a nuptial gift. Furthermore, there was evidence of discrimination among mates because males sometimes approached a number of females before mating with one.

Fig. 2.11 A newly mated female of the Mormon cricket, *Anabrus simplex*, with the male's spermatophore attached at the base of her ovipositor. The female will bend her head under her body to consume this nutritious packet of materials as the sperm contained within enter her genital tract.

Is there anything about the biologies of the water bug, the katydids and the dance fly that might explain the reversal in the reproductive roles of males and females? Although these insects are very different in most respects, they have one element in common. Males provide a costly investment: male *A. herberti* provide parental care whereas the male katydids and dance fly provide nuptial meals to their mates. The two kinds of male effort appear to be analogous in that they constitute investment in offspring. In the giant water bug the male assists directly in the hatching success of offspring (Smith, 1976) and in the katydids the food provided to the female at mating later increases the survival of offspring (Gwynne, 1984a, 1988b).

It was the relative parental investment by the sexes that G. C. Williams (1966) and R. L. Trivers (1972) argued should control the intensity of sexual selection on the sexes and thus the degree to which the sexes compete. In the vast majority of animal species the number of descendants a male leaves is primarily a function of the number of females that mate with him and use his small, and inexpensive, sperm to fertilize their large, and physiologically costly, eggs. Females produce expensive eggs, and in many groups nurture their offspring. As a result, females are usually a limited resource for males, and this should generate competition among males for access to females. Winners will succeed in fertilizing many eggs and leaving many descendants, shaping the next generation in their image. Females can afford the 'luxury' of choice because there will usually be many sexually active males for every willing female. The main prediction from the Williams–Trivers theory is that in cases in which it is the male who invests in offspring the sexual differences in behaviour should be decreased; if the male contribution to the next generation of offspring exceeds that of the female, the reproductive roles should be reversed (Thornhill and Gwynne, 1986). It was with the Williams–Trivers theory in mind, and the knowledge that males of giant water bugs and certain katydids invest greatly in reproduction, that the 'altered roles' in the mating behaviour of these insects were predicted and confirmed.

2.6 WHY THEN DO MALES OF CERTAIN INSECTS FEED THEIR MATES?

The presentation of a nuptial gift by male katydids is similar to other insects we have already mentioned. Scorpionflies provide prey items to their mates (e.g. Thornhill, 1976) and fruit flies, *Drosophila subobscura*, regurgitate or egest a droplet of nuptial food (Steele, 1986a). Despite offering their mates these items, male scorpionflies, some *Drosophila* species and many other nuptial-gift giving insects (Fig. 2.12) still exhibit more-or-less typical sex roles with males competing for mates and

Fig. 2.12 Gift-giving males occur in a number of insect groups, including the thynnine wasps. Here a male gives his wingless mate a drop of regurgitated nectar while the two wasps copulate.

females doing the choosing. In these species, and unlike the katydids and dance fly discussed in the previous section, males (and their offerings) are apparently not a limiting resource for the reproduction of females.

Why do males of certain insects feed their mates? Proponents of the view that reproductive behaviour functions to benefit the group would probably argue that females of these insects are unable to collect sufficient food for reproduction and that males cooperate in food acquisition in order to assure species survival. The question becomes much more interesting when an individual selectionist perspective is taken because this approach predicts frequent conflict between the sexes when it comes to reproduction. Thus females should be selected to obtain nutritious courtship gifts from males whereas males might well be expected to try to obtain copulations without the concomitant food investment.

There are two apparent selective contexts for the evolution of nuptial feeding behaviour. Food gifts could be sexually selected in that they have evolved to obtain fertilizations or they could be a result of natural selection favouring male investment in the fitness of their offspring. Courtship feeding in groups such as scorpionflies appears to be mainly due to sexual selection in that sexual competition among males is mediated by their ability to capture or produce nutritious food items for the female; males donate food items in exchange for fertilizations (Thornhill, 1976, 1981). In contrast male parental care by giant water bugs appears to be a result of natural selection for increased offspring fitness because males have a high confidence of parentage in the eggs they care for (Smith, 1979b).

Interestingly, courtship feeding of spermatophores in katydids appears to be functionally more similar to parental care by water bugs than to courtship feeding in scorpionflies. The nutritious spermatophylax provided by male katydids (*Requena verticalis*) is on average more than twice the size predicted if males simply exchanged food gifts for fertiliz-ations of eggs (Gwynne, 1986). Studies of the role of the spermatophylax meal in this species showed that male's investment results in an increase in the fitness of the male's own offspring (Gwynne 1984a, 1988a, b). The high confidence of paternity of katydid males is due to the fact that males mating with virgin females fertilize virtually all of the eggs in the next clutch of eggs laid. The pattern of sperm competition in this species is quite different from that of most insects (Parker, 1970; Gwynne, 1984b).

2.7 CAN SELECTIVE THINKING BE USEFUL IN INSECT CONTROL PROGRAMMES?

Although evolutionary biologists may be generally enthusiastic about the range and power of individual selectionist thinking as a guide to under-standing insect reproductive behaviour, it remains to be demonstrated that the approach will yield benefits for those interested in the practical problem of dealing with insect pests. Nevertheless, there are grounds for supposing that an accurate analysis of the function of behavioural tactics used by such pests would be helpful in seeking to defeat them. Certainly any tendency to view pest species as composed of individuals that are committed to species-benefiting behaviour is likely to be mistaken and could subvert effective biological control programmes. Thus, for example, the widespread acceptance of the hypothesis that bark beetles communicate for the purpose of maximizing efficiency of utilization of their host plant resources in order to maximize population growth has several unfortunate consequences (Alcock, 1982). In particular, such a view induces the pest control manager to ignore the likelihood that the members of a bark beetle population are engaged in intense reproductive competition. The individual selectionist approach generates the hypoth-esis that cooperation exists within a population only to the extent that it raises an individual's chances of reproducing successfully. If true, we can expect to find bark beetles frequently working at cross-purposes or in opposition to each other. The inevitability of reproductive competition within species as a result of Darwinian natural selection raises the possibility that pest managers may be able to tap such competition for their own ends—but only if they are aware of its existence.

In addition, an appreciation of the operation of individual selection can offer a guide to some problems likely to arise in the course of biological control programmes. For a case study, let us consider the attempt to control the infamous 'medfly' *Ceratitis capitata* in Western Australia. The area around Carnarvon, Western Australia, is an important fruit-growing centre, which was infested with the medfly in the 1970s. Tephritid fruit flies lay their eggs on fruit, and the larvae feed upon and damage the produce. The state government of Western Australia funded a biological control scheme using the sterile insect technique in which masses of lab-reared males are sterilized using radiation and then released (Fisher, 1981). The aim of this technique is to swamp the release area with sterile males so that most mating females will lay non-viable eggs.

The geography of the Carnarvon fruit-growing area was such that it seemed a perfect testing ground for sterile male release. One of the most important factors affecting the success of such a programme is the degree of immigration of wild flies into the control area (Ito and Kawamoto, 1979); Carnarvon is surrounded by the Indian Ocean to the west and semi-desert on the other three sides, a ring of obstacles to the immigration of fruit flies. The sterile release programme was successful in that the numbers of wild flies were greatly reduced after several years of fly release. However, although the population of wild flies was eventually reduced to a few per cent of its former size, it remained at this level with no further reduction. The explanation for the persistence of wild flies was not obvious, but knowledge of the mating biology of the medfly and some 'selectionist thinking' may provide an answer.

The reproductive success of a female mating only with a sterile male is zero. Therefore, the theory that forms the basis of this chapter predicts that there would be a large selective advantage to females that discriminated against sterile males in favour of the wild males. Furthermore, males that are descended from generations of laboratory reared flies may well be adapted to mate under artificial conditions so that in nature wild males would have the edge in sexual competition.

Is there any evidence that sexual selection may favour wild males over laboratory-raised individuals, thus explaining why wild individuals persist in the population? The only study of medflies in nature revealed exactly the sort of mating system in which a high degree of sexual selection on males is expected. Most matings result from the attraction of females to groups of displaying males in leks located on leaves (Prokopy and Hendrichs, 1979). Each male performs a wing-waving display and emits a pheromone that attracts females. Although a fruit resource is necessary for the reproduction of medflies, the mating system appears to be mainly non-resource based because oviposition locations are not predictable: some 253 species of fruit hosts have been recorded (Burk and Calkins, 1983). Mate choice in medflies has not been studied but females of another tephritid, the Carib fly *Anastrepha suspensa*, are known to discriminate among lekking males (Burk and Webb, 1983).

Receptive female medflies respond strongly to the displaying males (Burk and Calkins, 1983) and there is some evidence that laboratory-raised males are less responsive than wild males to pheromones produced by lekking conspecific males (Supastian, 1986). Male–male competition may also influence mating success in medfly populations containing released sterile males and wild males. A detailed study of competition between laboratory-reared and wild males of the screw worm fly (*Cochliomyia hominivorax*, Calliphoridae), a pest of cattle and the subject of the first major programme of sterile male release, found that laboratory conditions selected for alleles favouring flight at warmer temperatures. As a result, when released, sterilized males lost in competition with wild males because the wild individuals were better able to locate females in the early, cooler part of the day (summarized in Bush, 1978).

Burk and Calkins (1983) reported how an understanding of the medfly mating system provided a solution to previously unexplained results on the attraction of the flies to pheromones. Sticky traps that attract flies are baited with a synthetic male pheromone. However, more males than females are caught in these traps, presumably because males exploit pheromone cues to locate leks that they might join. All males must find a lek every day whereas females will respond once or only a few times in their lifetimes. Therefore, male–male competition may underlie the responsiveness of males to male sex pheromone in these flies as well as in certain bark beetles (Alcock, 1982). Burk and Calkins remark, 'that sexual selection can provide a convincing explanation of a previously puzzling pattern such as this is strong justification for using it in studies of other aspects of medfly mating behaviour.'

In conclusion, a full understanding of natural selection and adaptation, which has already made very important contributions to the study of animal social and sexual systems, can also be a useful tool in approaching economic problems. Already 'selective thinking' is involved in research into the use of recombinant DNA and genetic engineering to design pest-resistant crops (see Gould, 1988). G. C. Williams, who initiated the era of critical thinking about natural selection, recently suggested (1985) that biologists have been tardy in making full use of the Mendelian formulation of natural selection, stating 'what is sometimes called modern Darwinism is a field in its infancy at best.' As we have seen, many useful (testable) predictions have been derived from this body of theory. It is intriguing to consider what future pest managers might do with models derived from natural selection theory like the one generated by Parker (1985), a model predicting that under certain conditions competitive behaviour among males of the same species could lead to the extinction of that species.

References

Alcock, J. (1979) The evolution of intraspecific diversity in male reproductive strategies in some bees and wasps. In *Sexual Selection and Reproductive Competition in Insects* (eds M. S. Blum and N. A. Blum), Academic Press, New York, 381–402.

Alcock, J. (1982) Natural selection and communication among bark beetles. *Fla. Entomol.*, **65**, 17–32.

Alcock, J. (1987) The effects of experimental manipulation of resources on the behavior of two calopterygid damselflies that exhibit resource-defense polygyny. *Can. J. Zool.*, **65**, 2475–2482.

Alcock, J. and Houston, T. F. (1987) Resource defense and alternative mating tactics in the banksia bee, *Hyaleus alcyoneus* (Erichson). *Ethology*, **76**, 177–189.

Alexander, R. D. (1961) Aggressiveness, territoriality, and sexual behavior in field crickets (Orthoptera: Gryllidae). *Behaviour*, **171**, 130–223.

Birch, M. C. (1978) Chemical communication in pine bark beetles. *Amer. Sci.*, **66**, 409–419.

Boggs, C. I. and Gilbert, L. E. (1979) Male contribution to egg production in butterflies: evidence for transfer of nutrients at mating. *Science*, **206**, 83–84.

Borgia, G. (1979) Sexual selection and the evolution of mating systems. In *Sexual Selection and Reproductive Competition in Insects* (eds M. S. Blum and N. A. Blum), Academic Press, New York, 19–80.

Bradbury, J. W. and Andersson, M. B. (eds) (1987) *Sexual Selection: Testing the Alternatives*, John Wiley & Sons Ltd, Chichester.

Burk, T. and Calkins, C. O. (1983) Medfly mating behavior and control strategies. *Fla. Entomol.*, **66**, 3–18.

Burk, T. and Webb, J. C. (1983) Effect of male size on calling propensity, song parameters, and mating success in Caribbean fruit flies, *Anastrepha suspensa* (Loew) (Diptera: Tephritidae). *Ann. Entomol. Soc. Amer.*, **76**, 678–682.

Bush, G. L. (1978) Planning a rational quality control program for the screw worm fly. In *The Screw-worm Problem. Evolution of Resistance to Biological Control* (ed. R. H. Richardson), University of Texas Press, Austin, 37–47.

Cade, W. (1980) Alternative male reproductive strategies. *Fla. Entomol.*, **63**, 30–45.

Cade, W. H. (1984) Genetic variation underlying sexual behavior and reproduction. *Amer. Zool.*, **24**, 355–366.

Campbell, B. (ed.) (1972) *Sexual Selection and the Descent of Man 1871–1971*. Aldine Publishing, Chicago.

Carayon, J. (1964) Un cas d'offrande nuptiale chez les Heteroptères. *Comp. Rend. Acad. Sci.*, **259**, 4815–4818.

Cook, D. (1988) Sexual selection in dung beetles. II. Female fecundity as an estimate of male reproductive success in relation to horn size and alternative behavioural strategies in *Onthophagus binodis* Thunberg (Scarabaeidae: Onthophagini). *Austr. J. Zool.*, **36**, 521–532.

Cook, D. (1990) Differences in courtship, mating and precopulatory behaviour between male morphs of the dung beetle *Onthophagus binodis* Thunberg (Coleoptera: Scarabaeidae) *Anim. Behav.*, **40**, 428–36.

Darwin, C. (1859) *On the Origin of Species*, John Murray, London.

Darwin, C. (1871) *The Descent of Man and Selection in Relation to Sex*, John Murray, London.

Davies, N. B. and Houston, A. I. (1984) Territory economics. In *Behavioural Ecology, An Evolutionary Approach* (eds J. R. Krebs and N. B. Davies), Sinauer Associates, Sutherland, MA, 148–169.

Dawkins, R. (1980) Good strategy or evolutionarily stable strategy? In *Sociobiology: Beyond Nature/Nurture?* (eds G. W. Barlow and J. Silverberg), Westview Press, Boulder, CO, 331–367.

Dominey, W. J. (1984) Alternative mating tactics and evolutionarily stable strategies. *Amer. Zool.*, **24**, 385–396.

Downes, J. A. (1970) The feeding and mating behaviour of the specialized Empidinae (Diptera): observations on four species of *Rhamphomyia* in the arctic and a general discussion. *Can. Entomol.*, **102**, 769–791.

Eberhard, W. G. (1979) The function of horns in *Podischnus agenor* (Dynastinae) and other beetles. In *Sexual Selection and Reproductive Competition in Insects* (eds M. S. Blum and N. A. Blum), Academic Press, New York, 231–258.

Eberhard, W. G. (1982) Beetle horn dimorphism: making the best of a bad lot. *Amer. Nat.*, **119**, 420–426.

Eberhard, W. G. (1983) Behavior of adult bottle brush weevils (*Rhinostomus barbirostris*) (Coleptera: Curculionidae). *Rev. Biol. Trop.*, **31**, 233–244.

Emlen, S. T. and Oring, L. W. (1977) Ecology, sexual selection and the evolution of mating systems. *Science*, **197**, 215–222.

Fisher, K. (1981) Fruit fly under attack . . . from the sterile-insect-technique. *J. Agric. West. Austr.*, **22**, 51–52.

Forsyth, A. and Montgomerie, R. D. (1987) Alternative reproductive tactics in the territorial damselfly *Calopteryx maculata:* sneaking by older males. *Behav. Ecol. Sociobiol.*, **21**, 73–82.

Goldsmith, S. K. (1987a) Resource distribution and its effect on the mating system of a longhorned beetle, *Perarthrus linsleyi* (Coleoptera: Cerambycidae). *Oecologia*, **73**, 317–320.

Goldsmith, S. K. (1987b) The mating system and alternative reproductive behaviors of *Dendrobias mandibularis* (Coleoptera: Cerambycidae). *Behav. Ecol. Sociobiol.*, **20**, 111–115.

Gould, F. (1988) Evolutionary biology and genetically-engineered crops. *Bioscience*, **38**, 26–33.

Gwynne, D. T. (1981) Sexual difference theory: Mormon crickets show role reversal in mate choice. *Science*, **213**, 779–780.

Gwynne, D. T. (1982) Mate selection by female katydids (Orthoptera: Tettigoniidae, *Conocephalus nigropleurum*). *Anim. Behav.*, **30**, 734–738.

Gwynne, D. T. (1984a) Courtship feeding increases female reproductive success in bushcrickets. *Nature*, **307**, 361–363.

Gwynne, D. T. (1984b) Male mating effort, confidence of paternity and insect sperm competition. In *Sperm Competition and the Evolution of Animal Mating Systems* (ed. R. L. Smith), Academic Press, New York, pp. 117–149.

Gwynne, D. T. (1985) Role-reversal in katydids: habitat influences reproductive behaviour (Orthoptera: Tettigoniidae: *Metaballus* sp.). *Behav. Ecol. Sociobiol.*, **16**, 355–361.

Gwynne, D. T. (1986) Courtship feeding in katydids (Orthoptera: Tettigoniidae): investment in offspring or in obtaining fertilizations? *Amer. Nat.*, **128**, 342–352.

Gwynne, D. T. (1988a) Courtship feeding and the fitness of females in katydids. *Evolution*, **42**, 545–555.

Gwynne, D. T. (1988b) Courtship feeding in katydids benefits the mating male's offspring. *Behav. Ecol. Sociobiol.*, **23**, 373–377.

Hughes, A. L. and Hughes, M. K. (1982) Male size, mating success, and breeding habitat partitioning in the whitespotted sawyer *Monochamus scutellatus* (Say) (Coleoptera: Cerambycidae). *Oecologia*, **55**, 258–263.

Huxley, T. H. (1910) *Lectures and Lay Sermons*, E. P. Dutton, New York.

Ito, Y. and Kawamoto, H. (1979) Number of generations necessary to attain eradication of an insect pest with sterile insect release method. *Res. Popul. Ecol.*, **20**, 216–226.

Jacobs, M. E. (1955) Studies on territorialism and sexual selection in dragonflies. *Ecology*, **36**, 566–586.

Kirkpatrick, M. (1987) Sexual selection by female choice in polygamous animals. *Ann. Rev. Ecol. Syst.*, **18**, 43–70.

Koenig, W. D. and Albano, S. S. (1985) Patterns of territoriality and mating success in the white-tailed skimmer *Plathemis lydia* (Odonata: Anisoptera). *Amer. Midl. Nat.*, **114**, 1–12.

Lack. D. (1966) *Population Studies of Birds*, Clarendon Press, Oxford.

Lloyd, J. E. (1971) Bioluminescent communication in insects. *Ann. Rev. Entomol.*, **16**, 97–122.

Lloyd, J. E. (1979) Sexual selection in luminescent beetles. *Sexual Selection and Reproductive Competition in Insects* (eds M. S. Blum and N. A. Blum), Academic Press, New York, 293–342.

McLain, D. K. (1984) Host plant density and territorial behavior of the seed bug, *Neocoryphus bicrucis* (Hemiptera: Lygaeidae). *Behav. Ecol. Sociobiol.*, **14**, 181–187.

Meyer-Holzapfel, V. M. (1984) Zur ethologie des Männchens der Trugbiene (*Panurgus banksianus* Kirby) (Hymenoptera, Apidae). *Z. Tierpsychol.*, **64**, 221–252.

Mousseau, T. and Roff, D. (1987) Natural selection and the heritability of fitness components. *Heredity* **59**, 181–197.

O'Neill, K. M. and Evans, H. E. (1983) Body size and alternative mating tactics in the beewolf *Philanthus zebratus* (Hymenoptera, Sphecidae). *Biol. J. Linn. Soc.*, **220**, 175–184.

Orians, G. H. (1969) On the evolution of mating systems in birds and mammals. *Am. Nat.*, **103**, 589–603.

Otte, D. (1980) Theories of flash synchronization in fireflies. *Amer. Nat.*, **116**, 587–590.

Otte, D. and Stayman, K. (1979) Beetle horns: some patterns in functional morphology. In *Sexual Selection and Reproductive Competition in Insects* (eds M. S. Blum and N. A. Blum), Academic Press, New York, 259–292.

Parker, G. A. (1970) Sperm competition and its evolutionary consequences in the insects. *Biol. Rev.*, **45**, 525–567.

Parker, G. A. (1985) Population consequences of evolutionarily stable strategies. In *Behavioural Ecology. Ecological Consequences of Adaptive Behaviour* (eds R. M. Sibley and R. H. Smith), Blackwell Scientific, Oxford, 33–58.

Partridge, L. (1980) Mate choice increases a component of offspring fitness in fruit flies. *Nature*, **283**, 290–291.

Partridge, L., Ewing, A. and Chandler, A. (1987) Male size and mating success in *Drosophila melanogaster*: the roles of male and female behaviour. *Anim. Behav.*, **35**, 555–562.

Perrins, C. (1964) Survival of young swiftlets in relation to brood-size. *Nature*, **201**, 1147–1148

Peschke, K. (1987) Male aggression, female mimicry and female choice in the rove beetle, *Aleochara curtula* (Coleoptera, Staphylinidae). *Ethology*, **75**, 265–284.

Phelan, P. L. and Baker, T. C. (1986) Male-size-related courtship success and intersexual selection in the tobacco moth, *Ephestia elutella. Experientia*, **42**, 1291–1293.

Prokopy, R. J. and Hendrichs, J. (1979) Mating behavior of *Ceratitis capitata* on a field-caged host tree. *Ann. Entomol. Soc. Amer.*, **72**, 642–648.

Sakaluk, S. K. (1984) Male crickets feed females to ensure complete sperm transfer. *Science*, **223**, 609–610.

Searcy, W. A. (1982) The evolutionary effects of mate selection. *Ann. Rev. Ecol. Syst.*, **13**, 57–85.

Shelly, T. E., Greenfield, M. D. and Downum, K. R. (1987) Variation in host plant quality: influences on the mating system of a desert grasshopper. *Anim. Behav.*, **35**, 1200–1209.

Smith, R. L. (1976) Male brooding behavior of the water bug *Abedus herberti* (Hemiptera: Belostomatidae). *Ann. Entomol. Soc. Amer.*, **69**, 740–747.

Smith, R. L. (1979a) Paternity assurance and altered roles in the mating behaviour of a giant water bug, *Abedus herberti* (Heteroptera: Belostomatidae). *Anim. Behav.*, **27**, 716–725.

Smith, R. L. (1979b) Repeated copulation and sperm precedence: paternity assurance for a male brooding water bug. *Science*, **205**, 1029–1031.

Smith, R. L. (ed.) (1984) *Sperm Competition and the Evolution of Animal Mating Systems*, Academic Press, New York.

Steele, R. H. (1986a) Courtship feeding in *Drosophila subobscura*. I. The nutritional significance of courtship feeding. *Anim. Behav.*, **34**, 1087–1098.

Steele, R. H. (1986b) Courtship feeding in *Drosophila subobscura*. II. Courtship feeding by males influences female mate choice. *Anim. Behav.*, **34** 1099–1108.

Supastian, C. (1986) The mating behaviour of the Mediterranean fruit fly *Ceritatis capitata* (Weid.): Behavioural changes with laboratory rearing. MSc Thesis, University of Western Australia.

Svensson, B. G. and Petersson, E. (1987) Sex-role reversed courtship behaviour, sexual dimorphism and nuptial gifts in the dance fly, *Empis borealis* (L.). *Ann. Zool. Fenn.*, **24**, 323–334.

Thornhill, R. (1976) Sexual selection and nuptial feeding behavior in *Bittacus apicalis* (Insecta: Mecoptera). *Amer. Nat.*, **110**, 529–548.

Thornhill, R. (1980) Competitive, charming males and choosy females: was Darwin correct? *Fla. Entomol.*, **63**, 5–30.

Thornhill, R. (1981) *Panorpa* (Mecoptera: Panorpidae) scorpionflies: systems for understanding resource-defense polygyny and alternative male reproductive effort. *Ann. Rev. Ecol. Syst.*, **12**, 355–386.

Thornhill, R. and Alcock, J. (1983) *The Evolution of Insect Mating Systems*, Harvard University Press, Cambridge, MA.

Thornhill, R. and Gwynne, D. T. (1986) The evolution of sexual differences in insects. *Amer. Sci.*, **74**, 382–389.

Trivers, R. L. (1972) Parental investment and sexual selection. In *Sexual Selection and the Descent of Man, 1871–1971* (ed B. Campbell), Aldine, Chicago, 136–179.

Verner, J. and Willson, M. F. (1966) The influence of habitats on mating systems of North American passerine birds. *Ecology*, **47**, 143–147.

Waage, J. K. (1973) Reproductive behavior and its relation to territoriality in *Calopteryx maculata* (Beauvois) (Odonata: Calopterygidae). *Behaviour*, **47**, 240–256.

Waage, J. K. (1979) Dual function of the damselfly penis: sperm removal and transfer. *Science*, **203**, 916–918.

Waage, J. K. (1986) Evidence for widespread sperm displacement ability among Zygoptera (Odonata) and the means for predicting its presence. *Biol. J. Linn. Soc.*, **28**, 285–300.

Waage, J. K. (1987) Choice and utilization of oviposition sites by the damselfly, *Calopteryx maculata* (Odonata: Calopterygidae). I. Influence of site size and the presence of other females. *Behav. Ecol. Sociobiol.*, **20**, 439–446.

Watt, W. B., Carter, P. A. and Donohue, K. (1986) Female's choice of 'good genotypes' as mates is promoted by an insect mating system. *Science*, **233**, 1187–1190.

Wendelken, P. W. and Barth Jr, R. H. (1985) On the significance of pseudofemale behavior in the neotropical cockroach genera of *Blaberus*, *Archimandrita*, and *Byrostria*. *Psyche*, **92**, 493–504.

West, M. J. and Alexander, R. D. (1963) Sub-social behavior in a burrowing cricket *Anurogryllus muticus* (DeGeer): Orthoptera: Gryllidae. *Ohio J. Sci.*, **63**, 19–24.

Williams, G. C. (1966) *Adaptation and Natural Selection*, Princeton University Press, Princeton, NJ.

Williams, G. C. (1975) *Sex and Evolution*, Princeton University Press, Princeton, NJ.

Williams, G. C. (1985) A defence of reductionism in evolutionary biology. In *Oxford Surveys in Evolutionary Biology*, Vol. 2 (eds R. Dawkins and M. Ridley), Oxford University Press, Oxford, 1–27.

Windsor, D. M. (1987) Natural history of a subsocial tortoise beetle, *Acromis sparsa* Boheman (Chrysomelidae: Cassidinae) in Panama. *Psyche*, **94**, 127–150.

Wittenberger, J. F. (1981) *Animal Social Behavior*, Duxbury Press, Boston.

Wynne-Edwards, V. C. (1962) *Animal Dispersion in Relation to Social Behaviour*, Oliver and Boyd, Edinburgh.

3

Mate finding: selection on sensory cues

Winston J. Bailey

3.1 INTRODUCTION

Once an insect has emerged as an adult it must either call or search for a mate. Although in some cases both sexes may call, it is more usual for one sex to do the calling and the other the searching. The 'distance' cue for mate location should provide the searching insect with information with regard to species' identity—through pattern, colour, odour or sound—as well as direction. Vision, although giving the most accurate directional information, and a range of cues of both colour and shape for identity, is restricted to day-time behaviour (except in those animals using luminescence). Communication requires a line-of-sight between signaller and receiver and, not surprisingly, most insects that use vision as part of their social behaviour are highly mobile. Odour as a pheromone relies on unique molecular structures for identity as well as an ability by the searching insect to zig-zag upwind to find the odour source. Here, although specificity may be achieved through specialized receptors, directional information over any distance is subject to the vagaries of air movements. Hearing has an undoubted advantage over both these sensory modalities in that it may be used during the day or night, and further, its temporal pattern or frequency structure can provide the necessary uniqueness for identity. The sound source can be located with some precision with a binaural system of hearing and because sound is only partially impeded by intervening obstacles, it may be used in quite densely vegetated habitats.

This chapter is about the evolution of insect calling and searching, and although drawing on a range of signalling modalities, it is primarily concerned with acoustic cues. I adopt such a narrow line for three reasons. First, the acoustic system has been widely studied over a range of functional disciplines (Bennet-Clark, 1984; Michelsen and Larsen 1985) and their results suggest the limits of selection. The questions generated are essentially 'how' (Mayr, 1982). Second, because acoustic behaviour has been the focus of evolutionary biologists in a variety of taxa (e.g. Gerhardt, 1982, 1987; Arak, 1983; Gwynne and Morris, 1983; Payne 1983; Searcy & Andersson, 1986; Gerhardt *et al.*, 1987) we may, both by comparative studies and by experiment, address a number of 'why' questions; those concerned with evolution. Lastly, as W. Somerset Maugham writes in the preface to '*Of Human Bondage*', 'I (have) learnt that it is easier to write of what you know than of what you don't.'

3.1.1 Group and individual phenomena

In the preceding chapter, Alcock and Gwynne presented a case for viewing behaviour that to many would appear as 'group' phenomena, as evolution acting at the level of the individual. They use as their straw man Wynne-Edward's (1962) thesis of group selection, where for example social signalling might benefit the population as an 'epidiectic diplay'. The *idée fixe* on group and 'species' behaviour is not restricted to the domain of social behaviour, and is the axiom of most physiologists: the species' distinct patterns are set at a 'species' level, and any variation around this 'average' is statistical noise. In other words the survival of the species through its unique mechanisms for species isolation is housed within the insect's neural pathways for no other reason than the maintenance of species' distinctiveness. Littlejohn, (1988) by referring to 'homogamy', deflects discussion from 'species'' patterns (a group event) to the recognition of one individual by another (an individual event). With the individual's response to its conspecific partner in mind, it is only recently that workers have begun to look at variation in female searching behaviour to different song patterns, and to consider that this variation may have a significant role in the individual's mating success (e.g. Crankshaw, 1979; Hedrick, 1986; Partridge *et al.*, 1987; Simmons, 1988ab; Gwynne and Bailey, 1988).

Displays of insects aggregating for reproduction are spectacular—the din of screeching cicadas in a neotropical forest, the spiralling column of lake flies rising hundreds of metres from Lake Victoria or the beacon trees of fireflies in Southeast Asia. All conjure up immediate thoughts of cooperative behaviour. And although the simplest concept of group behaviour by acoustical insects is the 'chorus', our views of this phenomenon have required considerable readjustment in the light of the group versus individual controversy of the late 1960s (Alexander, 1975). Hence, where swarming results in mating, evolutionary theory would predict there to be some advantage to each individual in remaining within the group. The logic of this is simply set out by Lloyd (1979b), who describes the theoretical scenario of non-adaptive altruism in fireflies.

Male and female fireflies congregate in a single tree, producing a display of synchronous flashing that acts as a beacon to other fireflies. If there is a gene for aggregating but not flashing, a form of energy cheating, the non-flashing males could devote their time to mating rather than flashing. These males would provide more genes in the following generation, and the gene for flashing would eventually become extinct. Since this has not occurred, Lloyd concludes that there must be a competitive advantage to males in flashing over not flashing, and although the reasons for this advantage may not be immediately apparent, it is highly unlikely that group activity of this nature will be based entirely on a gene for cooperation!

3.1.2 Flexibility and generalizations

It is not only the call that is under selection from different quarters to enhance reproductive success, but also the associated strategies of when to call and for how long, and for the searching insect, the most propitious moment to begin the search. When examining the behaviour of individuals in nature we are left with a complex array of strategies. Although Maynard Smith (1976) was the first to use the term 'evolutionary stable strategy', Dawkins (1979) popularized the notion of a mixture of behavioural strategies simultaneously occurring within a population: how each evolved and how each could be maintained. He considered a strategy to be genetically determined and judged only in relation to the natural selection of alternatives. He used terms such as 'pure', 'mixed' or 'conditional', and although such a classification provided a useful base for theoretical argument, animals seldom obey the rules (Cade, 1980). Alcock (1979) recognized flexibility in 'male' behaviour as falling between a range from genetically fixed behavioural polymorphisms to a graded flexibility within an individual. For example, genetically fixed polymorphisms may exist where a horned scarab beetle follows a particular mating pattern in a dung pad, and its small and hornless conspecific follows an entirely different mating strategy. Even here each morph may exhibit a range of behaviours dependent on the quantity of the dung and the degree of male competition (Cook, 1988; Ridsdill-Smith, Chapter 10, this volume). Field crickets adopt an aggressive role within male aggregations until densities reach a level where alternative strategies of female interception may become more successful (Cade, 1981; Evans, 1983).

3.1.3 Calling for a mate—time and distance

Time between emergence as an adult to the first attempts at copulation is an opportunity to gain valuable food, but with each day there will be a decrease in the likelihood of mating due to mortality caused by accident, disease and predation. Similarly, the distance over which each sex must move to locate a mate will again decrease the chances of survival and reduce the opportunity of a successful mating. Reducing both time and distance reaches one extreme in protandrous male hymenopteran parasitoids; they remain on their emergence site with the expectation of encountering a female from the same host—the risks to the male seem minimal and to the emerging female near zero. At the other end of the spectrum, a male noctuid moth may fly over several kilometres in response to a female pheromone trail, risking capture by bats, or in a rare though spectacular fashion by the deceit of bolas spiders (Eberhard, 1977).

Justification for the 'expense' involved in finding a mate will be the number of eggs a male is able to fertilize, or the number of fertilized eggs the female can deposit in a way that maximizes their survival. Expense will be the metabolic energy diverted from egg production to mate location, risks of damage and predation, as well as the effort in searching for a suitable oviposition site. For the male, although sperm may be cheap, there will be energetic costs in competitively advertising to a limited number of receptive females and this advertising not only attracts the searching partner, but also competing conspecifics, and in some cases predators. Thus, metabolic costs and predation risks will rise with any protraction of the time between emergence and mating, and will increase with the physical distance over which both sexes have to move before they meet and the female can oviposit.

3.1.4 Investment and costs—who calls whom?

There is presumably a cost to mating (broadly seen as the entire process, or specifically as the behaviour of physically pairing (McCauley and Lawson, 1986; see Gwynne, 1989 for some recent comment), and in a chapter on mate location, it is necessary to at least consider the context of the single item of location. Daly (1978) suggested six primary costs of mating: (1) those energetic costs of the sexual mechanisms (structure building); (2) sexual behaviour; (3) escape from unwanted sexual attention; (4) predation; (5) disease transmission; and (6) injury inflicted by the male. Of these, predation risks provided the weakest data set, and Gwynne (1989) suggests from three studies that the act of mating indicated no additional risks to the participants: perhaps all the risk is in getting there.

Given that calling for mates is costly, there are a number of generalizations across insect taxa in respect of calling strategies that may provide leads as to the manner in which selection operates on the caller and receiver. In other words, given the conditions under which selection may operate (reviewed by Alcock and Gwynne, Chapter 2, this volume) we may make certain predictions as to who should do the calling and who the searching.

Darwin (1874), in tune with the attitudes of the time, quoted the Greek poet Xenarchus' rejoicings in the sexual differences of cicadas, with the remark that 'Happy the Cicadas live, since they all have voiceless wives'. An observation that is not entirely true since duetting occurs in a number of species! Despite this sexist remark, there do appear to be established patterns of sexual differences in calling behaviour that require some explanation. For example, why do most females use pheromones to attract their mates (Jacobson, 1972) and most males visual or acoustic cues (Otte, 1977)? Why should these rules be changed when the mating system shifts from high maternal investment to high male investment (Thornhill, 1979)?

As a consequence of these sexual differences, Thornhill and Alcock (1983) observe that the fixation on sexual roles may be mirrored in structures associated with signal detection and searching. Where there is wing dimorphism, males have better developed wings than females, implying that the female is the calling sex and the male does the searching. And where there is antennal dimorphism, the male antennae are more plumose than females, where again the male has the job of locating the calling female. Responses of the social signal to the effects of predation (natural or non-sexual selection (Endler, 1986)) will inevitably reflect the investment in the offspring by each sex. We may predict that the high-investment sex will tend to be less exposed than its lower-investment counterpart, and this sex (usually the female) will take fewer risks. Where males provide a degree of parental care or a nuptial gift to the female, a reversal of a male calling role may be evident (Thornhill, 1979). Are the risks from predation really more severe for visual and acoustic signallers compared to those using pheromones? Is there really such a distinction between pheromone and acoustic signals in terms of energetic costs? Can environmental changes induce a role reversal within a single system?

Greenfield (1981) suggests that predation based on pheromonal cues is unlikely to evolve as most sexual signals using chemicals are highly species specific, and most predators tend to be generalists. Thus, it would be a high risk for a predator to evolve a suite of specialist recognition sites to track down one species' odour. The exceptions to this rule should be those insect parasitoids that have evolved systems of host attraction through kairomones (see Wellings, Chapter 4, this volume), but the number of cases of sex pheromone attraction is limited (Vinson, 1985). A simple explanation of this may be that in most cases the host female is already too old for the parasitoid to complete its development: the majority are egg or larval parasites.

3.1.5 Reversal of signalling roles

Role reversal, where the male becomes the limiting sex and females compete for an opportunity to mate, has been demonstrated in a number of 'specialized' mating systems. For example, female Mormon crickets, during times of high density and short supplies of food, will compete for the opportunity to mate with males in order to acquire its protein-rich spermatophylax. Here the male hardly needs to call to attract a mate (Gwynne, 1984). Male bitticids (Mecoptera) produce a pheromone to attract the female, but as with the Mormon cricket it is the male that has a high investment, perhaps in excess of that provided by the female (Thornhill and Gwynne, 1986). More commonly it is the female that is the limiting sex, and so where searching becomes more hazardous than calling females may take on a signalling role, leaving the searching to males. In these cases males assume both risks of calling and searching and so patrolling males will call intermittently from different perches with the expectation of an answer from the female. If there is an appropriate answer, he will continue to call moving in on the responding female. The female for her part may remain stationary, answering only the calls of close conspecifics. A duet of this nature may be a halfway house to a full reversal of calling roles. Such a system is illustrated in many male phaneropterine bushcrickets where the male will call, evoking an acoustic answer in the female to which he will orient (Robinson, 1990). A similar reversal appears in cicadas (Gwynne, 1987), Homoptera (Claridge, 1987) and Neuroptera (Henry, 1985). And using another sensory system, patrolling male fireflies induce a flash response in the females to which they are attracted (Lloyd, 1979a).

In the event that role reversal does not take place, selection will act on males to reduce their conspicuousness, yet at the same time maintain a competitive advantage over their neighbours; however, they still have to produce an effective call for female location. Swords are usually two-edged! Hence, it may pay males to alter their calling strategies in order to reduce the risk of detection, and one common way out of this dilemma is to call from within a group of males.

3.2 THE ACOUSTIC SIGNAL

Darwin (1874) considered the songs of insects as being produced like many other seemingly exaggerated structures that are used by females to assess males. Indeed, the temporal structures of many insect songs show remarkable ornateness (Fig. 3.1) that is hard to justify in terms of reproductive isolation or signalling efficiency. This complexity may be explained by a parallel argument to that of Eberhard (1985), who suggested that overly complex genitalia have been arrived at, not as mechanisms enhancing reproductive isolation (lock and key), but through sexual selection. Hence one view of the complexity of mating signals is that they

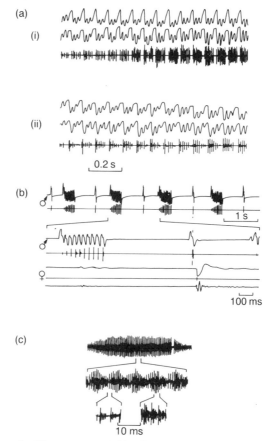

Fig. 3.1 A range of calling songs. (a) *Chorthippus biguttulus* (Acrididae) The upper two traces show the movements of the two legs and the lower trace the sound pattern of (i) the male and (ii) the female (after Helversen and Helversen, 1983). (b) Male calling song and female response of *Ancistrura nigrovittata* (Tettigoniidae). The upper pair of traces showing the wing movement (upper) and sound (lower) of a solitary male. The lower expanded trace shows the male's two-part song with the female's response following the short click section of the song (after Josephson and Halversen, 1986). (c) The complex song pattern of the cicada (*Psaltoda harrisii*). The time mark of the lower trace is 10 ms. After Young and Josephson (1983).

are embellishments of simple courtship patterns inducing a supernormal response in the receptors (West-Eberhard, 1984); females choose them because they are more intense and perhaps easier to locate. Of course, equally plausible is that the apparent complexity is merely an effect of the mechanics of signal production. This is certainly hard to explain when one examines the strange wing movements producing sound in some phaneropterine bushcrickets (Heller, 1990) (Figure 3.1). Where signals are repetitive, receptor systems may accommodate the stimuli and hence novelty within the signal may bring the attention of the receptor back into focus (West-Eberhard, 1984). I develop these ideas to the point that selection may have taken hold of receptor pathways evolved primarily for predator detection as attention-grabbers within the social signal.

When confronted with a complex array of species distinct signals, the most obvious place to start is with the evolution of call structure through processes of speciation. In fact I relegate this to the 'back-page', not that it is unimportant, but rather that it has so dominated our thinking that we seem almost surprised when other modifiers are suggested. Attempts to redress the balance have been slow to emerge (but see West-Eberhard, 1983, 1984). One reason is that it is difficult to design experiments that effectively separate the components of sexual selection (see Searcy and Anderssen, 1986), male competition or intrasexual selection and epigamic selection (Huxley, 1938; Thornhill and Alcock, 1983).

In summary, selection on the call should:

1. reduce the vulnerability of the caller to predation;
2. optimize the energy consumed in signalling;
3. optimize the time of calling;
4. optimize processes that improve the signal's localization;
5. optimize those processes that improve the resolving power of male conflicts;
6. optimize any differential preference of the attracted female; and
7. ensure recognition of the caller by conspecifics.

3.2.1 Reducing predation risks

As the male has to find the female, he requires organs of sense and locomotion, but if these organs are necessary for the other purposes of life, as is generally the case, they will have been developed through natural selection.

(Darwin, 1874)

In this section, which forms the central part of my thesis, I trace three evolutionary scenarios in orthopteroid (*sensu latro*) communication— pheromones to substrate vibration (cockroaches to crickets), vision to sound (acridids), and the separation of ultrasonic bat cries from the calls of conspecifics (crickets and tettigoniids). I suggest that coyness on the part of a male or female may have its roots in predator avoidance.

The acoustic signal as a sex attractant is incidentally available to predators, and because of this we may predict that predators will have a strong influence on signal structure (Morris, 1980; Morris and Beier, 1982; see review by Burk, 1988). Visually obvious swarms of male nematoceran flies are not only attractive to the females, but also to a range of insect and

Table 3.1 Examples of predators using acoustic signals to track their insect prey.

Prey	Predator	Reference
Crickets	Cat	Walker (1964)
	Herons	Bell (1979)
	Tachiniid Flies	Cade (1975)
	Birds	Rost and Honneger (1987)
Bushcrickets	Cat	Walker (1964)
	Bats	Belwood and Morris (1987), Tuttle *et al.* (1985), Belwood (1990), Buchler and Childs (1981)
	Tachiniid flies	Burk (1982), Bailey and McCrae (1978), W. J. Bailey (unpublished data)
	Geckos	Gwynne and Bailey (1988), Sakaluk and Belwood (1984), Nutting (1953)
	Birds	Bailey and McCrae (1978)
Orthoptera	Tachiniid flies	Sabrosky (1953)
Cicadas	Birds	Simmons *et al.* (1970), Nigamine and Ito (1980)
	Bats	Bornaccorso (1979)
	Sarcophagid flies	Soper *et al.* (1976)
	Spiders (indirect)	Gwynne (1987)

avian predators that gather at dusk at marker points for swarm formation (Downes, 1970). Thus, selection for those components of the signal that are effective in mate attraction will be balanced by the requirement to avoid predators. Despite Belwood and Morris' (1987) convincing argument that calling male neotropical tettigoniids are under selection from predators, and the elegant example of prey-specific bats locating tettigoniids by their call (Tuttle *et al.*, 1985; Belwood, 1990), there are few well-documented cases of prey location on acoustic cues (Table 3.1). Perhaps the most convincing and near-natural experiment was by Bell (1979), who described the behaviour of herons locating ground-dwelling crickets. Herons were observed feeding in late evening on calling crickets, and the same herons were attracted to speakers producing cricket calls. The majority of cases refer to the attraction of parasitic tachinids to calling crickets where the fly lays its larvae on or close to the male. The parasite, once feeding within the body of the host, will reduce the effective calling time of a male from 2–3 months to 1–2 weeks (Burk, 1982).

Various authors have noted the reduction in calling by male bush-crickets in neotropical forests (Rentz, 1975; Morris, 1980; Morris and Beier, 1982; Morris *et al.*, 1989; Belwood, 1990), but here it is as a strategy to reduce the caller's exposure to the remarkably high predation by foliage gleaning bats rather than as a consequence of parasitism. It may also be that signalling in humid dense foliage is energetically expensive (Griffin, 1971): the transmission of high frequencies in such a vapour laden habitat could be counter-productive.

(a) Pheromones, wing flicks and substrate vibration

In all insects, the mechanoreceptors are available for both social and non-social signals. For example in cockroaches and crickets the anal cerci, although highly adept at perceiving air movements created by predators (Camhi and Tom, 1978), may also detect the modulation patterns produced by the insect's own wing movements (Dambach *et al.*, 1983; Dambach and Rausche, 1985), and undoubtedly the patterned motion of the wings of other insects calling nearby (Kamper and Dambach, 1981). In some phalangopsid crickets, the wings are used in silent flicks, producing pulsed patterns of air that have a clearly social function (Kamper and Dambach, 1981; Boake and Capranica, 1982; Boake, 1983; Heinzel and Dambach 1987). The cerci in this case will be interpreting a social signal rather than one produced by a predator.

Similar ambivalence will occur where substrate-borne vibrations are detected by the sensitive chordotonal receptors known as subgenual organs situated beneath the knees of an insect. These organs are present on the legs of all insects, and therefore like the equally ubiquitous mechanoreceptive hairs, would have a primary function of detecting the presence of other intruding organisms. As with the specialized reception of social signals by cercal hairs, these receptors also have a demonstrated role in social signalling in a wide variety of groups (e.g. Orthoptera (Otte, 1974; Morris, 1980), Neuroptera (Henry, 1985), and Hemiptera (Cokl, 1983; Claridge, 1987)). In fact Kalmring and Kuhne (1980) were able to show that input from the subgenual organ augmented that of the airborne signal, amplifying the song's temporal pattern in the insects nervous system. An insect using this channel for the recognition and location of its mate must differentiate between environmental and social noises. Pattern recognition must therefore, under certain conditions, override one of the insect's primary prey-detection systems. A diagram illustrating the conflict between predator avoidance and male attraction in the ensiferan Orthoptera is given in Fig. 3.2.

(b) Shift from visual to acoustic signals

For diurnal creatures the most developed recognition system is vision. However, vision is the least species-specific sensory modality of all: the same light frequencies are available to nearly all animals (Endler, 1986). By contrast, the acoustic signal requires highly specialized receptors for the entire range of the auditory ranges—there are no general purpose

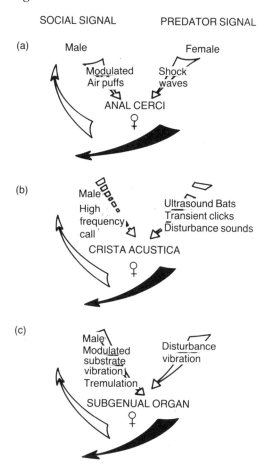

SOCIAL SIGNAL   PREDATOR SIGNAL

(a)   Male   Female

Modulated   Shock
Air puffs   waves

ANAL CERCI
♀

(b)   Male

High   Ultrasound Bats
frequency   Transient clicks
call   Disturbance sounds

CRISTA ACUSTICA
♀

(c)   Male   Disturbance
Modulated   vibration
substrate
vibration
Tremulation

SUBGENUAL ORGAN
♀

Fig. 3.2 A schematic diagram illustrating the conflict between predator avoidance and male attraction in the ensiferan Orthoptera. (a) Low frequency air movements detected by the anal cerci evoking the avoidance response from predators and attraction when modulated at the species pattern in phalangopsid crickets. (b) The detection of high frequency airborne sound from a calling male (attraction) or from bats or the disturbance sounds of approaching predators (avoidance). (c) The response to substrate vibration detected by the subgenual organ. Attraction to modulated tremulation occurs in certain tettigoniids and an avoidance reaction follows substrate disturbance by an approaching predator.

broad-band receptors. Above all, mate finding may take place under the blanket of night when visual signals, except where luminescence is used, no longer function.

Otte (1977: after Alexander, 1962) considers that the tympanum of acridids is monophyletic, and argues that its function may not have been primarily for the detection of social signals in that not all families use acoustic signals in courtship and yet all but a few possess tympana. Further, the auditory centres within the thoracic ganglia show remarkable constancy between acoustic and non-acoustic species (Riede and Kamper, (*pers. comm.*)). Then how did the hearing system acquire a social function? The scenario was set by Otte (1977) and can be summarized in the following way. Diurnal acridids use patterned visual signals in mate recognition (Otte, 1972; Riede, 1987), and where these patterned signals were supplemented by sound, the signal could be effective over a longer range and in complex vegetation.

Unlike the tympana, the methods of sound production are polyphyletic, as evidenced by the range of body parts used in both visual and acoustic signalling; the legs, wings, head, antennae and abdomen. In some taxa, and by no means universally (Riede, 1987), pattern recognition through the auditory route must have taken over from visual signalling with sound becoming the principal distant cue. This switch has evolved to a level in certain gomphocerines where individuals may recognize the highly complex song structure of their own species (Fig. 3.1) (e.g. Perdeck, 1958; Helversen and Helversen, 1983), and yet in others there is no acoustic signal at all. Riede (1987) points out that only four subfamilies of South American grasshoppers produce sound; the majority are silent. Could the acoustic signal have evolved under relaxed selection of predation in one part of the nearctic and not in the neotropics?

(c) Differentiation of ultrasound

One group where social communication is dominated by sound is the ensiferan Orthoptera, perhaps because they were derived from a primitively nocturnal ancestor (long legs and long antennae). Certainly the mechanism for social communication by ensiferan males, unlike the acridids, appears to be monophyletic, as does the sound-receiving system—the tympana under the knees of the forelegs in all ensiferan groups (Alexander, 1962; Elsner, 1983; Bailey, 1990), and the uniformly present tegmenal stridulation. As further support of the association between sound production and reception, the two appear to be genetically coupled (see below).

Crickets, with their shrill calls at around 3–5 kHz, are able to detect and respond behaviourally to the ultrasound of bats (Popov and Shuvalov, 1977; Moiseff *et al.*, 1978). And like many night-flying insects they show avoidance behaviour when they hear the bat's cries. There has been no comparable work on night-flying tettigoniids, although they are able to hear much higher frequencies than crickets can. Strangely, unlike crickets and despite the fact that there are areas within the auditory neuropile that are dedicated to high and low frequencies (Römer, 1983), there is no clear behavioural separation of high and low frequencies showing positive and negative movements to the sound source. It could be argued that the tettigoniid system at one stage was evolved for predator detection and later took on the function of a social receptor (Bailey, 1991). Regardless of its origin or fate, a receptor system covering such a wide frequency spectrum as that of the tettigoniids suffers in the same way as the visual receptor—it is non-specialized. And, like the cruder hair receptors on the anal cerci of crickets and cockroaches, the animal must face the ambivalence of fleeing in the face of predation or being attracted to the high-pitched calls of a mate. The effect (Williams, 1966) may be coyness —a behaviour more normally considered as an evolved consequence of female choice.

(d) Coyness in responding to mates

If social signalling evolved around receptors used primarily for predator detection, this may leave us with a proximate basis for coyness—defined as a reluctance to accept the advances of the partner. Coyness is usually a feature of female choice, the vacillation between two or more courting males. In evolutionary terms, the mechanism inducing coyness is irrelevant as long as the outcome is for females to exhibit reluctance for a long enough period for there to be ambivalence; there is therefore opportunity to choose, but the observed coyness may be a reflection of a neural pathway that was once used in predator detection.

If the same communication channel is used for both behaviours then there should be a neural switch at some point in the circuit, perhaps similar to that proposed between locomotion and stridulation in bifunctional wing muscles of acridids (Elsner, 1983), the grasshopper uses the same muscles for its complex stridulation at one moment and then moments later for an entirely different pattern in locomotion.

As the female approaches the male, she must remain aware of predators until she is at a sufficient distance to forego caution and accept the signal as of social value. If the neural circuits are indeed separate we may expect there to be a time difference between predator avoidance and mate attraction, and in fact there is. The latency (the time between stimulus and effect) for bat avoidance is significantly shorter than for mate attraction (Moiseff *et al.*, 1978; Moiseff and Hoy, 1983; Doherty and Hoy, 1985).

Unlike crickets and bushcrickets, the sound receptor system of moths appears to be almost entirely dedicated to predator detection (Roeder, 1966), and where species use sound in a social context, extreme ambivalence should be evident. The discoveries of Lepidoptera using sound for mate attraction seems to increase yearly (Spangler, 1988), and given this conflict of interest moths should either call during the day or under conditions that are less exposed to predation by bats. The agaristid moths fulfil one of these requirements in switching to diurnal calling (Bailey,

Fig. 3.3 The calling of the moth *Symmoracma*. This pyralid moth has highly modified genitalia which are rubbed against the ventral part of the abdomen. Song frequencies are close to 50 kHz. (Photograph by W. J. Bailey.)

1978; Alcock *et al.*, 1989). The noctuid *Heliothis zea* produces a series of wing clicks that are reputed to have a function in forming male assemblies (Kay, 1969), and again this behaviour may be distinct from that shown when exposure to bat predation is highest. Gallerine moths such as *Achroia grisella* and *Corcyra cephalonica* call from within the beehive on the honeycomb (Spangler, 1984), and *Galleria mellonella* only from within an enclosure used by cavity-nesting bees (Spangler, 1986), again away from the predatory signals of bats. A recently described ultrasonic pyralid goes against this prediction in that it calls from the exposed tops of trees and bushes in xeric woodlands of southwestern Australia (Gwynne and Edwards, 1986); however an Old World tropical forest species, *Symmoracma minoralis* from southern Thailand, calls from the under-surface of leaves within 1.5 m of the floor of forest clearings (W. J. Bailey, unpublished data) (Fig. 3.3).

3.2.2 Energy consumption

Although calling in insects is about as energetically expensive as walking and flying, compared with other forms of social signalling it is an expensive way to attract mates (Heath and Josephson, 1970; Stevens and

Josephson, 1977). However there appears to be no real evidence that this in any way affects calling time or duration (K. N. Prestwich, personal communication). Hence, in terms of selection on the calling time, insufficient energy reserves may be only a weak selective factor compared to others such as sexual selection and predator avoidance. The maximum observed increase in energy consumption over the resting state was that for a copiphorine tettigoniid, *Neoconocephalus robustus*, in which energy consumption was increased 14 times (Heath and Josephson, 1970). Prestwich and Walker (1981) observed a ten-fold increase over resting levels in the cricket *Anurogryllus* (Table 3.2). Walker (1983a) suggested that species with a high calling rate should have a shorter calling duration, and although this may be the case for some crickets, it does not appear to be a consistent feature of other groups. Conversion of muscle energy to acoustic power is highly inefficient at close to 3%, and given such a low conversion, selection is unlikely to have operated towards a greater muscle mass for the production of more intense signals.

Table 3.2 Metabolic costs of calling*.

Insect	Metabolic increase (Calling/resting rate based on O_2 consumed)	Reference
Crickets		
Anurogryllus arboreus	10.0–15.8	Prestwich and Walker (1981)
Oecanthus celerinictus	6.25–12.0	Prestwich and Walker (1981)
Oecanthus quadripunctatus	6.5–8.0	Prestwich and Walker (1981)
Teleogryllus commodus	3.9	Kavanagh (1987)
Gryllotalpa australis	13.4	Kavanagh (1987)
Bushcrickets		
Euconocephalus nasutus	14.2	Stevens and Josephson (1977)
Neoconocephalus robustus	15.3	Stevens and Josephson (1977)
Cicada		
Cystosoma saundersii	18.4	MacNally and Young (1981)
Frogs		
Physalaemus pustulosus	2.1–4.3	Ryan (1985)
Hyla versicolor	5–22	Taigon and Wells (1985)

*In part after Burk (1988).

3.2.3 Calling time

Males should call most when it is reproductively profitable, and a factor influencing this will be female availability (Walker, 1983a). It is, however, debatable where selection is occurring in that male behaviour may drive the female to respond when males are most active, or alternatively, females with their higher investment may be selecting on male singing activity. If movement attracts a predator, females may prefer to take this risk under the cover of night, and where body temperature affects the energy used in locomotion, it may also pay females to move early in the evening rather than later while ground and air temperatures are still high. In addition, selection may operate on females to be available for the 'best' males that are most able to hold their territory early rather than later in the evening. Late-calling males may have already mated, or be those that were unable to hold a territory during the peak of male activity.

A similar series of arguments may apply to the male, notably advantages through lower metabolic requirements, and sexual selection through female choice and male competition. Hence, if males are required to increase muscle temperature before calling commences (Josephson and Helverson, 1971; Heller, 1987), calling at dusk before air and body temperatures fall may allow stridulation to take place without a period of warm-up. As the evening progresses, air temperatures may fall to levels that make singing too costly (Walker, 1983a). Some species will even change their diel pattern of calling as seasonal ambient temperatures fall (Nielsen and Dreisig, 1970). Early-calling males have the opportunity to dominate and outcompete late risers, and if virgin females reach sexual maturity during the previous 24 hours in a random manner throughout the diel cycle, there will be a greater probability that more females will be responsive to males in the early evening, increasing selection on early singers (Walker, 1983a). Many nocturnally active species of tettigoniid call for much of the day while they establish territories, and mating only occurs after sunset (e.g. *Requena verticalis* and *Metaballus* spp. unpublished data; Gwynne, 1985). Although using non-acoustic cues, the costs of searching earlier than one's competitors has a parallel in the emergence of forest tent caterpillar moths (*Malacosoma disstria*), where mate competition forces males to search for females emerging from their cocoons during daylight hours, resulting in the capture of moths by birds (D. N. Bieman; MSc Thesis, Michigan State University, cited by Burk, 1982).

Calling at dusk coincides with the peak of predator activity—air temperatures are still high and metabolic costs for both poikilotherms and small bats will be lower. Data on these risks are scant although Belwood (in press) has described a species of neotropical tettigoniid that calls for a brief period before dawn. She correlates this with bat activity, which is at its lowest during this time. For most, however, the risks may be worth it.

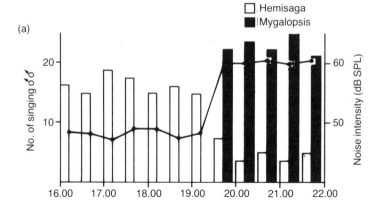

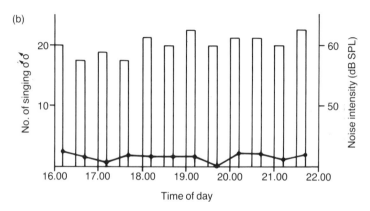

Fig. 3.4 Heterospecific masking in the Tettigoniidae. (a) The number of singing male *Hemisaga denticulata* at the site where *Mygalopsis marki* commences calling at dusk (20.00 h). (b) The number of calling male *H. denticulata* between 16.00 h and 22.00 h at a nearby population where *M. marki* was absent. After Römer *et al.* (1989).

Finally, species with similar calling times may indirectly compete for calling sites, which results in time-sharing. Bailey and Morris (1986) describe the possibility of one syntopic Canadian conocephaline masking the call of a second. Greenfield (1988) showed how the call of *Neoconocephalus spiza* inhibited the call of three congenerics, and a similar situation exists between quite different subfamilies where the short call of *Hemisaga dendiculata* is inhibited by the almost continuous call of *Mygalopsis marki* (Fig. 3.4). Removal of *M. marki* allowed *H. denticulata* to recommence singing, and in habitats where there are no *M. marki*, *H. denticulata* called throughout both day and night (Römer *et al.*, 1989).

3.2.4 Improving signal localization

The essential element of any acoustic signal is that it must be heard and recognized. Being heard relates to both the intensity of the signal and to the threshold of the receptors, whereas recognition may involve a temporal pattern within a fixed frequency range. But small animals have problems in determining the direction of sound when the wavelength is greater than the distance between the ears, and one solution to this is to broadcast at high frequencies where the wavelength is short, and/or employ devices in the receptor system that increase binaural differences; effectively increasing the 'distance' between the ears (Lewis, 1983). Calling with high frequency signals is not simple, as high frequency sounds require far more energy to propagate than low frequencies over an equivalent distance. More importantly, the same high frequencies are more easily lost in the environment through absorption and scattering (Wiley and Richards, 1978). And thus many insects appear to trade off the efficiency of sound production for the requirements of both auditory acuity and signal loss.

The loss of signal clarity in terms of its temporal character has been shown to decrease female preference in *Gryllus bimaculatus* (Simmons, 1988b). Pulse structure affects the manner in which both the acridid and tettigoniid hearing system responds to syllable patterns (Roessler *et al.*, 1989). Further, males producing high frequencies facilitate location of the sound source (Rheinlaender and Römer, 1980; Latimer and Lewis, 1986; Bailey and Yeoh, 1988). The ease with which sound is localized may be of greater relevance to a female deciding between two or more calling males than any other character of the song. For example, the female bushcricket *Requena verticalis* prefers males whose songs contain high frequencies: her path toward the sound source is far more direct than her path toward an alternative sound in which the high frequency part of the song is suppressed. Rheinlaender and Römer (1990) provide a review of mate location in the Orthoptera and illustrate from the point of view of the receptor how frequency structure is an essential component of auditory acuity in both ensiferan and caeliferan species.

Finally, calling posture and perch position affect the loudness of the sound field. Where an insect is producing sounds with its wings, as in crickets and bushcrickets, sound levels at the ventral surface may be up to 6 dB lower compared to those from the dorsum (Bailey, 1985). Males may select vantage sites within the habitat from which to project their call; thus bladder cicadas call from the sides of small rather than large bushes (Doolan and MacNally, 1981), short-tailed crickets call from 1 m up the trunk of a tree (Paul and Walker, 1979), and in the bushcricket *Mygalopsis marki* males will compete within the canopy of low shrubs for perch advantage (Bailey, 1985; Dadour and Bailey, 1985).

3.2.5 Aggregations and male conflict

Male insects frequently call from within aggregations (Alexander, 1975; for a review see Greenfield and Shaw, 1983). The advantages of this behaviour have variously been considered as: (1) providing protection (Simmons *et al.*, 1970; Cade, 1979); (2) producing a larger or louder signal source (this is an equivocal argument for acoustic insects, but see Forrest (1983) and Morris and Fullard (1983)); (3) requiring limited calling sites within a patchy habitat; or (4) allowing males to clump near 'hot spots' through which females may pass (Greenfield and Shaw, 1983; Feaver, 1983; Shelly *et al.*, 1987; Gwynne and Bailey, 1988). These are male-initiated reasons for aggregating, whereas Bradbury and Gibson (1983) add to the list female-initiated reasons such as: (5) reduction of predation; (6) preference of females for clumped males in less desirable habitats, thereby reducing the competition between the sexes for resources; and (7) opportunity for female choice. Examples of acoustic insects that clearly support these hypotheses are minimal, and in some cases entirely lacking.

One expectation is that males calling within an aggregation should gain more mates than males calling alone. The male advantage of calling in aggregations in terms of the expected number of copulations appears less than for a male calling alone. For example, chorusing bladder cicadas can only attract 0.61 females per calling male (Doolan and MacNally, 1981), and the per capita male success of *Anurogryllus arboreus* was no better than for solo males (0.42 versus 0.45) (Walker, 1983b). Forrest (1983) estimates that a male mole cricket would have to call on average for 4.2 nights (*Scapteriscus acletus*: average number of females per male 0.79) or 6.5 nights (*S. vicinus*: average number of females per male 0.52) to achieve a mating. This is more than compensated by the observed preference of females for groups rather than for solitary males under field conditions (Doolan and McNally, 1981; Walker, 1983b). Cade (1981) was able to show that *Gryllus integer* females were more attracted to groups of speakers than to solitary speakers. Under arena conditions, *Conocephalus nigropleurum* prefers the calls of a group of males to that of a single male (Morris and Fullard, 1983).

This observed preference may mean that females do prefer groups rather than individuals. However, the alternative explanation is that 'passive attraction' merely leads the female to the most 'intense' (in many cases the most continuous) signal (Parker, 1983). The curious feature of the experiments in which females are attracted to the unstructured noise of a chorus is that females no longer appear to require a coherent species-specific song, and indeed random noise is even preferred to the call of a single male (Morris and Fullard, 1983). The loss of signal structure through interference from either the same or different species calling around the listening insect is unequivocal (Römer and Bailey, 1986; Römer *et al.*, 1989), and this interference affects the ease with which females may find a single male (Bailey and Morris, 1986). Hence, group calling still poses numerous theoretical problems, particularly in terms of female advantage. The phenomenon of masking within the aggregation may be viewed from the receptor level, but direct observation of searching females within aggregations appears to belie the effects seen in the nervous system; she is still able to locate a specific calling male.

3.3 INDIVIDUAL VARIATION

3.3.1 Improving female choice

Where there is a difference in parental investment, resulting in more males available for mating than females, patterns that enhance male competitive status in either attracting a female, or retaining a female required and limited resource, are likely to be subject to intense selection. Where these components are identifiable, they may be independent of the species-specific call pattern, and although females will be attracted to those parameters of the song closest to the mean of the population (e.g. *Chorthippus brunneus*; Butlin *et al.*, 1985), variation in the way the basic species' signal is produced may convey to a searching female information relating to the quality of the male.

If selection on any one of these 'better-male' attributes provides advantage to the caller, then these traits will become rapidly fixed in the population, there will be a concomitant reduction in variance. However, where variation is provided in an arbitrary manner, as for example through larval nutrition, then traits related to the size of the signalling adult are unlikely to be fixed, but traits in the receiver that take advantage of male size may become fixed in the searching sex: females may always choose larger males in a population. Where the female's choice appears to be governed by direct benefit, such as a nuptial gift, two competing strategies may be operable. Where adult nutrition is low and densities are high, males may be reluctant to give females their hard won protein resource (e.g. the spermatophylax of many tettigoniids) and a role-reversed situation may develop with the male becoming the choosey sex (see above). However, where female density is low and males are competing for females, this protein source may be sufficient to increase female choice (for a discussion of role reversal, see Thornhill and Gwynne, 1986). Given this possibility it is tempting to predict that females would use size-related acoustic cues to locate males with large nuptial gifts: there is a good correlation between male size and spermatophylax size (Gwynne *et al.*, 1984) and lower carrier frequencies or louder songs are produced by larger males (Latimer, 1981).

There are a number cases suggesting that females are able to differentiate between song parameters that vary outside the pattern-specific template. Thus ensiferan females will choose males with higher calling rates (Hedrick, 1986), signals of dominant males (Crankshaw, 1979), more intense calls (Forrest, 1983; Bailey, 1985; Bailey and Yeoh, 1988), those with emphasized lower frequencies (Latimer and Sippel, 1986), or with emphasized high frequency components (Latimer and Sippel, 1986; Gwynne and Bailey, 1988; Bailey and Yeoh, 1988). I have already referred to females preferring a clear signal to one that indicates a more distant male (Simmons, 1988b), and also the calls of older males (Zuk, 1987). Finally there is indirect evidence that call structure may indicate heavier males (Gwynne, 1982).

Of the two song characters correlated to male size, frequency is the most reliable in that it is dependent on the structure of the sound-producing apparatus: longer violin strings will produce lower tones. Intensity, however, is compounded by posture and the density of the intervening vegetation. Gerhardt (1987) argues that, at least in the Anura, cues based on unique elements in the song such as frequency are unlikely to influence female choice in nature, since within aggregations these factors will be obscured by both the habitat and interference of calling neighbours. Although frequency discrimination has been demonstrated in two-speaker trials in a number of tettigoniid species, preference is always reversed by increasing the intensity of the non-preferred signal (Latimer and Sippel, 1986; Gwynne and Bailey, 1988; Bailey and Yeoh, 1988; Bailey *et al.*, 1990). A similar situation occurs in the creosote bushes of the Sonoran deserts, where again female *Ligurotettix conquilletti* select males on call intensity (M. D. Greenfield and T. E. Shelly, unpublished data).

3.3.2 The recognition of conspecifics

The species distinctive call allows like to recognize like and this paradigm, like no other, has remained central to proximate studies of acoustic signalling. It has not only formed the basis of experiments directed at unravelling problems of female attraction, but more significantly has been the founding rationale for most major studies of the neural control of both sound production and reception (e.g. Huber, 1962, 1988; Elsner and Popov, 1978; Doherty and Hoy, 1985). For example, the search for a neural template governing species-specific signal recognition has been at the centre of the neurophysiologist's stage for many decades (Boyan, 1984; Huber, 1985; Schildberger, 1985). It has a demonstrated role in terms of species recognition in a wide range of taxa (e.g. crickets: Walker, 1957; Zaretsky, 1972; bushcrickets: Bailey and Robinson, 1971; grasshoppers: Perdeck, 1958; Helversen and Helversen, 1983; planthoppers: Claridge,

1987; lacewings: Henry, 1985; fruit flies: Ewing and Bennet-Clark, 1968).

What are the creative forces producing these distinct and recognizable patterns? The calls of most species within a population are conservative in their structure and in terms of its temporal character, females prefer the population's mean value (e.g. Butlin *et al.*, 1985). Changes are presumed to occur when populations become separated and each is subject to different selective pressures. And thus, in the simplest of terms, we can imagine a population of calling insects at first isolated by some geological event from its parent species. The call of the insect in this novel environment will be influenced by: (1) the habitat's structure; (2) the bioacoustic environment including inter- and intra-specific interference (Ryan and Brenowitz, 1985; Littlejohn, 1988); (3) pleiotropy; and (4) mutation and finally the effects of sexual selection (West-Eberhard, 1983; Thornhill and Alcock, 1983, p. 412; Gwynne and Morris, 1986). (For a review of these processes in crickets see Otte, 1989.) Just what triggers off song changes during periods of isolation, or indeed what maintains the momentum of change, would appear to be a lottery: no single avenue has been given any demonstrable role. Although attractive theoretically, the influence of sexual selection through the Fisherian run-away model could lead to character extremes, but Otte (1989) is more cautious as to this 'unproven' process. However, if change through sexual selection were either the catalyst or its continuing drive, the process would inevitably be checked by physics and the price of predation.

Does change always take place when species are isolated, or more importantly do sister species consistently show song differences when they have returned to sympatry? Otte (1989) observes that sympatric species are 'virtually always' distinctive in their calling patterns, whereas Dadour and Johnson (1983) note that a species pair of tettigoniids on the coast of Western Australia have indistinguishable songs and are isolated through post-zygotic mechanisms: the F1 progeny are sterile. By contrast, Claridge (1987) has found the song of the brown planthopper to be highly labile, changing with each new 'bio-race'. If the song does change due to this array of selective pressures, or even by random mutation, then not only must the output of the song change in response to these pressures, but also the input: the receptor system and the recognition of pattern. The pattern generator and template receiver were considered to be genetically linked in crickets (see review by Doherty and Hoy, 1985). Contrary to these conclusions Otte (1989) points out that, as these systems are under polygenic and multichromosomal control (Hoy and Paul, 1973), it would be expected that hybridization would produce the intermediate call patterns and preference for hybrid siblings: the sender and receiver may also be under separate genetic control. This separation has been clearly established in grasshoppers (Bauer and Helversen, 1987), where undoubtedly the two systems have co-evolved.

3.4 INDIVIDUALS AND POPULATIONS—BEYOND SPECIATION

As the behavioural physiologist becomes more sensitive to within-population variation, so the experimental paradigm may shift. For example, Doherty (1985) suggests that at least two levels of response may occur to different song parameters of male *Gryllus bimaculatus*. Females searching for a male's call in a two-choice situation may 'trade off' attraction to one temporal character, say chirp duration, against a second character such as chirp period. In other words, by allowing the female cricket to select between two alternatives, quite different perceptive mechanisms are exposed that are not revealed by a single-choice paradigm. Elsner and Popov (1978) suggested a similar split in approach between a system geared in a stereotyped manner to species-specific parameters, and a second less tangible feature of graded responses that may be regarded as motivation. If the neuroethologist uses the logic that two 'motivational' levels are being applied within this system, then these may be switched under different experimental regimes. This is the insect's context of the experiment.

Selectionist's arguments may similarly propose different conditional strategies in the same female, or perhaps fixed polymorphic strategies in a population where one female will prefer one song type at a certain time to another. What happens if these strategies are also linked to locomotion and each strategy dictates a different response to the male's call? Thus, characters associated with, for example, chirp period in *G. bimaculatus*, may be attractive under one strategy and this may be associated with phonotaxis. Hence, at least theoretically, one could use females within a choice experiment that were pre-disposed to one extreme of a range of stimuli, and by overlooking females that are reluctant to run, miss a genotype that did not fit the experimental paradigm. Crickets show a pronounced change in phonotaxis when presented with visual and acoustic targets (Weber *et al.*, 1987; Stout *et al.*, 1987), and this may indicate that the acoustic cue is only part of a complex of inputs, each able to modulate the female's searching behaviour. Huber and Thornsen (1985) comment on individualist crickets running on a Kramer treadmill under highly controlled acoustic conditions—'one gets to know individual females' in the qualitative way they run to the target speaker. Butlin and Hewitt (1986) imply that variation in female response should be an integral part of experimental design. For example they were able to show significant variation in female phonoresponse of *Chorthippus parallelus* to the calls of males associated with female mating history and exposure to males.

Thus, once a female has located a male quite different stimuli will become important in its courtship behaviour: there is no longer a need for species recognition. For example, unmodulated white noise is attractive to female *Conocephalus nigropleurum* (Morris and Fullard, 1983), and whereas the fruit fly *Zapronius* uses a highly specific calling song, the close courtship call is similar to those used by related species (Lee, 1986). Again many cricket species use a contact courtship call that differs from the species-specific calling pattern (e.g. Harrison *et al.*, 1988).

As evolutionary biologists begin to propose wider hypotheses that relate to the individual's sensory perception, these too should filter down to the physiologist and thereby create even more demands on the ingenuity of the experimenter. However, the gap remains between physiologists and evolutionary biologists, and was aptly summarized back in 1985 by Franz Huber:

At present, neurobiologists are confronted with and suffer from a plethora of speculation and hypotheses formulated by behavioural ecologists, often without a thought for an experimental solution. On the other hand, what neurobiologists could gain from behavioural ecologists is to become open to the diversities in strategies and tactics animals have evolved, and to be confronted with field work and the comparative method.

Or in verse:

> Those who neglect the mechanistic approach
> and happily ride their philosophical coach
> may be reminded that the WHY of a thing
> will gain more with the HOW coming in

(one stanza from a two stanza poem: F. Huber, unpublished)

ACKNOWLEDGEMENTS

Acknowledgements are due to Tom Walker and Jim Lloyd, who suffered the early explorations of this chapter, and to Jackie Belwood, whose work on tropical forest tettigoniids inspired me to continue along the line of selection through predation. Franz Huber, while encouraging me to explore the interface between the proximate and evolutionary biologist, provided a moderating influence to my initial extremes. Thanks are also due to Mike Greenfield, Darryl Gwynne, Andrea Schatral and Dale Roberts and James Ridsdill-Smith. Any errors of interpretation are still my own.

References

Alcock, J. (1979) The evolution of intraspecific diversity in male reproductive strategies in some bees and wasps. In: *Sexual Selection and Reproductive Competition in Insects* (eds M. Blum and N. Blum), Academic Press, New York.

Alcock, J., Gwynne, D. T. and Dadour, I. R. (1989) Territoriality and mating in acoustically-signalling moths (Lepidoptera: Agaristidae). *Insect Behav.*, in press.

Alexander, R. D. (1962) Evolutionary change in cricket acoustical communication. *Evolution*, **16**, 443–467.

Alexander, R. D. (1975) Natural selection and specialized chorusing behaviour in acoustical insects. In *Insects, Science and Society* (ed. D. Pimentel), Academic Press, New York, 35–77.

Arak. A. (1983) Male–male competition and mate choice in anuran amphibians. In *Mate Choice* (ed. P. Bateson), Cambridge University Press, Cambridge, 181 –210.

Bailey, W. J. (1978) Resonant wing systems in the Australian whistling moth *Hecatesia* (Agaristidae, Lepidoptera). *Nature*, **272**, 444–446.

Bailey, W. J. (1985) Acoustic cues for female choice in bushcrickets (Tettigoniidae). In *Acoustic and Vibrational Communication in Insects* (eds K. Kalmring and N. Elsner), Paul Parey, Berlin, 101–110.

Bailey, W. J. (1990) The bushcricket ear. In *The Tettigoniidae: Behaviour, Systematics and Evolution* (eds W. J. Bailey and D. C. F. Rentz), Crawford House Press, Bathurst, 217–47.

Bailey, W. J. and McCrae, A. W. R. (1978). The general biology and phenology of swarming in the East African tettigoniid *Ruspolia differens* (Serville) (Orthoptera). *J. Nat. Hist.*, **12**, 259–288.

Bailey, W. J. and Morris, G. K. (1986) Confusion of phonotaxis by masking sounds in the bushcricket *Conocephalus brevipennis* (Tettigoniidae: Conocephalinae). *Ethology*, **73**, 19–28.

Bailey, W. J. and Robinson, D. (1971) Song as a possible isolating mechanism in the genus *Homorocoryphus* (Tettigonioidea, Orthoptera). *Anim. Behav.*, **19**, 390–397.

Bailey, W. J. and Yeoh, P. B. (1988) Female phonotaxis and frequency discrimination in the bushcricket *Requena vertiaclis* (Tettigoniidae: Listroscelidinae). *Physiol. Entomol.*, **13**, 363–372.

Bailey, W. J., Cunningham, R. and Lebel, L. (1990) Song power, spectral distribution and female phonotaxis in the Bush cricket *Requena verticalis* (Tettigoniidae: Orthoptera): Active female choice or passive attraction. *Anim. Behav.*, **40**, 33–42.

Bauer, M. and Helversen, O. von (1987) Separate localisation of sound recognising and sound producing neural mechanisms in a grasshopper. *J. Comp. Physiol.*, **161**, 95–101.

Bell, P. D. (1979) Acoustic attraction of herons by crickets. *J. N. Y. Entomol. Soc.*, **87**, 126–127.

Belwood J. (1990) Anti-predator defences and ecology of neotropical forest katydids, especially the Pseudophyllinae. In *The Tettigoniidae: Behaviour, Systematics and Evolution* (eds W. J. Bailey and D. C. F. Rentz), Crawford House Press, Bathurst, 8–26.

Belwood, J. and Morris, G. K. (1987) Bat predation and its influence on calling behaviour in Neotropical katydids. *Science*, **238**, 64–67.

Bennet-Clark, H. C. (1984) Insect hearing: acoustics and transduction. In *Insect Communication* (ed. T. Lewis), Academic Press, London, 49–82.

Boake, C. R. B. (1983) Mating systems and signals in crickets. In *Orthopteran Mating Systems. Sexual Competition in a Diverse Group of Insects* (eds D. T. Gwynne and G. K. Morris), Westview Press, Boulder, CO, 25–44.

Boake, C.R.B. and Capranica, R. R. (1982) Aggressive signal in 'courtship' chirps of a gregarious cricket. *Science*, **218**, 580–582.

Bornaccorso, F. J. (1979) Foraging and reproductive ecology in a panamanian bat community. *Bull. Fla. State Mus.*, **24**, 359–408.

Boyan, G.S. (1984) Neural mechanisms of auditory information processing by identified interneurons in Orthoptera. *J. Insect Physiol.*, **30**, 27–41.

Bradbury, J. W. and Gibson, R. M. (1983) Leks and mate choice. In *Mate Choice* (ed. P. Bateson), Cambridge University Press, Cambridge, 109–138.

Buchler, E. R. and Childs, S. B. (1981) Orientation to distant sounds by big brown bats (*Eptescus fuscus*). *Anim. Behav.*, **29**, 428–432.

Burk, T. (1982) Evolutionary significance of predation on sexually signalling males. *Fla. Entomol.*, **65**, 90–104.

Burk, T. (1988) Acoustic signalling, arms races and the costs of honest signalling. *Fla. Entomol.*, **71**, 400–409.

Butlin, R. K. and Hewitt, G. M. (1986) The response of female grasshoppers to male song. *Anim. Behav.*, **34**, 1896–1899.

Butlin, R. K., Hewitt, G. M. and Webb, S. F. (1985) Sexual selection for intermediate optimum in *Chorthippus brunneus* (Orthoptera: Acrididae). *Anim. Behav.*, **33**, 1282–1292.

Cade, W. H. (1975) Acoustically orienting parasitoids: fly phonotaxis to cricket song. *Science*, **190**, 1312–1313.

Cade, W. H. (1979) The evolution of alternative male reproductive strategies in field crickets. In *Sexual Selection and Reproductive Competition in Insects* (eds M. S. Blum and N. M. Blum), Academic Press, New York, 343–349.

Cade, W. H. (1980) Alternative male reproductive strategies. *Fla. Entomol.*, **63**, 30–44.

Cade, W. H. (1981) Alternative male strategies: Genetic differences in crickets. *Science*, **212**, 563–564.

Camhi, J. and Tom, W. (1978) The escape behaviour of the cockroach *Periplaneta americana*. I. Turning response to wind puffs. *J. Comp. Physiol.*, **128**, 193–201.

Claridge, M. F. (1987) Acoustic signals in the Homoptera: behavior, taxonomy, and evolution. *Ann. Rev. Entomol.*, **30**, 297–317.

Cokl, A. (1983) Functional properties of vibroreceptors in the legs of *Nezara viridula* (L) (Heteroptera, Pentatomidae). *J. Comp. Physiol.*, **150**, 261–269.

Cook, D. (1988) Sexual selection in dung beetles. II. Female fecundity as an estimate of male reproductive success in relation to horn size: An alternative behavioural strategy in *Onthophagus binodis*. *Austr. J. Zool.*, **36**, 521–532.

Crankshaw, O. S. (1979) Female choice in relation to calling and courtship songs in *Acheta domesticus*. *Anim. Behav.*, **27**, 1274–1275.

Dadour, I. R. and Bailey, W. J. (1985) Male agonistic behaviour of the bushcricket *Mygalopsis marki* Bailey in response to conspecific song (Orthoptera: Tettigoniidae). *Z. Tierpsychol.*, **70**, 320–330.

Dadour, I. R. and Johnson, M. S. (1983) Genetic differentiation, hybridization and reproductive isolation in *Mygalopsis marki* Bailey (Orthoptera: Tettigoniidae). *Austr. J. Zool.*, **31**, 353–360.

Daly, M. (1978). The cost of mating. *Amer. Nat.*, **112**, 771–774.

Dambach, M. and Rausche, G. (1985) Low-frequency airborne vibrations in crickets and feedback control of calling song. In *Acoustic and Vibrational Communication in Insects* (eds K. Kalmring and N. Elsner), Paul Parey, Hamburg, 177–182.

Dambach, M., Rausche, H.-G., and Wendler, G. (1983) Proprioceptive feedback influences the calling song of the field cricket. *Naturwissenschaften*, **70**, 417.

Darwin, C. (1874) *The Descent of Man, and Sexual Selection in Relation to Sex* , John Murray, London.

Dawkins, R. (1979) Good strategy or evolutionarily stable strategy? In *Sociobiology: Beyond Nature/Nurture?* (eds C. W. Barlow and J. Silverberg), Westview Press, Boulder, CO, 331–367.

Doherty, J. A. (1985) Trade-off phenomena in calling song recognition and phonotaxis in the cricket, *Gryllus bimaculatus* (Orthoptera, Gryllidae). *J. Comp. Physiol.*, **156**, 787–801.

Doherty, J. and Hoy, R. R. (1985) Communication in insects. III. The auditory behaviour of crickets: some views of genetic coupling, song recognition, and predator detection. *Q. Rev. Biol.*, **60**, 457–472.

Doolan, J. M. and MacNally, R. C. (1981) Spatial dynamics and breeding ecology in the cicada *Cytosoma saundersii*: the interactions between distributions of resources and intraspecific behaviour. *J. Anim. Ecol.*, **50**, 925–940.

Downes, J. A. (1970) The feeding and mating behaviour of the specialized Empidinae (Diptera): observations of four species of *Rhamphomyia* in high arctic and a general discussion. *Can. Entomol.*, **102**, 769–791.

Eberhard, W. G. (1977) Aggressive mimicry by a bolas spider. *Science*, **198**, 1173–1175.

Eberhard, W. G. (1985) *Sexual Selection and Animal Genitalia*, Harvard University Press, Cambridge, MA.

Elsner, N. (1983). A neuroethological approach to the phylogeny of leg stridulation in gomphocerine grasshoppers. In *Neuroethology and Behavioural Physiology* (eds F. Huber and H. Markl), Springer-Verlag, Berlin and Heidelberg, 54–68.

Elsner, N. and Popov, A. V. (1978) Neuroethology of acoustic communication. *Adv. Insect. Physiol.*, **13**, 229–355.

Endler, J. A. (1986) *Natural Selection in the Wild*, Princeton University Press, Princeton, NJ.

Evans, A. R. (1983) A study of the behaviour of the Australian field cricket *Teleogryllus commodus* (Walker) (Orthoptera: Gryllidae) in the field and in habitat situations. *Z. Tierpsychol.*, **62**, 269–290.

Ewing, A. W. and Bennet-Clark, H. C. (1968) The courtship songs of *Drosophila*. *Behaviour*, **31**, 288–301.

Feaver, M.N. (1983) Pair formation in the katydid *Orchelimum nigripes* (Orthoptera: Tettigoniidae). In *Orthopteran Mating Systems: Sexual Selection in a Diverse Group of Insects* (eds D. T. Gwynne and G. K. Morris), Westview Press, Boulder, CO, 205–239.

Forrest, T. G. (1983) Calling songs and mate choice in male crickets. In *Orthopteran Mating Systems: Sexual Competition in a Diverse Group of Insects* (eds D. T. Gwynne and G. K. Morris), Westview Press, Boulder, CO, 185–204.

Gerhardt, H. C. (1982) Sound pattern recognition in some North American treefrogs (Anura: Hylididae): implications for mate choice. *Amer. Zool.*, **22**, 581–595.

Gerhardt, H. C. (1987) Evolutionary and neurobiological implications of selective phonotaxis in the green treefrog, *Hyla cinerea. Anim. Behav.*, **35**, 1479–1489.

Gerhardt, H. C., Daniel, R. E., Stephen, A. and Schramm, S. (1987) Mating behaviour and male mating success in the green treefrog. *Anim. Behav.*, **35**, 1490–1503.

Greenfield, M. D. (1981) Moth sex pheromones: an evolutionary perspective. *Fla. Entomol.*, **64**, 4–17.

Greenfield, M. D. (1988) Interspecific acoustic interactions among katydids (*Neoconocephalus*): inhibition-induced shifts in diel periodicity. *Anim. Behav.*, **36**, 684–695.

Greenfield, M. D. and Shaw, K. C. (1983) Adaptive significance of chorusing with special reference to the Orthoptera. In *Orthopteran Mating Systems: Sexual Competition in a Diverse Group of Insects* (eds D. T. Gwynne and G. K. Morris), Westview Press, Boulder, CO, 1–27.

Griffin, D. R. (1971) The importance of atmospheric attenuation for the echolocation of bats (Chiroptera). *Anim. Behav.*, **19**, 55–61.

Gwynne, D. T. (1982) Mate selection by female katydids (Orthoptera: Tettigoniidae, *Conocephalus nigropleurum*). *Anim. Behav.*, **30**, 734–738.

Gwynne, D. T. (1984) Sexual selection and sexual differences in mormon crickets (Orthoptera: Tettigoniidae, *Anabrus simplex*). *Evolution*, **38**, 1011–1022.

Gwynne, D. T. (1985) Role-reversal in katydids: habitat influences reproductive behaviour (Orthoptera: Tettigoniidae, *Metaballus* sp.). *Behav. Ecol. Sociobiol.*, **16**, 355–361.

Gwynne, D. T. (1987) Sex-biased predation and the risky mate-locating behaviour of male tick-tock cicadas (Homoptera: Cicadidae). *Anim. Behav.*, **35**, 571–576.

Gwynne, D. T. (1989). Does copulation increase the risk of predation? *Tree*, **4**, 54–56.

Gwynne, D. T. and Bailey, W. J. (1988) Mating system, mate choice and ultrasonic calling in a Zaprochiline katydid (Orthoptera: Tettigoniidae). *Behaviour*, **105**, 202–223.

Gwynne, D. T. and Edwards, E. D. (1986) Ultrasound production by genital stridulation in *Syntonarcha iriastis* (Lepidoptera: Pyralidae): long-distance signalling by male moths? *Zool. J. Linn. Soc.*, **88**, 363–376.

Gwynne, D. T. and Morris, G. K. (eds) (1983) *Orthopteran Mating Systems: Sexual Competition in a Diverse Group of Insects*, Westview Press, Boulder, CO.

Gwynne, D. T. and Morris, G. K. (1986). Heterospecific recognition and behavioural isolation in acoustic Orthoptera (Insecta). *Evol. Theory*, **8**, 33–38.

Gwynne, D. T., Bowen, B. J. and Codd, C. G. (1984) The function of the katydid spermatophore and its role in fecundity and insemination (Orthoptera: Tettigoniidae) *Austr. J. Zool.*, **32**, 15–22.

Harrison, L., Horseman, G. and Lewis, D. B. (1988) The coding of the courtship song by an identified auditory neurone in the cricket *Teleogryllus oceanicus* (Le Guillou). *J. Comp. Physiol.*, **163**, 215–225.

Heath, J. E. and Josephson, R. K. (1970) Body temperature and singing in the katydid *Neoconocephalus robustus* (Orthoptera: Tettigoniidae). *Biol. Bull.*, **138**, 272–285.

Hedrick, A. V. (1986) Female preferences for calling bout duration in a field cricket. *Behav. Ecol. Sociobiol.*, **19**, 73–77.

Heinzel, H.-G. and Dambach, M. (1987) Travelling air vortex rings as potential communication signals in a cricket. *J. Comp. Physiol.*, **160**, 79–88.

Heller, K.-G. (1987) Warm-up and stridulation in the bushcricket *Hexacentrus unicolor* Serville (Orthoptera, Conocephalidae, Listroscelidinae). *J. Exp. Biol.*, **126**, 97–109.

Heller, K.-G. (1990) The evolution of song pattern between the sexes in bush-crickets In *The Tettigoniidae: Behaviour, Systematics and Evolution* (eds W. J. Bailey and D. C. F. Rentz), Crawford House Press, Bathurst, 130–151.

Heller, K.-G. and Helversen, O. von (1986) Acoustic communication in phaeropterid bushcrickets: species-specific delay of female stridulatory response and matching male sensory time window. *Behav. Ecol. Sociobiol.*, **18**, 189–98.

Helversen, D. von and Helversen, O. von (1983) Species recognition and acoustic localization in acridid grasshoppers. A behavioural approach. In *Neuroethology and Behavioural Physiology* (eds F. Huber and H. Markl), Springer-Verlag, Berlin, Heidelberg, New York, Tokyo, 95–102.

Henry, C. S. (1985) The proliferation of cryptic species in *Chrysoperla* green lacewings through song difference. *Fla. Entomol.*, **68**, 18–38.

Hoy, R. R. and Paul, R. C. (1973) Genetic control of song specificity in crickets. *Science*, **180**, 82–83.

Huber, F. (1962) Central nervous control of sound production in crickets and some speculations on its evolution. *Evolution*, **16**, 429–442.

Huber, F. (1985) Approaches to insect behaviour of interest to both neurophysiologists and behavioural ecologists. *Fla. Entomol.*, **68**, 52–78.

Huber, F. (1988) Invertebrate neuroethology: guiding principles. *Experimentia*, **44**, 428–431.

Huber, F. and Thornsen, J. (1985) Cricket auditory communication. *Sci. Amer.*, **253**, 60–68.

Huxley, J. (1938) Darwin's theory of sexual selection and the data subsumed by it, in the light of recent research. *Amer. Nat.*, **72**, 416–433.

Jacobson, M. (1972) *Insect Sex Pheromones*, Academic Press, New York.

Josephson, R. K. and Halverson, R. C. (1971) High frequency muscles used in sound production by a katydid. I. Organisation of the motor system. *Biol. Bull.*, **141**, 411–433.

Kalmring, K. and Kuhne, R. (1980) The coding of airborne-sound and vibration signals in bimodal ventral-nerve cord neurons of the grasshopper *Tettigonia cantans*. *J. Comp. Physiol.*, **139**, 267–275.

Kamper, G. and Dambach, M. (1981) Response of the cercus-to-giant interneuron system in crickets to species-specific song. *J. Comp. Physiol.*, **141**, 311–317.

Kavanagh, M. W. (1987) The efficiency of sound production in two cricket species, *Gryllotalpa australis* and *Teleogryllus commodus* (Orthoptera: Grylloidea). *J. Exp. Biol.*, **130**, 107–119.

Kay, R. E. (1969) Acoustic signalling and its possible relationship to assembling and navigation in the moth *Heliothis zea*. *J. Insect Physiol.*, **15**, 989–1001.

Latimer, W. (1981) Variation in the song of the bushcricket *Platycleis albopunctata* (Orthoptera, Tettigoniidae). *J. Nat. Hist.*, **15**, 245–263.

Latimer, W. and Lewis, B. (1986) Song harmonic content as a parameter determining acoustic orientation behaviour in the cricket *Teleogryllus oceanicus* (Le Guillou). *J. Comp. Physiol.*, **158**, 535–591.

Latimer, W. and Sippel, M. (1986) Acoustic cues for female choice and male competition in *Tettigonia cantans*. *Anim. Behav.*, **35**, 887–910.

Lee, C. P. (1986) The role of male song in sex recognition in *Zaprionus tuberculatus* (Diptera, Drosophilidae). *Anim. Behav.*, **34**, 641–648.

Lewis, B. (1983) *Bioacoustics: A Comparative Approach*, Academic Press, London.

Littlejohn, M. J. (1988) The retrograde evolution of homogamic acoustic signalling systems in hybrid zones. In *The Evolution of the Amphibian Auditory System* (ed. B. Fritsch), John Wiley, New York, 613–635.

Lloyd, J. (1979a) Sexual selection in luminescent beetles. In *Sexual Selection and Reproductive Competition in Insects* (eds M. A. Blum and N. A. Blum), Academic Press, New York, 293–342.

Lloyd, J. E. (1979b) Mating behaviour and natural selection. *Fla. Entomol.*, **62**, 17–34.

MacNally, R. C. and Young, D. (1981) Song energetics of the bladder cicada *Cystosoma saundersii*. *J. Exp. Biol.*, **90**, 185–196.

Maynard Smith, J. (1976) Sexual selection and the handicap principle. *J. Theoret. Biol.*, **57**, 239–242.

Mayr, E. (1982) *The Growth of Biological Thought: Diversity, Evolution and Inheritance*. Harvard University Press. Cambridge, MA.

McCauley, D. E. and Lawson, E. C. (1986) Mating reduces predation on male milkweed beetles. *Amer. Nat.*, **127**, 112–117.

Michelsen, A. and Larsen, O.N. (1985) Hearing and sound. In *Comprehensive Insect Physiology, Biochemistry and Pharmacology* (eds G. A. Kerkut and L. I. Gilbert), Vol. 6, Pergamon Press, Oxford, 496–556.

Moiseff, A. and Hoy, R. R. (1983) Sensitivity to ultrasound in an identified auditory interneuron in the cricket: a possible neural link in phonotactic behavior. *J. Comp. Physiol.*, **152**, 155–167.

Moiseff, A., Pollack, G. S. and Hoy, R. R. (1978) Steering responses of flying crickets to sound and ultrasound: mate attraction and predator avoidance. *Proc. Nat. Acad. Sci. USA*, **75**, 4052–4056.

Morris, G. K. (1980) Calling display and mating behaviour of *Copiphora rhinoceros* Pictet (Orthoptera: Tettigoniidae). *Anim. Behav.*, **28**, 42–51.

Morris, G. K. and Beier, M. (1982) Song structure and description of some Costa Rican katydids (Orthoptera: Tettigoniidae). *Trans. Amer. Entomol. Soc.*, **108**, 287–314.

Morris, G. K. and Fullard, J. H. (1983) Random noise and congeneric discrimination in *Conocephalus* (Orthoptera: Tettigoniidae) In *Orthopteran Mating Systems: Sexual Selection in a Diverse Group of Insects* (eds D. T. Gwynne and G. K. Morris), Westview Press, Boulder, CO, 73–96.

Morris, G. K., Klimas, D. and Nickle, D. A. (1989) Acoustic signals and systematics of false leaf katydids from Ecuador (Orthoptera: Tettigoniidae: Pseudophyllinae). *Trans. Amer. Entomol. Soc.*, **114**, 215–264.

Nagimine, M. and Ito, Y. (1980) 'Predator-foolhardiness' in an epidemic cicada population. *Res. Popul. Ecol.*, **22**, 89–92.

Nielsen, E. T. and Dreisig, H. (1970) The behaviour of stridulation in Othoptera

Ensifera. *Behaviour*, **37**, 205–252.

Nutting, W. L. (1953) The biology of *Euphasiopteryx brevicornis* (Townsend) (Diptera, Tachinidae), parasitic in the cone-headed grasshoppers (Orthoptera, Copiphorinae). *Psyche*, **60**, 69–81.

Otte, D. (1972) Simple vs elaborate behaviour in grasshoppers: an analysis of communication in the genus *Syrbula*. *Behaviour*, **42**, 292–322.

Otte, D. (1974) Effects and function in the evolution of signalling systems. *Ann. Rev. Ecol. Syst.*, **5**, 385–417.

Otte, D. (1977) Communication in Orthoptera. In *How Animals Communicate* (ed. T. Sebeok), Indiana Press, Bloomington, IN, 334–361.

Otte, D. (1989) Speciation in Hawaiian crickets. In *Speciation and its Consequences* (eds D. Otte and J. A. Endler), Sinauer Press, Sunderland, MA, 482–526.

Parker, G. A. (1983) Mate quality and mating decisions. In *Mate Choice* (ed. P. Bateson), Cambridge University Press, Cambridge, 141–164.

Partridge, L., Ewing, A. and Chandler, A. (1987). Male size and mating success in *Drosophila melanogaster*: the roles of male and female behaviour. *Anim. Behav.*, **35**, 555–562.

Paul, R. C. and Walker, T. J. (1979) Arboreal singing in a burrowing cricket, *Anurogryllus arboreus*. *J. Comp. Physiol.*, **132**, 217–223.

Payne, R. B. (1983) Bird songs, sexual selection and female mating strategies. In *Social Behaviour of Female Vertebrates* (ed. S. K. Wasser), Academic Press, New York, 55–90.

Perdeck, A. C. (1958) The isolating value of specific song patterns in two sibling species of grasshoppers (*Chorthippus brunneus* Thunb. and *C. biguttulus* L.). *Behaviour*, **12**, 1–75.

Popov, A. V. and Shuvalov, V. F. (1977) Phonotactic behaviour of crickets. *J. Comp. Physiol.*, **119**, 111–126.

Prestwich, K. N. and Walker, T. J. (1981) Energetics of singing in crickets: effect of temperature in three trilling species (Orthoptera: Gryllidae). *J. Comp. Physiol.*, **143**, 199–212.

Rentz, D. C. (1975) Two new katydids of the genus *Melanonotus* from Costa Rica with comments on the life history strategies (Tettigoniidae: Pseudophyllinae). *Entomol. News*, **86**, 129–140.

Rheinlaender, J. and Römer, H. (1980) Bilateral coding of sound direction in the CNS of the bushcricket *Tettigonia viridissima* L. (Orthoptera, Tettigoniidae). *J. Comp. Physiol.*, **140**, 101–111.

Rheinlaender, J. and Römer, H. (1990) The neuroethology of sound reception in bushcrickets In *The Tettigoniidae: Behaviour, Systematics and Evolution* (eds W. J. Bailey and D. C. Rentz), Crawford Press, Bathurst (in press).

Riede, K. (1987) A comparative study of mating behaviour in some neotropical grasshoppers (Acridoidea). *Ethology*, **76**, 265–296.

Robinson, D. (1990) Acoustic communication between the sexes in bushcrickets. In *The Tettigoniidae: Behaviour, Systematics and Evolution* (eds W. J. Bailey and D. C. Rentz), Crawford Press, Bathurst, 248–64.

Roeder, K. D. (1966) Acoustic sensitivity of the noctuid tympanic organ and its range for the cries of bats. *J. Insect. Physiol.*, **12**, 843–859.

Roessler, W., Bailey, W. J., Schroeder, J. and Kalmring, K. (1990) Resolution of time and frequency patterns in the tympanal organ of tettigoniids. I. Synchronisation and oscillation in the activity of receptor populations. *Zool. Jb. Physiol.*, **94**, 83–9.

Römer, H. (1983) Tonotopic organization of the auditory neuropile in the bush-cricket *Tettigonia viridissima*. *Nature*, **306**, 60–62.

Römer, H. and Bailey, W. J. (1986) Insect hearing in the field. II. Male spacing behaviour and correlated acoustic cues in the bushcricket *Mygalopsis marki*. *J. Comp. Physiol.* **159**, 627–638.

Römer, H., Bailey, W. A. and Dadour, I. R. (1989) Insect hearing in the field. III. Masking by noise. *J. Comp. Physiol.*, **164**, 609–620.

Rost, R. and Honegger, H. W. (1987) The timing of premating and mating behaviour in a field population of the cricket *Gryllus campestris* L. *Behav. Ecol. Sociobiol.*, **21**, 279–290.

Ryan, M. K. (1985) *The Tungara Frog: A Study in Sexual Selection and Communication*, University of Chicago Press, Chicago, IL.

Ryan, M. J. and Brenowitz, E. A. (1985) The role of body size, phylogeny, and ambient noise in the evolution of bird song. *Amer. Nat.*, **126**, 87–100.

Sabrosky, C. W. (1953) Taxonomy and host relationships of the tribe Ormüni in the western hemisphere (Diptera, Larvaevoridae). *Proc. Entomol. Soc. Washington*, **55**, 167–183.

Sakaluk, S. K. and Belwood, J. J. (1984) Gecko phonotaxis to cricket song: a case of satellite predation. *Anim. Behav.*, **32**, 659–62.

Schildberger, K. (1985) Recognition of temporal patterns by identified auditory neurons in the cricket brain. In *Acoustic and Vibrational Communication in Insects* (eds K. Kalmring and N. Elsner), Parey-Verlag, Hamburg, 41–49.

Searcy, W. A. and Andersson, M. (1986) Sexual selection and the evolution of song. *Ann. Rev. Ecol. Syst.*, **17**, 507–533.

Shelly, T. E., Greenfield, M. D. and Downum, K. R. (1987) Variation in host plant quality: influences on the mating system of a desert grasshopper. *Anim. Behav.*, **35**, 1200–1209.

Simmons, J. A., Wever, E. G. and Pylka, J. M. (1970) Periodical cicada: sound production and hearing. *Science*, **171**, 212–213.

Simmons, L. W. (1988a) Male size, mating potential and lifetime reproductive success in the field cricket, *Gryllus bimaculatus* (De Geer). *Anim. Behav.*, **36**, 372–379.

Simmons, L. W. (1988b) The calling song of the field cricket, *Gryllus bimaculatus* (De Geer): constraints on transmission and its role in intermale competition and female choice. *Anim. Behav.*, **36**, 380–394.

Soper, R. S., Shewell, G. E. and Tyrrell, D. (1976) *Colcondamyia auditrix* nov. sp. (Diptera: Sarcophagidae), a parasite which is attracted by the mating song of its host, *Okanagana rimosa* (Homoptera: Cicadidae). *Can. Entomol.*, **108**, 61–68.

Spangler, H. G. (1984) Attraction of female lesser wax moths (Lepidoptera: Pyralidae) to male-produced and artificial sounds. *J. Econ. Entomol.*, **77**, 346–349.

Spangler, H. G. (1986) Further observations on sound production by the lesser wax moth, *Achroia grisella* (F.) (Lepidoptera: Pyralidae). *J. Kansas Entomol. Soc.*, **59**, 555–557.

Spangler, H. G. (1988) Moth hearing, defense, and communication. *Ann. Rev. Entomol.*, **33**, 59–81.

Stevens, S. K. and Josephson, R. K. (1977) Metabolic rate and body temperature in singing katydids. *Physiol. Zool.*, **50**, 31–42.

Stout, J. F., Atkins, G., Weber, T. and Huber, F. (1987) The effect of visual input on calling song attractiveness for female *Acheta domesticus*. *Physiol. Entomol.*, **12**, 135–140.

Taigon, T. L. and Wells, K. D. (1985) Energetics of vocalization by an anuran amphibian (*Hyla versicolor*). *J. Comp. Physiol.*, **155**, 163–170.

Thornhill, R. (1979) Male and female sexual selection and the evolution of mating systems in insects. In *Sexual Selection and Reproductive Competition in Insects* (eds M. S. Blum and N. A. Blum), Academic Press, New York, 81–121.

Thornhill, R. and Alcock, J. (1983) *The Evolution of Insect Mating Systems*, Harvard University Press, Cambridge, MA.

Thornhill, R. and Gwynne, D. T. (1986) The evolution of sexual differences in insects. *Amer. Sci.*, **74**, 382–389.

Tuttle, M. D., Ryan, M. J. and Belwood, J. (1985) Acoustical resource partitioning by two species of phyllostomatid bats (*Trachops cirrhosus* and *Tonatia silvicola*). *Anim. Behav.*, **33**, 1369–1371.

Vinson, S. D. (1985) The behaviour of parasitoids. In *Comprehensive Insect Physiology, Biochemistry and Pharmacology* (eds G. A. Kerkut and L. I. Gilbert), Pergamon Press, Oxford, 417–69.

Walker, T. J. (1957) Specificity in the response of female tree crickets (Orthoptera: Gryllidae: Oceanthinae) to calling songs of the males. *Ann. Entomol. Soc. Amer.*, **50**, 626–636.

Walker, T. J. (1964) Experimental demonstration of a cat locating orthopteran prey by the prey's calling song. *Fla. Entomol.*, **47**, 163–165.

Walker, T. J. (1983a) Diel patterns of calling in nocturnal Orthoptera. In *Orthopteran Mating Systems: Sexual Selection in a Diverse Group of Insects* (eds D. T. Gwynne and G. K. Morris), Westview Press, Boulder, CO, 45–72.

Walker, T. J. (1983b) Mating modes and female choice in short-tailed crickets (*Anurogryllus arboreus*). In: *Orthopteran Mating Systems: Sexual Selection in a Diverse Group of Insects* (eds D. T. Gwynne and G. K. Morris), Westview Press, Boulder, CO, 240–267.

Weber, T., Atkins, G., Stout, J. F. and Huber, F. (1987) Female *Acheta domesticus* track acoustical and visual targets with different walking modes. *Physiol. Entomol.*, **12**, 141–147.

West-Eberhard, M. J. (1983) Sexual selection, social competition, and speciation. *Rev. Biol.*, **58**, 155–183.

West-Eberhard, M. J. (1984) Sexual selection, competitive communication and species-specific signals in insects. In *Insect Communication* (ed. T. Lewis), Academic Press, London, 283–324.

Williams, G. C. (1966) *Adaptation and Natural Selection*, Princeton University Press, Princeton, NJ.

Wiley, R. H. and Richards, D. G. (1978) Physical constraints on acoustic communication in the atmosphere: implications for the evolution of animal vocalisation. *Behav. Ecol. Sociobiol.*, **3**, 69–94.

Wynne-Edwards, V. C. (1962) *Animal Dispersion in Relation to Social Behaviour*, Oliver & Boyd, Edinburgh.

Young, D. and Josephson, R. K. (1983) Mechanisms of sound-production and muscle kinetics in cicadas. *J. Comp. Physiol.*, **152**, 183–195.

Zaretsky, M. D. (1972) Specificity of the calling song and short term changes in the phonotactic response by female crickets *Scapsipidus marginatus* (Gryllidae). *J. Comp. Physiol.*, **79**, 153–172.

Zuk, M. (1987) Variability in attractiveness of male field crickets (Orthoptera: Gryllidae) to females. *Anim. Behav.*, **35**, 1240–1248.

4

Host location and oviposition on animals

Paul W. Wellings

4.1 INTRODUCTION

Price (1980) has provided a general analysis of the feeding habits of 16 929 species of British insects. He categorized these species into five groups: predators; non-parasitic herbivores and carnivores (e.g. grasshoppers, bees, mosquitos); saprophages (e.g. dung beetles); parasites on plants; and parasites on animals (Fig. 4.1). The commonest feeding habits were insects parasitic on plants (35.1%) (see Jones, Chapter 5, this volume) and those parasitic on animals (35.6%). Three orders dominate the 6031 species classified in the last category. Diptera and Phthiraptera each represented at 5% of the species and about 89% were Hymenoptera. The remaining 1% were comprised of Siphonaptera, some Coleoptera and some Homoptera. While our knowledge of the world's insect fauna is limited, it is clear that the parasitic Hymenoptera are a remarkably diverse and abundant group. As a result, it seems likely that insects that are parasitic on animals are extraordinarily common.

The vast majority of insects that are animal parasites exhibit complex life cycles: defined by Istock (1967) as those involving two or more ecologically distinct phases. Each phase has its own set of resource and competitive interactions which do not overlap with the set of interactions limiting abundance in the other phase(s). Istock (1967) also pointed out that:

> In species with complex life-cycles the ecologically distinct phases evolve independently of one another to a large degree. . . . Such independence is incomplete to the degree that some behavioral, developmental, and morphological characteristics necessary for making the transitions between the phases must be maintained.

This suggests that we should ask what factors influence the degree of linkage between phases. One major factor may be the influence that environmental heterogeneity has in shaping life histories. Levins' (1968)

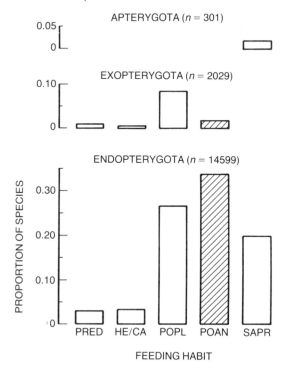

Fig. 4.1 The feeding habits of British insects classified in five groups: PRED, predators; HE/CA, herbivores and carnivores; POPL, parasites on plants; POAN, parasites on animals; SAPR, saprophages. *n* is the number of species in each Division (based on data in Price, 1980).

concept of environmental graininess provides an important insight into the way the characteristics necessary for shifting between phases in complex life cycles may have been shaped. The idea of graininess stems from the size of patches in the environments experienced by the organism. So, if an individual spends its whole life in a single patch, the environment is thought to be coarse-grained (C). This contrasts with fine-grained environments (F); in which an individual may move between patches during its life. In complex life cycles with two phases, there could be three categories of life history as both phases have the potential of existing in fine- and/or coarse-grained environments (i.e. C–C, F–F, C–F/F–C).

When both phases exist in fine-grained environments the coupling between the phases may be minimal as the fitness of any one phase is maximized by those traits that determine the within-stage patch selection. These complex life cycles should contrast with those in which one stage exists in a fine-grained environment and the other in a coarse grain.

Here, the phase living in the fine-grained environment may have a marked effect on the fitness of the other stage because it selects the environment in which the other stage spends its life. If there is any variation in the quality of coarse grains, it is highly probable that the stage occupying the fine-grained environment should evolve a set of behavioural and ecological adaptations to maximize the fitness of the other stage. The latter, once placed in/on/by a coarse grained environment, is constrained to utilize only this resource. The final category in this classification, in which both phases occupy coarse-grained environments, seems to be biologically implausible.

Some insects are parasitic on their hosts in all stages (e.g. lice, polyctenids, some hippoboscids and a single species of flea). In others the immature stages exist well away from the hosts' body (e.g. pyralids, some scarabaeids, many hippoboscids and a few fleas) (Marshall, 1981). Movement between hosts of the former group, usually occurs when the hosts are in contact with one another, for example in nests and during copulation whereas in the later group, adults of these organisms tend to be winged or very good jumpers. Both phases in these complex life cycles exist in fine-grained environments and movement onto hosts or between hosts comes about through recognition of a number of environmental cues associated with the host or host habitat. For example, Osbrink and Rust (1985) demonstrate that visual and thermal cues are responsible for orientation and attraction towards hosts in the cat flea. Marshall (1981) reviews other examples of factors influencing host seeking and recognition: in some of these examples environmental cues appear to activate the host-seeking process but, after that, host location appears to be random. Individuals in the host population may differ in their susceptibility to parasites. For example, Hopkins' (1985) data suggest that flea burdens on small mammals are greater on reproductively active males than on either females or non-reproductively active male conspecifics. Similarly, seasonal variation in parasite burdens between sexes in rock dassies (*Procavia capensis*) appear to depend on physiological condition and social organization (Fourie *et al.*, 1987). The effect of variations in host location abilities and host susceptibilities may result in the characteristic non-random distributions of parasites on their hosts (Fig. 4.2).

The mechanisms influencing host location and oviposition in insect parasites that spend both phases in fine-grained environments are relatively well described, while functional explanations of different behaviour patterns and the consequences of the behaviour of parasites on host and parasite populations are poorly understood. In contrast, the behaviour of some parasites that make the transition from fine- to coarse-grained environments has been studied from both a mechanistic and functional point of view. One large group, the hymenopterous insect parasitoids, has been the subject of many studies, probably for two

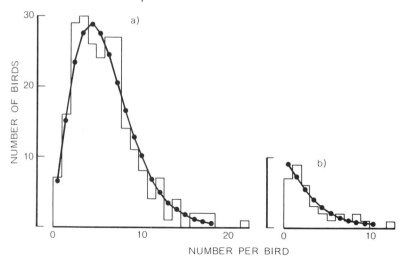

Fig. 4.2 Frequency distributions of ectoparasites on seabirds: (a) *Halipenus pelagicus* on Storm Petrels and (b) *Perineus nigrolimbatus* on Fulmars. Observed frequencies are illustrated with histograms and the joined circles are calculated negative binomial frequencies (after Fowler and Miller, 1984).

reasons. First, they are excellent organisms for laboratory studies: the adult females exhibit a suite of behavioural and life history traits that are used in host location and oviposition. Thus, carefully designed experiments can quantify the consequences of different behaviours. Second, these insects are economically important and have been extensively used in biological control programmes and integrated pest management. As a result, researchers have examined behavioural factors influencing aspects of parasitoid fitness in mass rearing programmes, augmentation of female performance in the field and have studied the impact of parasitoids on host populations. The rest of this chapter is devoted to insect parasitoids and provides an overview of the proximate causes for changes in behaviour and the adaptive significance of particular behaviour patterns.

4.2 HOST SELECTION

Host selection has been viewed as a sequence of behavioural responses to environmental or habitat cues which bring the adult female parasitoid into an appropriate habitat and, from there, into contact with the host. Doutt (1964) suggested that this sequence could be divided into a number of steps: host habitat location, host location and host acceptance. In addition, Doutt recognized a fourth step which he called host suitability. This classification has proved to be a useful framework in which to consider the factors leading to successful parasitism by insects. Individual elements of this sequence have been the subject of major reviews, for example Lewis *et al.* (1976), Vinson (1976, 1981), Weseloh (1981) and van Alphen and Vet (1986).

There is often marked spacial and temporal variability in the abundance of a parasitoid's host and it would not be unusual for newly emerged adult parasitoids to encounter a limited number of hosts or a cohort of hosts with an age structure unsuitable for parasitization. In these circumstances the adaptive strategy may be to disperse into new host habitats. This process appears to be highly directional and non-random. The mechanisms underlying this behaviour are very poorly understood in most parasitoids. Lewis *et al.* (1976) outlined a generalized model of habitat location. They suggested that there is a transition from random parasitoid movement to scanning selected habitats. This transition results in orientation to habitat or plant cues. Kitano (1978) hypothesized that volatiles from plants are the first cue in host selection and showed that food plants of the host, such as *Brassica*, *Cleome* and *Chaenomeles* were attractive to the parasitoid, *Cotesia* (= *Apanteles*) *glomerata*. Similarly, Camors and Payne (1972) demonstrated that a larval parasitoid of bark beetles was attracted to a terpene in a food plant of its host. In contrast, Dicke *et al.* (1984) found that *Leptopilina heterotoma* did not use olfactory cues from the host plant in long distance habitat location. They suggested that such cues cannot be used to distinguish between potential host habitats and habitats *containing* hosts.

The role of plant and other cues characteristic of habitats containing hosts have been shown to be important in host habitat location (Spradbery, 1970, 1974; Mueller, 1983; Odell and Godwin, 1984). For example, Monteith (1964) found that the tachinids *Drino bohemica* and *Bassa harveyi* made between two and four times as many attacks on sawfly larvae exposed on unhealthy trees as on those exposed on healthy trees. Olfactory stimuli from unhealthy trees were more attractive to the tachinids than those from healthy trees. Further, *D. bohemica* preferred the odour of old to that of new tree growth (Monteith, 1966). Such responses may be adaptive when host plants subjected to herbivore attack grow less vigorously than uninfested host plants. More recently a manipulative experiment, conducted by Faeth and Bultman (1986), explored the effect of painting tannins on leaves containing leafminers. Leafminers in control leaves experienced 35% parasitization, while those in tannin-painted leaves experienced 57% parasitization. The marked increase in parasitism could be due to adult parasitoids being attracted to volatile products of tannin degradation. Alternatively, the increase could be a by-product of a reduction in the development rate of hosts in tannin-painted leaves; thereby exposing the leafminers to parasitoids for a longer period (Faeth and Bultman, 1986).

Within habitats, host-associated chemical, auditory and visual cues appear to mediate the searching process. The functional role of these cues is clear: by evolving directed responses to cues associated with the host, female parasitoids may minimize the time spent in areas devoid of hosts and restrict themselves to searching areas containing high densities of hosts. Most recent work has focused on chemical cues. These kairomones may act in two ways; some are searching stimulants, others are important in host recognition. For example, Clement *et al.* (1986) found that a kairomone in the frass of *Agrotis ipsilon* triggered larviposition activity in the tachinid parasitoid, *Bonnetia comta*. Parasitoids were found to be responsive on moth-damaged corn seedlings but not on seedlings subjected to mechanical damage. Similarly, van Leerdam *et al.* (1985) found that the host-finding behaviour of *Cotesia flavipes* was influenced by a water-soluble substance occurring in the fresh frass of its host. On encountering frass, *C. flavipes* responded by intensive palpitation of the frass with its antennae and subsequently moved at a decreased rate. Bouchard and Cloutier (1984) showed that *Aphidius nigripes* spends more time searching on potato plants previously infested with its aphid host than on uninfested plants and that this was due to the presence of aphid feeding excretions (honeydew). Female parasitoids were initially arrested in areas containing honeydew, and exhibited abdominal projection. They also reduced walking speed and increased the rate of turning while searching. On reaching the perimeter of an affected area, parasitoids often exhibited a klinotactic response causing them to return into the patch. Bouchard and Cloutier (1985) also demonstrated that only female *A. nigripes* were attracted to host and honeydew odours. Males did not respond to these cues but were attracted to conspecific females.

Chemical cues are present in many other host and host-plant products. Prokopy and Webster (1978) found that *Opius lectus*, a parasitoid of the fruit fly *Rhagolitis pomonella*, was retained on the fruit as a result of fruit odour and *R. pomonella*'s own oviposition-deterring pheromone (see Fletcher and Prokopy, Chapter 6, this volume). The pheromone induced changes in the searching behaviour of *O. lectus* but did not direct it to the host. Host oviposition products can also influence host recognition and acceptance. For example, the parasitoids *Telenomus remus* and *Trichogramma pretiosum* respond to the accessory gland material associated with the eggs of their hosts, *Spodoptera fruigiperda* and *Heliothis zea*, respectively (Nordlund *et al.*, 1987). Vinson (1975) also demonstrated that host searching in the egg-larval parasitoid *Chelonus taxanus* was influenced by chemicals associated with the host egg. Extracts of these chemicals promoted intensive local searching behaviour. Other research has found oviposition stimuli in, for example, the wax of scale insects (Takabayashi and Takahashi, 1985), the wing scales of moths (Chiri and Legner, 1982) and cocoons, silk and pupae of moths (Sandlan, 1980).

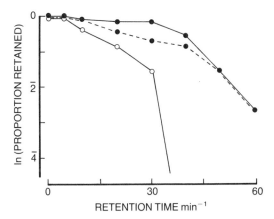

Fig. 4.3 Retention rate of adult female *Microplitis croceipes* searching on plants with ●—● 12 h feeding by *Heliothis zea*. Larvae and frass were left on these plants. ●--● 12 h feeding by *H. zea* but larvae removed. ○—○ No feeding damage or larvae present. It appears that kairomones in the frass prolong retention (based on data in Lewis *et al.*, 1976).

The general effect of chemical cues is to alter parasitoid behaviour. These cues lead to parasitoids spending greater periods of time in host-infested areas (Fig. 4.3) as a result of changes in the rate of movement and frequency of turning. Host recognition appears to be, in part, a function of non-volatile chemical cues. Chiri and Legner (1982) found that moth wing scales induced changes in the searching behaviour of a species of *Chelonus*. They hypothesized that shed scales were likely to signal the presence of hosts and found that parasitoids responded to chemical extracts from these wing scales. The proportion of female parasitoids responding to these extracts was dependent on the concentration of the chemicals (Fig. 4.4). Chiri and Legner (1982) found that this parasitoid also responded to wing scale extracts of non-host moth species. This suggests that, when host and non-host species occur concurrently, the field performance of parasitoids could be reduced due to time wasted in responding to non-hosts. The spatial dynamics and niche requirements of closely related host and non-host species may have a marked effect on shaping host-location behaviour and searching patterns of parasitoids.

Polyphagous parasitoids face a similar set of problems. They may prefer some species in the suite of potential hosts and the suitability of these host species may vary, but outweighing these factors may be large spatial and temporal changes in the abundance of different hosts. Under these circumstances, host habitat location may be difficult, especially as many volatile cues appear to be perceived in a very general fashion: in evolutionary terms, it may be too expensive for polyphagous parasitoids to have sets of receptor systems dedicated to the cues from each host

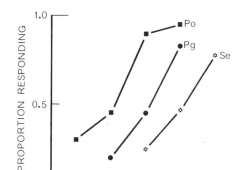

Fig. 4.4 Responses of the parasitoid *Chelonus.* sp. nr. *cirvimaculatus* to hexane extracts of the wing scales of three moths: the natural host is *Pectinophora gossypiella* (Pg), while *Phthorimaea operculella* (Po) and *Spodoptera exigua* (Se) are factitious hosts. 0.1 g of scales from each host were washed with 100 ml of hexane and concentrated to a stock solution of 10 mg equivalent of scales per 1 ml hexane. Bioassays were conducted using serial dilutions of this stock (based on data in Chiri and Legner, 1982).

species in their repertoire. However, some parasitoids exhibit plasticity in behavioural responses which appear to maximize their encounters with locally abundant hosts. Arthur (1966, 1971) investigated the host-searching behaviour in *Itoplectis conquisitor* (a polyphagous parasitoid) and *Nemeritis canescens* (an oligophagous species). He demonstrated that *I. conquisitor* could be conditioned to associate visual stimuli (coloured tubes) with the presence of hosts and subsequently used these cues for long periods. Similarly, *N. canescens* learned to associate the odour of geraniol (an alien substance in the parasitoid's habitat) with the presence of hosts (Arthur, 1971). In these two species the duration of conditioning to novel visual and chemical cues was much longer in the polyphagous species. Arthur (1971) suggested that associative learning is a significant factor in the success of polyphagous species foraging in a range of habitats.

More recently, Dmoch *et al.* (1985) showed that experience influenced subsequent host-selection behaviour in *Cotesia* (= *Apanteles*) *marginiventris*. Previous experience with hosts or a range of host products caused an increase in the searching time of these individuals compared with inexperienced parasitoids. Host products frequently contain volatile recognition cues and it appears that some parasitoids may learn to associate with non-volatile kairomones. For example, Lewis and Tumlinson (1988) showed that when *Microplites croceipes* encountered *H. zea* frass it antennated this material and, in doing so, recognized a non-volatile kairomone. At the same time, this parasitoid was able to detect and associate volatiles present in the frass with the contact kairomone and subsequently used the volatile cue as a tracking kairomone. It seems that associative learning can occur even without direct contact with the host and this plasticity in behavioural responses may give an adaptive advantage to parasitoids searching for cryptic hosts (Lewis and Tumlinson, 1988). Other examples of learning in parasitoids are reviewed by van Alphen and Vet (1986).

Learned behaviour is not persistent and the ability to learn may be an age-specific trait (van Alphen and Vet, 1986). There is little information on the mechanisms influencing learned behaviour. However, because of the potential importance of augmentation of parasitoid performance in the field through the use of kairomones and disruption of herbivore populations through the use of synthetic mating or oviposition pheromones, is such that one aspect of learning, namely habituation, is central to the development of these management practices. Habituation is thought to occur if repeated applications of a stimulus results in decreased responsiveness in the parasitoid. Gardner and van Lenteren (1986) investigated the responses of *Trichogramma evanescens* on encountering contact kairomones and/or eggs of *Pieris brassicae*. They presented a sequence of seven arenas, each containing a single patch of contact kairomone. A single egg was also present in these patches in arenas 1 and 6. The time spent

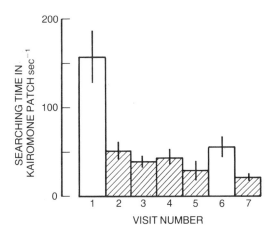

Fig. 4.5 The searching time (s ± s.e.) of experienced *Trichogramma evanescens* in arenas containing strips impregnated with kairomones of *Pieris brassicae* over a series of seven visits. Single host eggs were placed in the centre of the kairomone patches (and encountered by the parasitoids) on visits 1 and 6 (based on data in Gardner and van Lenteren, 1986).

searching in these kairomone areas decreased in the absence of eggs (which elicit arrestment behaviour) and also between the fourth and fifth encounter with kairomone patches, indicating that habituation may occur in this parasitoid (Fig. 4.5). Other laboratory studies have not been able to find evidence for habituation in parasitoids (Dmoch *et al.*, 1985).

The usefulness of applications of kairomones as tools in pest management may depend on the rate at which habituation occurs. Laboratory experiments have recorded increased rates of parasitization (Takabayaski and Takahashi, 1988); however, results from field studies using contact kairomones have been inconclusive (Lewis *et al.*, 1975; Gross *et al.*, 1984). Rates of parasitization at low densities do not appear to be influenced by applications of contact kairomones. In these cases, parasitoids may be retained by the kairomones but an increased proportion may spend their time searching in patches without hosts.

Applications of synthetic pheromones to disrupt mating in pest species may influence the searching behaviour of parasitoids, and this is important if both pheromones and parasitoids are thought to be integral components of pest management. Parasitoids do respond to these volatiles in laboratory trials. For example, *T. evanescens* was attracted to various synthetic and natural moth sex pheromones (Zaki, 1985). The consequences of this behaviour in the field are still to be resolved. It may be that rates of parasitization are increased. On the other hand, false odour trails may lead to wasted searching time and habituation, although field applications of a synthetic sex pheromone against the Douglas fir tussock moth had no detrimental effects on the performance of two species of egg parasitoid (Sower and Torgersen, 1979).

4.3 HOST ACCEPTANCE, SUITABILITY AND PREFERENCE

Once a biotope is located and hosts are encountered, individual parasitoids respond to physical and chemical stimuli associated with the host, and these cues may lead to acceptance of the host. In laboratory studies, host acceptance is usually assayed by offering hosts of different stages or species, in no-choice experiments, to individual parasitoids. Those classes of host that elicit the required behavioural responses are thought to be acceptable.

Wilson *et al.* (1974) investigated the factors which cause larvae of *Heliothis virescens* to be acceptable to the parasitoid *Campoletis sonorensis*. On encountering a host, the parasitoid examined the larva by antennation. Host movement and size were relatively unimportant in determining acceptability. In contrast, host shape was a major stimulus, with straight cylindrical shapes being more acceptable than round flat shapes.

Kairomones associated with the host then provide the ovipositional stimulus. Similarly, Strand and Vinson (1983) investigated host recognition factors of *Telenomus heliothidis* encountering eggs of *H. virescens*. They found that both physical and chemical cues influence host recognition. Host size and shape were particularly important: spherical hosts with diameters of 0.5–0.63 mm were the most acceptable. Host colour did not appear to be critical. Highest responses were recorded when some part of the egg was coated with kairomone. In *T. heliothidis* recognition of the kairomone and host shape was achieved by antennation.

In laboratory assays, many parasitoids will accept a range of classes of host but variation in the biology of hosts may cause asymmetrical patterns of acceptance when the parasitoids encounter more than one type of host (e.g. van Veen and van Wijk, 1987). Also, within the same host species different host genotypes may show different susceptibilities to parasitism. For example, Carton and David (1985) demonstrated a polygenically determined polymorphism in larval foraging strategies of *Drosophila melanogaster*, with one strain exhibiting deep digging behaviour and superficial digging behaviour. These two strains differed in their susceptibility to the parasitoid *Leptopilina boulardi* in that deep diggers were encountered at lower rates by this parasitoid than were superficial diggers. Similarly, Sokolowski and Turlings (1987) worked with two strains of *D. melanogaster* exhibiting a polymorphism in larval locomotory behaviour (movers and non-movers). They offered both forms to two species of parasitoid with differing within patch host location strategies: *Leptopilina heterotoma* rhythmically probes the surface of a patch with its ovipositor, while *Asobara tabida* uses vibrotaxis to detect hosts. The two strains of *Drosophila* were located and parasitized at different rates. The 'movers' were more susceptible than 'non-movers', when encountered by *A. tabida*, but there were no significant differences in susceptibility of either form to *L. heterotoma*.

Parasitization can also influence the susceptibility of hosts. For example, the song of the male field cricket (*Gryllus integer*) attracts conspecific females but also attracts adult female tachinids (*Euphasiopteryx ochracea*) which deposit larvae on the calling males (Cade, 1984). Parasitism influences the duration of the calling song: prior to parasitization, crickets called for an average of about 4.3 h per night, while after parasitization the duration of the call was reduced to 2.2 h per night. This suggests that parasitized individuals are less likely to be encountered by other tachinids than non-parasitized hosts. In addition, Cade (1975) also found that some males in the unparasitized cohort did not call at all. These non-calling males are infrequently parasitized, indicating that polymorphism in calling behaviour influences susceptibility.

Even if a host is both acceptable and susceptible to the adult parasitoid, it may vary in its degree of suitability for the development of offspring.

Vinson and Iwantsch (1980) have reviewed host suitability for insect parasitoids and note four major factors that influence individual fitness:

1. whether the parasitoid is able to evade (or defend itself against) the host's internal defence systems;
2. the degree of interspecific and intraspecific competition experienced by the parasitoid in the host;
3. whether the host contains toxins detrimental to the parasitoid's eggs/larvae;
4. whether the nutritional quality of the host is suitable for development.

Variation in host suitability can affect a wide range of parasitoid life history parameters; for example, pre-emergence survivorship, development rate, adult size, adult fecundity and adult longevity. Research on *Trichogramma* has demonstrated that the performance of these egg-parasitoids is influenced by host age. In old hosts the proportion of hosts parasitized, the number of progeny oviposited and the proportion of parasitized hosts in which development was successfully completed declined (Juliano, 1982). The type of host can also influence the subsequent performance of adults. Boldt (1974) showed that adults emerging from eggs of *Trichoplusia ni* moved faster than did those reared from eggs of *Sitotroga cerealella*.

Intraspecific variation in hosts can also have significant effects on parasitoid performance. For example, Chabora (1970a, b) examined the suitability of pupae of *Musca domestica* from two populations (Florida and New York) as hosts for the parasitoid *Nasonia vitripennis*. Parasitoids exhibited marked differences in the rates of oviposition, pre-emergence survivorship and adult life span on the two strains.

The relative rates of parasitization may be asymmetrical when a range of host types are offered simultaneously to parasitoids. For example, the encyrtid parasitoid, *Anagyrus indicus*, was able to complete development and emerge, in no-choice experiments, from 1st, 2nd and 3rd instar and adult females of its mealybug host, *Nipaecoccus vastator*. However, in choice tests significantly more adult females were parasitized than 3rd instars whereas 1st and 2nd instars were not parasitized at all (Nechols and Kikuchi, 1985). Preferences between hosts of different species may also occur. Wright *et al.* (1989) examined the preferences of an *Aleochara* sp. (a staphylinid beetle whose larvae are parasitoids of fly pupae). These parasitoids accepted hosts widely different in mass (range ~0.5–45.0 mg) in no-choice experiments. However, in choice experiments, intermediate-sized hosts were preferred.

Hopper and King (1984) examined the preferences of *Micropletis croceipes* attacking various instars of *Heliothis* spp. and suggested that preference did not depend on the total host density or the density of the most preferred stage. They also examined the relationship between preference for a particular stage and the suitability of that stage for the parasitoids' offspring. Cock's (1978) review of preference indices suggests that the design of an experimental study will dictate the selection of the preference index. Hopper and King (1984) used Manly's index (Manly *et al.*, 1972) which allows for the changing availability of non-parasitized hosts over the course of the experiment. Development rate (a measure of suitability) and instar preference are positively correlated in *Micropletis* (Fig. 4.6). An analysis of the relationship between progeny sex-ratio and preference in Nechols and Kikuchi's (1985) study of *Anagyrus* shows a similar trend: progeny emerging from non-preferred hosts are predominantly male, while those emerging from preferred hosts are predominantly female (Fig. 4.6).

Preference for particular classes of host in insect parasitoids may have evolved as a result of the marked differences in the fitness of progeny emerging from these hosts, and this may be a functional explanation of the observed relationships between single measures of suitability and preference. Recent studies on host preference indicate that preferences should also be a function of time (e.g. Ward, 1987; see also Ward, Chapter 8, this volume). In addition within-patch exploitation will alter the availability and/or suitability of the most preferred hosts. Some current habitat selection models predict that under these circumstances the next preferred class of resource should then be exploited: this is in marked contrast to the observations made by Hopper and King (1984). The patterns of preference may also depend on the life history strategies of adult parasitoids (e.g. whether they are egg-limited, whether they can

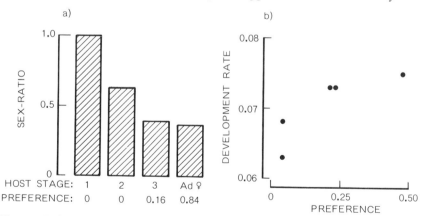

Fig. 4.6 Relationships between single measures of host suitability and preference: (a) *Anagyrus indicus* attacking *Nipaecoccus vastator*—sex ratio (proportion males) of the offspring is female biased in preferred host stages (based on data in Nechols and Kikuchi, 1985); (b) *Micropletis croceipes* attacking different instars of *Heliothis* species—development rate per day is greatest in preferred instars (based on data in Hopper and King, 1984).

disperse to other patches) and the spatial differences in the age structure of host populations. Clearly, more research is needed here in order to answer these problems.

4.4 PATCH-TIME ALLOCATION

In nature a parasitoid may encounter a series of areas occupied by differing number of hosts with a range of suitabilities. The way in which female parasitoids exploit these host patches can have marked effects on fitness and, as a result, selection should have shaped behaviours caus- ing individuals to move between patches as well as behaviour within patches.

Some experimental data suggest that patches with high host densities intercept proportionally more parasitoids and retain them for longer periods. The net effect is that hosts in high host-density patches experi- ence higher rates of parasitization (e.g. Gross *et al.*, 1984). However, detailed studies reveal that this pattern can come about for a wide variety of reasons. Some laboratory studies have identified methodological prob- lems. For example, the form of the relationship between time in patch and patch density can be a function of experimental design and may change when parasitoids are free to leave patches (e.g. Collins *et al.*, 1981; Hertlein and Thorarinsson, 1987). Other patterns can be attributed to specific behavioural mechanisms.

Nealis (1986), in a laboratory study, showed that *Cotesia rubecula* spent most time on leaves that were subject to the greatest levels of feeding

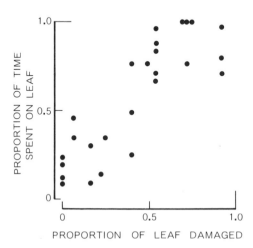

Fig. 4.7 Proportion of time spent by *Cotesia rubecula* on leaves with given levels of host feeding damage (after Nealis, 1986).

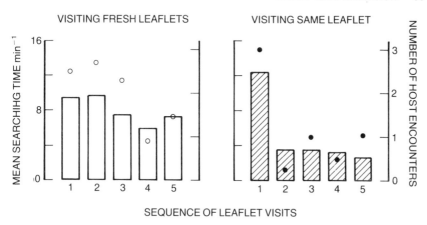

Fig. 4.8 Discrimination of previously searched leaflets by the leafminer parasitoid *Dapsilarthra rufiventris*. Searching times (bars) and number of hosts encountered (circles) are greater when this parasitoid is offered fresh leaflets rather than leaflets previously encountered in a series of visits (after Sugimoto *et al.*, 1986).

damage by their host, *Pieris rapae* (Fig. 4.7). The parasitoids moved slowly in areas with feeding damage and responded to a kairomone in the oral secretions of the hosts. There appeared to be no evidence to suggest that the amount of time spent by C. *rubecula* on a leaf depended on the number of previous encounters with non-parasitized hosts and, therefore, the profitability of the patch. Rather, the increased time spent in high density patches arose as a result of behaviours exhibited after attacking a host (e.g. grooming). Thus, this parasitoid spends more time on plants where hosts have been previously encountered. Field studies indicate that C. *rubecula* moves over the whole of the plant, hovering in flight near plants that have been damaged before and landing on these leaves with damage. Similar observations demonstrating the importance of foraging flights within habitats containing hosts are presented by Ayal (1987).

Within-patch variables can also influence the time spent in each patch. For example, Sugimoto *et al.* (1986) examined the behaviour of the parasitoid *Dapsilarthra rufiventris* searching for larvae of the leaf-mining fly, *Phytomyza ranunculi*. These larvae are non-randomly distributed on the leaflets of their host plants. Effective searching by this parasitoid comes about when they avoid intensive searching of previously visited leaflets. These authors showed that searching time and encounter rate decrease rapidly when parasitoids are presented with the same leaflet to search on five successive occasions. In contrast, the decrease was less when fresh leaflets were available on each searching occasion (Fig. 4.8). *D. rufiventris* allocated less time to previously searched patches than unsearched cues by discrimination at the *leaflet* level. This change in behaviour resulted from detection of a volatile pheromone left on the leaflet by the ovipositing female rather than from encountering a previously parasitized host (Sugimoto *et al.*, 1986). In addition, the amount of pheromone deposited on a leaflet by a female parasitoid influences the time that she spends in that patch (Sugimoto *et al.*, 1987).

4.5 HOST DISCRIMINATION AND SUPERPARASITISM

Within patches, parasitoids search for and encounter potential hosts, and some hosts may already have been parasitized. This presents a problem for individual parasitoids as oviposition in a parasitized host may have a substantial influence on fitness. Under these circumstances host discrimination, that is the ability of parasitoids to recognize and respond differentially to hosts which are already parasitized by conspecifics or themselves, may be a selective advantage. Bakker *et al.* (1985) recognize three processes which are essential in the evolution of discrimination: that parasitized hosts should be marked in some way; that these marks should be detectable and recognizable; and that the recognition of marks leads to a reduction in the willingness to parasitize the hosts. These authors also outline some possible advantages of host discrimination:

1. if oviposition takes longer than host checking then the ability to discriminate may save time;
2. when repeated parasitization increases the mortality rate of hosts, then discrimination may prevent the wastage of hosts;
3. if supernumerary larvae are eliminated in competition then discrimination prevents wastage of eggs; a factor that may be important in egg-limited parasitoids;
4. the changing rates of encounter with parasitized and non-parasitized hosts may be used by the female to assess the profitability of a patch.

Many parasitoids have the ability to discriminate and this frequently arises through the detection of kairomones deposited on the surface of the host or by substances injected into the host by conspecific females (e.g. Greany and Oatman, 1972a,b; Vinson and Guillot, 1972; Weseloh, 1976). Discrimination is not found in all parasitoids. For example, species in the same genus may vary in their ability to discriminate (Legaspi, 1986). Similarly, species with very similar life histories may vary in ability: the aphid parasitoids *Ephedrus californicus* and *E. cerasicola* can discriminate between parasitized and unparasitized hosts (Chow and Mackauer, 1986; Hofsvang and Hagvar, 1986) while *Aphidius rhopalosiphi* is unable to do so (Gardner *et al.*, 1984).

The surface kairomone source may be removed by washing the host and these hosts may then be subject to further attacks where the parasitoid will attack by probing but will not lay any eggs. If a host is already parasitized it may show a substantial physiological change detectable to a second female (e.g. Lawrence, 1981; Wylie, 1983). The kairomones used by parasitoids appear to be species specific: most parasitoids will readily attack hosts that are already parasitized by other species.

Discrimination is not absolute and failure to discriminate comes about for a wide variety of reasons. Van Lenteren (1976) suggested that parasitoid experience may be important and observed that naive individuals do not respond to kairomones. Similarly, Chow and Mackauer (1986) demonstrated that experienced, but not naive females of *E. californicus*, respond to an external marker. Superparasitism, the product of failure to discriminate absolutely, is not merely a result of inexperience. For example, it increases as the density of conspecific parasitoids increases (Lawrence, 1981). Host density also influences superparasitism rates. *Cotesia* (= *Apanteles*) *glomerata* varies the number of eggs laid per host depending on the frequency with which hosts are presented. This appears to maximize the fitness of individual females throughout life (Ikawa and Susuki, 1982; Ikawa and Okabe, 1985).

Intervals between successive contacts with hosts are also important in solitary species. Cloutier *et al.* (1984) show that females of *Aphidius nigripes* re-encountering hosts at intervals of less than 2 h cannot discriminate. If, however, the time interval between attacks is greater than 2 h then the probability of discrimination increases (Fig. 4.9). Temporal changes in the rate of oviposition restraint in parasitized hosts are frequently associated with the changing physiological status of the hosts as the offspring of the first female hatch and begin development within the host.

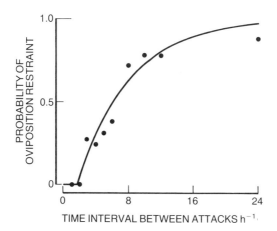

Fig. 4.9 Host discrimination in the aphid parasitoid *Aphidius nigripes*: oviposition restraint is a function of time elapsed since parasitization (after Cloutier *et al.*, 1984).

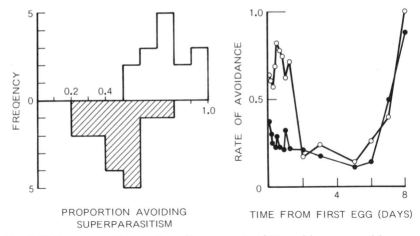

Fig. 4.10 Superparasitism in the solitary parasitoid *Nemeritis canescens*: (a) frequency distributions of female parasitoids avoiding self (open histograms) and conspecific (hatched histograms) superparasitism; (b) the relationships between rates of avoidance of self (open circles) and conspecific (closed circles) superparasitism and time elapsed between first oviposition and re-encounter with individual hosts (after Hubbard *et al.*, 1987).

Hubbard *et al.* (1987) also observed time-dependent changes in the ability to discriminate. These authors, however, found major differences between the responses of *Nemeritis canescens* (a solitary parasitoid of the larvae of flour moths) encountering hosts parasitized by conspecifics and hosts parasitized by themselves. Females avoided both types of host if they had been previously parasitized seven or eight days prior to encounter. Females of this species were also able to detect hosts which they *themselves* had encountered, but only if the encounter had taken place in the previous two days. In other words, they were able to avoid self-superparasitism (Fig. 4.10) and this short-term differential discrimination seemed to depend on a specific chemical marker (Hubbard *et al.*, 1987). In contrast, a study on self and conspecific superparasitism in a gregarious parasitoid (*T. evanescens*) found no evidence of differential discrimination (van Dijken and Waage, 1987). There may be some fundamental differences between solitary and gregarious parasitoids in the advantages of avoiding self-superparasitism. Active conspecific superparasitism may be adaptive in solitary parasitoids when unparasitized hosts are rare (Hubbard *et al.*, 1987), whereas avoidance of self-superparasitism in gregarious parasitoids may be less important; especially in species that are able to adjust patterns of progeny and sex allocation in response to host suitability (e.g. Waage and Godfray, 1985; Waage, 1986).

Age specific physiological changes in individual parasitoids can also influence patterns of discrimination and superparasitism. Some species reach adult life with a full complement of eggs, others emerge with almost no eggs, while still others mature additional eggs after oviposition begins. In the latter group, the rate of egg maturation may be a function of oviposition experience (Lawrence *et al.*, 1978). Ovipositional activity may depend on egg load and the rate at which further eggs are matured (Walter, 1988), and in some species this affects the way in which hosts are exploited. Collins and Dixon (1986) examined the way in which egg depletion influenced the foraging behaviour and host discrimination in the aphid parasitoid, *Monoctonus pseudoplatani*. They observed that the attack rate of this parasitoid declined over the course of experiments and showed that this could be partly attributed to the rejection of parasitized hosts as exploitation occurred. However, part of the change came about through a marked reduction in time spent searching for hosts at the end of the experiment compared with the start. There were also changes in the patterns of discrimination. Collins and Dixon (1986) offered these parasitoids sets of hosts at hourly intervals for six successive hours (Fig. 4.11

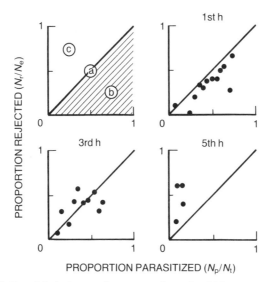

Fig. 4.11 Relationship between the proportion of aphids encountered by the parasitoid *Monoctonus pseudoplatani* that were rejected and the corresponding proportion of hosts which were parasitized. This figure illustrates data from three hours of six successive 1 h foraging periods. In this test, data points lying on the diagonal (a) indicate perfect discrimination. Data points in zone (b) show a shift towards superparasitism, while points in zone (c) show a tendency to reject hosts which are unparasitized. This aphid parasitoid is egg-limited and the data suggest that selectivity increases as egg-load decreases (after Collins and Dixon, 1986).

illustrates 3 hours). At the start of the experiment parasitoids rejected fewer hosts than expected, indicating a tendency towards superparasitism when egg loads were high. In contrast, in the fifth and sixth hours, the parasitoids rejected more hosts than expected; in these cases the parasitoids were encountering but not ovipositing in unparastized hosts (Fig. 4.11). The degree of discrimination shown by *M. pseudoplatani* appears to be related to egg load, which can be used as a measure of motivation to oviposit. As Collins and Dixon (1986) point out, motivation plays a crucial role in shaping foraging behaviour.

4.6 PROGENY AND SEX ALLOCATION

The previous sections of this chapter have outlined the ways in which ovipositing insects search for, locate and utilize hosts. This section is concerned with the decisions that arise after locating a suitable host. How many eggs should be laid into each host, and, because many parasitoids have haplo-diploid sex determination, what sex should their offspring be? Individual hosts represent a limited resource for the offspring of female parasitoids. As a result, interest has focused on the idea that female parasitoids may lay their eggs in clutch sizes that maximize the fitness of their offspring (see Iwasa *et al.*, 1984; Parker and Courtney, 1984; Charnov and Skinner, 1984, 1985; Waage and Godfray, 1985; Waage, 1986). The probability of surviving from egg to adult may decline with increased clutch size and the size of emerging adults may also depend on the number of conspecific competitors in each host. Charnov and Skinner (1985) and Waage and Godfray (1985) have made initial approximations of the fitness of clutches by calculating the product of the number of eggs laid, the probability of surviving to adulthood, and the number of eggs in the female offspring of the clutch. The last term takes into account the effects of within-clutch density dependence influencing fecundity through adult size. Their reviews of the empirical evidence suggest that:

1. within species, pre-emergence survivorship declines with increasing clutch size;
2. species with lower levels of within brood mortality are characterized by larger brood sizes;
3. the clutch size in nature is usually smaller than that which maximizes fitness per host.

Maximizing fitness per host may not maximize the life time fitness of the ovipositing female; although Charnov and Skinner (1985) point out that the above approximation of clutch fitness may be incomplete. In addition, optimal brood size may depend on whether the parasitoid is egg limited, on the probability of finding another host, and whether there is a cost of reproduction (i.e. a negative correlation between reproductive effort and adult female survival prior to producing the next clutch). Thus, clutch size in the field may be influenced by host density, the rate of search of the parasitoids and the time taken to handle a host. Laboratory experiments indicate that some species can modify clutch size in response to the rate of encounter with unparasitized hosts (e.g. Ikawa and Suzuki, 1982; Ikawa and Okabe, 1985).

Progeny allocation per host depends on host quality and the number of hosts available. Schmidt and Smith (1985, 1987) demonstrate that *Trichogramma minutum* modifies clutch size in response to the exposed surface area of host eggs, and this suggests that in some species the spatial arrangement of hosts may influence parasitoid fitness. Similarly, the pupal parasitoid *Pteromalus puparum* can detect differences in host size through exposed surface area and make corresponding changes to clutch size (Takagi, 1986). However, clutch size is not adjusted in direct proportion to the exposed surface area of hosts and this suggests that other qualitative factors are involved in this decision.

For gregarious parasitoids, host suitability may also depend on whether the host already contains the eggs of a conspecific. The value of the host for the superparasitizing female may depend on the original quality of the host, the number of eggs laid by the first female and the time elapsed since these eggs were laid (Charnov and Skinner, 1985). Models of these double oviposition situations predict that the clutch size of the second female should be smaller than that of the first, that high probabilities of superparasitism have only a marginal effect on the difference in clutch size, and that the differences in clutch sizes increases as the searching costs rise (Parker and Courtney, 1984). Experimental studies on *Nasonia vitripennis* (Holmes, 1972), *Cotesia glomeratus* (Ikawa and Suzuki, 1982) and *P. puparum* (Takagi, 1986) have shown that superparasitizing females produce smaller clutches.

Host quality has additional effects on the fitness of ovipositing females with haplo-diploid sex determination. This mode of reproduction is extremely common in the parasitic hymenoptera: in these species fertilized eggs give rise to female offspring, while unfertilized eggs give rise to male offspring. So, there may be mechanisms for controlling the production of fertilized eggs and thus the sex ratio of offspring produced in different ecological circumstances. Mechanistic and functional aspects of variation in sex ratios are reviewed by Charnov (1982), Frank (1983), Waage (1986) and Hurlbutt King (1987). One of the commonest observations is that sex ratios of solitary parasitoids emerging from small hosts tend to be male biased. Assuming that female fitness is relatively greater than males when emerging from larger hosts and vice versa from small hosts, then selection will favour mothers placing male eggs in smaller hosts and female eggs in larger hosts (Charnov, 1979). Most studies have scored sex ratio on emergence whereas Charnov's (1979) predictions relate to primary sex-ratio (i.e. the sex ratio at oviposition). If there are any differential mortality patterns of sexes in different quality hosts, emergence sex ratio could show a directional bias. There is a clear need to correct emergence sex ratios for any differential mortality patterns (Wellings *et al.*, 1986).

Fisher (1930) suggested that when individuals mate at random and the offspring of each sex are equally costly, the optimal sex ratio should be 0.5 (proportion males). Any deviation from this sex ratio results in a frequency-dependent advantage to the rarer sex. However, the optimal sex ratio can vary from 0.5 when mating patterns are structured. Hamilton (1967) modelled the situation where offspring in discrete patches of resource mate with each other prior to female dispersal to new patches. Under these circumstances of local mate competition (LMC), the optimal sex ratio depends on the number of females colonizing the resource patches and this sex ratio will be female biased if the number of females is small. In general, the observed qualitative patterns of female-biased sex ratios in gregarious parasitoids support Hamilton's LMC model (see Waage, 1986).

Gregarious parasitoids encountering an unparasitized host (the equivalent of a resource patch in the LMC model) produce female-biased sex ratios. The optimal behaviour of a second female encountering this host is to adjust clutch size (see above) and theoretical studies suggest that she should also modify the sex ratio of her offspring. The optimal sex ratio of the second female may depend on the relative clutch sizes of the two females and, in most circumstances, will tend to be more male biased than that of the first female (Suzuki and Iwasa, 1980; Werren, 1980). Some experimental studies on *N. vitripennis* support this idea (Werren, 1980), while others indicate that sex ratio patterns are highly variable from genotype to genotype (Orzack and Parker, 1986). Further studies are needed to resolve these problems. There is a need to understand how the mechanisms controlling sex differ between genotypes in differing ecological circumstances and also to investigate the selective advantages of variation in this trait. Throughout this section the problems of progeny and sex allocation have been treated as separate processes. Theoretical and experimental studies have used this division to make research tractable. As Waage and Godfray (1985) point out, this is an artificial dichotomy and in nature both processes occur simultaneously. The challenge for the future lies in developing models incorporating both problems and devising experiments which highlight the rules used in making these decisions.

4.7 INDIVIDUAL BEHAVIOUR AND POPULATION DYNAMICS

The behaviours exhibited by parasitoids in locating and exploiting hosts may have a major effect on the population dynamics of both parasitoids and hosts (Hassell and May, 1985). The patchiness of host populations and the behavioural factors influencing the way these patches are exploited determine the rate of immigration and emigration between patches and the birth rates of parasitoid populations. These, in turn, influence the death rates of host populations. Earlier sections of this chapter show how some parasitoids use kairomone sources of various types during the searching process. The adaptive significance of this behaviour is that parasitoids increase the chance of encountering and ovipositing in hosts. Laboratory studies also suggest that within-patch behaviour has the net effect of causing individual parasitoids to spend proportionally more of their searching time in high density patches. Proximate behavioural models, describing the allocation of foraging time in patches of different host densities, incorporate a variety of simple decision-making rules. These include foraging by expectation, foraging for a fixed time in each patch, fixed searching time in each patch and encounter rate mechanisms in which the parasitoid leaves the patch when the rate of encounter with hosts falls below a critical threshold (Waage, 1979). These models are deterministic and do not account for the highly variable searching times allocated by parasitoids to patches of *equal* host densities observed in some studies (e.g. Morrison, 1986). They do, however, provide a starting point for examining the relationship between foraging behaviour and population dynamics.

Host populations are non-randomly distributed throughout habitats, and models describing the dynamics of host/parasitoid populations suggest that aggregation of host populations has a powerful influence on stabilizing these interactions (Hassell, 1980; May *et al.*, 1981). Models describing these types of interactions incorporate non-random host distributions and are suitable for examining the effect of different foraging models on the number of hosts attacked. Hassell and May (1974) give a general function for parasitism of aggregated hosts:

$$H_a = n\bar{H}\left\{1 - \frac{1}{\bar{H}}\sum_{j=0}^{\infty}\left[p(j)j\exp\left(\frac{-aT(j)}{T_0 + n\sum_{j=0}^{\infty}[p(j)T(j)]}\right)\right]\right\} \quad (1)$$

where H_a is the number of hosts attacked, $\bar{H}$ is the mean number of hosts per patch, n is the number of patches, T_0 is the transit time between patches, $T(j)$ is the time spent by the parasitoid in a patch containing j hosts, a is the area of discovery or searching efficiency, and the hosts are distributed according to some statistical distribution so that the probability of having j hosts in a patch is $p(j)$ and

$$\sum_{j=0}^{\infty} p(j) = 1$$

It is possible to incorporate simple behavioural models of time allocated to patches of different host density via the term $T(j)$, and to examine the consequences to the host population of the different generalized behaviour patterns of individual parasitoids foraging for non-randomly distributed hosts. Host aggregation is most easily captured using the negative binomial distribution. Variation in the coefficient k describes aggregation: when k is small, populations are aggregated and, when k is large, populations tend towards random distributions. The relationships between three models of individual foraging behaviour and parasitism rates in different density host populations with two levels of aggregation are illustrated in Fig. 4.12. The consequences of the behavioural models for host populations are very similar in each case: parasitism rates increase with increasing mean host population densities and increasing levels of host aggregation. So, while these models of behaviour may differ in terms of adaptive significance and optimal use of resources for indi-

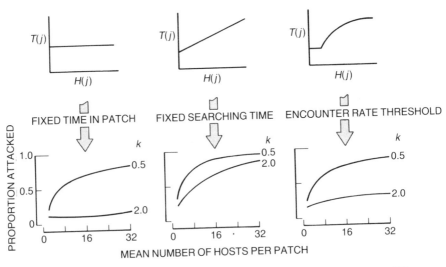

Fig. 4.12 The influence of three different behavioural models of the time (Tj) spent by the parasitoid in a patch of host density (Hj) on *the proportion of hosts parasitized. These models were included in Equation (1) (see text) and run for different mean numbers of hosts per patch ($\bar{H}$) and two levels of host aggregation (k)* (P. W. Wellings, unpublished data).

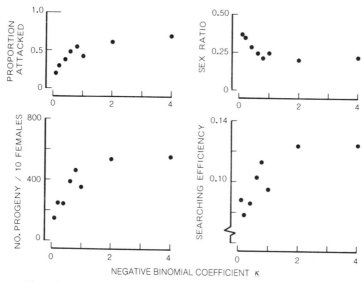

Fig. 4.13 The effect of host spatial distribution on parasitism by *Nasonia vitripennis*. Host density was held constant while spatial distribution was manipulated to conform with different values of *k* (after Jones and Turner, 1987).

vidual parasitoids, their consequences for host population dynamics are qualitatively similar.

Jones and Turner (1987) have examined the effects of host spatial distribution on the patterns of parasitism by the parasitoid *N. vitripennis*. In each treatment, they used 50 standardized host pupae and arranged them in various aggregated spatial distributions in arenas with 100 patches. Female parasitoids were then allowed to forage for and oviposite in hosts in these patches. These authors recorded the number of offspring produced, the proportion of hosts attacked, the searching efficiency of the parasitoids and the sex ratio of offspring for each of eight spatial distributions (Fig. 4.13). The results indicate that there are systematic changes in all these parameters with respect to host spatial distribution. However, some results are completely opposite to those predicted by the behavioural models used in Fig. 4.12. For example, in Jones and Turner's study the proportion of hosts attacked increased as host populations became *less* aggregated. The reasons for these differences are not clear, although they could be attributed to an assumption that searching efficiency is a constant, whereas experimental results indicate that this parameter declines with aggregation (Fig. 4.13). In addition, Jones and Turner used 10 female parasitoids simultaneously in each arena whereas the model (Equation (1)) does not incorporate any terms for mutual interference between females. In addition, it would be interesting to examine the way changes in mean host density per patch influence the observed patterns. The results presented in Fig. 4.13 are for a constant host density. My use of Equation (1) incorporates other limiting assumptions. New patches are assumed to be instanteneously discovered (i.e. T_0 = 0). In nature, travel time between patches may be large and very variable. There is a real need for detailed experimental studies on between patch movements in parasitoids. Equation (1) also assumes that host spatial distributions are independent of host density, whereas in the field, the spatial distributions of populations are density related (e.g. Wellings, 1987).

The differences between this experimental study and the theoretical models highlight avenues for future research. At present, there is a wealth of information, derived predominantly from laboratory studies, on the environmental cues influencing host location and oviposition decisions and this forms the basis for some models of the way female parasitoids exploit patchily distributed host populations (e.g. Charnov and Stephens, 1988). However, relatively few studies have investigated the assumptions built into these models and the sensitivity of the form of the results to experimental design. In addition, studies reporting field data on the behaviours of parasitoids exploiting host populations are very poorly represented in the literature.

References

van Alphen, J. J. M. and Vet, L. E. M. (1986) An evolutionary approach to host finding and selection. In *Insect Parasitoids*, (eds J. Waage and D. Greathead), Academic Press, London, 23–61.

Arthur, A. P. (1966) Associative learning in *Itoplectis conquisitor* (Say) (Hymenoptera: Ichneumonidae). *Can. Entomol.*, **98**, 213–223.

Arthur, A. P. (1971) Associative learning by *Nemeritis canescens* (Hymenoptera: Ichneumonidae). *Can. Entomol.*, **103**, 1137–1141.

Ayal, Y. (1987) The foraging strategy of *Diaeretiella rapae*. 1. The concept of the elementary unit of foraging. *J. Anim. Ecol.*, **56**, 1057–1068.

Bakker, K., van Alphen, J. J. M., van Batenburg, F. H. D., *et al.* (1985) The function of host discrimination and superparasitization. *Oecologia*, **67**, 572–576.

Boldt, P. E. (1974) Temperature, humidity, and host: effect on rate of search of *Trichogramma evanescens* and *T. minutum* auctt. (not Riley, 1871) *Ann. Entomol. Soc. Amer.*, **67**, 706–708.

Bouchard, Y. and Cloutier, C. (1984) Honeydew as a source of host-searching kairomones for the aphid parasitoid *Aphidius nigripes* (Hymenoptera: Aphidiidae). *Can. J. Zool.*, **62**, 1513.

Bouchard, Y. and Cloutier, C. (1985) Role of olfaction in host finding by aphid parasitoid, *Aphidius nigripes*. (Hymenoptera: Aphidiidae). *J. Chem. Ecol.*, **11**, 801–808.

Cade, W. H. (1975) Acoustically orienting parasitoids: fly phonotaxis to cricket song. *Science*, **190**, 1312–1313.

Cade, W. H. (1984) Effects of fly parasitoids in nightly calling duration in field crickets. *Can. J. Zool.*, **62**, 226–228.

Camors, F. B. and Payne, T. L. (1972) Responses of *Heydenia unica* (Hymenoptera: Pteromalidae) to *Dendroctonus frontalis* (Coleoptera: Scolytidae) pheromones and host tree terpene. *Ann. Entomol. Soc. Amer.*, **65**, 31–33.

Carton, Y. and David, J. R. (1985) Relation between the genetic variability of digging behaviour of *Drosophila* larvae and their susceptibility to a parasitic wasp. *Behav. Genet.*, **15**, 143–154.

Chabora, P. C. (1970a) Studies of parasite–host interaction. II. Reproductive and developmental response of the parasite *Nasonia vitripennis* (Hymenoptera: Pteromalidae) to strains of the house fly host, *Musca domestica*. *Ann. Entomol. Soc. Amer.*, **63**, 1632–1636.

Chabora, P. C. (1970b) Studies in parasite–host interaction. III. Host race effect on the life table and population growth statistics of the parasite *Nasonia vitripennis*. *Ann. Entomol. Soc. Amer.*, **63**, 1637–1642.

Charnov, E. L. (1979) The genetical evolution of patterns of sexuality: Darwinian fitness. *Amer. Nat.*, **113**, 465–480.

Charnov, E. L. (1982) *The Theory of Sex Allocation*, Princeton University Press, Princeton, NJ.

Charnov, E. L. and Skinner, S. W. (1984) Evolution of host selection and clutch size in parasitoid wasps. *Fla. Entomol.*, **67**, 5–21.

Charnov, E. L. and Skinner, S. W. (1985) Complementary approaches to the understanding of parasitoid oviposition decision. *Environ. Entomol.*, **14**, 383–391.

Charnov, E. L. and Stephens, D. W. (1988) On the evolution of host selection in solitary parasitoids. *Amer. Nat.*, **132**, 707–722.

Chiri, A. A. and Legner, E. F. (1982) Host-searching kairomones alter behaviour of *Chelonus* sp. nr. *curvimaenlatus*, a hymenopterous parasite of the pink bollworm, *Pectinophora gossypiella* Sanders. *Environ. Entomol.*, **11**, 452–455.

Chow, F. J. and Mackauer, M. (1986) Host discrimination and larval competition in the aphid parasite *Ephedrus californicus*. *Entomol. Exp. Appl.*, **41**, 243–254.

Clement, S. L., Rubink, W. L. and McCartney, D. A. (1986) Larviposition response of *Bonnetia comta* (Dipt., Tachinidae) to a kairomone of *Agrotis ipsilon* (Lep., Noctuidae). *Entomophaga*, **31**, 277–284.

Cloutier, C., Dohse, L. A. and Baudin, F. (1984) Host discrimination in the aphid parasitoid *Aphidius nigripes*. *Can. J. Zool.*, **62**, 1367–1372.

Cock, M. J. W. (1978) The assessment of preference. *J. Anim. Ecol.*, **47**, 805–816.

Collins, M. D. and Dixon, A. F. G. (1986) The effect of egg depletion on the foraging behaviour of an aphid parasitoid. *J. Appl. Entomol.*, **102**, 342–352.

Collins, M. D., Ward, S. A. and Dixon, A. F. G. (1981) Handling time and the functional response of *Aphelinus thomsoni*, a predator and parasite of the aphid *Drepanosiphum platanoidis*. *J. Anim. Ecol.*, **50**, 479–487.

Dicke, M., van Lenteren, J. C., Boskamp, G. J. F., *et al.* (1984) Chemical stimuli in host-habitat location by *Leptopilina heterotoma* (Thomson) (Hymenoptera: Eucoilidae), a parasite of *Drosophila*. *J. Chem. Ecol.*, **10**, 695–712.

van Dijken, M. and Waage, J. (1987) Self and conspecific superparasitism in *Trichogramma evanescens* Westwood. *Entomol. Exp. Appl.*, **43**, 183–192.

Dmoch, J., Lewis, W. J., Martin, B. P. and Nordlund, D. A. (1985) Role of host produced stimuli and learning in host selection behavior of *Cotesia* (= *Apanteles*) *marginiventris* (Cresson) (Hymenoptera, Braconidae). *J. Chem. Ecol.*, **11**, 453–463.

Doutt, R. L. (1964) Biological characteristics of entomophagous insects. In *Biological Control of Insect Pests and Weeds* (ed. P. DeBach), Chapman and Hall, London, 145–167.

Faeth, S. H. and Bultman, T. L. (1986) Interacting effects of increased tannin levels on leaf-mining insects. *Entomol. Exp. Appl.*, **40**, 297–300.

Fisher, R. A. (1930) *The Genetical Theory of Natural Selection*, Oxford University Press, Oxford.

Fourie, L. J., Horak, I. G. and Visser, E. (1987) Quantitative and qualitative aspects of the parasite burdens of rock dassies (*Procavia capensis* (Pallas, 1766)) in the Mountain Zebra National Park. *S. Afr. J. Zool.*, **22**, 101–106.

Fowler, J. A. and Miller, C. J. (1984) Non-haematophagous ectoparasite populations of procellariiform birds in Sheltland, Scotland. *Seabird.*, **7**, 23–30.

Frank, S. A. (1983) A hierarchical view of sex-ratio patterns. *Fla. Entomol.*, **66**, 42–75.

Gardner, S. M. and van Lenteen, J. C. (1986) Characterisation of the arrestment responses of *Trichogramma evanescens*. *Oecologia*, **68**, 265–270.

Gardner, S. M., Ward, S. A. and Dixon, A. F. G. (1984) Limitation of superparasitism by *Aphidius rhopalosiphi:* a consequence of aphid defensive behaviour. *Ecol. Entomol.*, 149–155.

Greany, P. D. and Oatman, E. R. (1972a) Demonstrations of host discrimination in the parasite *Orgilus lepidus* (Hymenoptera: Braconidae). *Ann. Entomol. Soc. Amer.*, **65**, 375–376.

Greany, P. D. and Oatman, E. R. (1972b) Analysis of host discrimination in the parasite *Orgilus lepidus* (Hymenoptera: Braconidae). *Ann. Entomol. Soc. Amer.*, **65**, 377–383.

Gross, H. R. Jr., Lewis, W. J., Beevers, M. and Nordlund, D. (1984) *Trichogramma pretiosum* (Hymenoptera: Trichogrammatidae). Effects of augmented densities and distributions of *Heliothis zea* (Lepidoptera: Noctuidae) host eggs and kairomones in field performance. *Environ. Entomol.*, **13**, 981–985.

Hamilton, W. D. (1967) Extraordinary sex ratios. *Science*, **156**, 477–488.

Hassell, M. P. (1980) *The Dynamics of Arthropod Predator–Prey Systems*, Princeton University Press, Princeton, NJ.

Hassell, M. P. and May, R. M. (1974) Aggregation of predators and insect parasites and its effect on stability. *J. Anim. Ecol.*, **43**, 567–594.

Hassell, M. P. and May, R. M. (1985) From individual behaviour to population dynamics. In *Behavioural Ecology: Ecological Consequences of Adaptive Behaviour* (eds R. M. Sibly and R. H. Smith), Blackwell Scientific, Oxford, 3–32.

Hertlein, M. B. and Thorarinsson, K. (1987) Variable patch times and the functional response of *Leptopilina boulardi* (Hymenoptera: Encoilidae). *Environ. Entomol.*, **16**, 593–598.

Hofsvang, T. and Hagvar, E. B. (1986) Oviposition behavior of *Ephedrus cerasicola* (Hym., Aphidiidae) parasitizing different instars of its aphid host. *Entomophaga*, **31**, 261–267.

Holmes, H. B. (1972) Genetic evidence for fewer progeny and a higher percent males when *Nasonia vitripennis* oviposits in previously parasitized hosts. *Entomophaga*, **17**, 119–126.

Hopkins, D. (1985) Host associations of Siphonaptera from central Oregon. *North West Sci*, **59**, 108–114.

Hopper, K. R. and King, E. G. (1984) Preference of *Microplitis croceipes* (Hymenoptera: Braconidae) for instars and species of *Heliothis* (Lepidoptera: Noctuidae). *Environ. Entomol.*, **13**, 1145–1150.

Hubbard, S. F., Marris, G., Reynolds, A. and Rowe, G. W. (1987) Adaptive patterns of superparasitism by solitary parasitic wasps. *J. Anim. Ecol.*, **56**, 387–401.

Hurlbutt King, B. L. (1987) Offspring sex ratio in parasitoid wasps. *Q. Rev. Biol.* **62**, 367–396.

Ikawa, T. and Okabe, H. (1985) Regulation of egg number per host to maximize the reproductive success in the gregarious parasitoid, *Apanteles glomeratus* L. (Hymenoptera: Braconidae). *Appl. Entomol. Zool.*, **20**, 331–339.

Ikawa, T. and Suzuki, Y. (1982) Ovipositional experience of the gregarious parasitoid *Apanteles glomeratus* (Hym., Braconidae) influencing her discrimination of the host larvae, *Pieris rapae crucivara*. *Appl. Entomol. Zool.*, **17**, 119–126.

Istock, C. A. (1967) The evolution of complex life cycle phenomena: an ecological perspective. *Evolution*, **21**, 592–605.

Iwasa, Y., Suzuki, Y. and Matsuda, H. (1984) Theory of oviposition strategy of parasitoids. 1. Effect of mortality and limited egg number. *Theor. Popul. Biol.*, **26**, 205–227.

Jones, T. H. and Turner, B. D. (1987) The effect of host spatial distribution on patterns of parasitism by *Nasonia vitripennis*. *Entomol. Exp. Appl.*, **44**, 169–175.

Juliano, S. A. (1982) Influence of host age on host acceptability and host suitability for a species of *Trichogramma* (Hymenoptera: Trichogrammatidae) attacking aquatic Diptera. *Can. Entomol.*, **114**, 713–720.

Kitano, H. (1978) Studies on oviposition activity and host searching behaviour of *Apanteles glomeratus* (Hym., Braconidae) a parasitoid of the Cabbage White Butterfly. *Kontyu*, **46**, 152–164.

Lawrence, P. O. (1981) Interference competition and optimal host selection in the parasitic wasp *Biosteres longicaudatus*. *Ann. Entomol. Soc. Amer.*, **74**, 540–544.

Lawrence, P. O., Greany, P. D. and Baranowski, R. M. (1978) Oviposition behavior of *Biosteres longicaudatus*, a parasite of the Caribbean fruit fly, *Anastrepha suspensa*. *Ann. Entomol. Soc. Amer.*, **71**, 253–256.

van Leerdam, M. B., Smith, J. W., and Fuchs, T. W. (1985) Frass-mediated, host-finding behavior of *Cotesia flavipes*, a braconid parasite of *Diatraea saccharalis* (Lepidoptera: Pyralidae). *Ann. Entomol. Soc. Amer.*, **78**, 647–650.

Legaspi, B. A. C. (1986) Host discrimination in two species of ichneumonid wasps, *Diadegma* spp., attacking larvae of *Plutella xylostella*. *Entomol. Exp. Appl.*, **41**, 79–82.

van Lenteren, J. C. (1976) The development of host discrimination and the prevention of superparasitism in the parasite *Pseudeucoila bochei* (Hym.: Cynipidae). *Neth. J. Zool.*, **26**, 1–83.

Levins, R. (1968) *Evolution in changing environments*. Princeton University Press, Princeton, NJ.

Lewis, W. J. and Tumlinson, J. H. (1988) Host detection by chemically mediated associative learning in a parasitic wasp. *Nature*, **331**, 257–259.

Lewis, W. J., Jones, R. L., Nordlund, D. A. and Sparks, A. N. (1975) Kairomones and their use for management of entomophagous insects: 1. Evaluation for increasing rates of parasitization by *Trichogramma* spp. in the field. *J. Chem. Ecol.*, **1**, 343–347.

Lewis, W. J., Jones, R. L., Gross, H. R. and Nordlund, D. A. (1976) The role of kairomones and other behavioral chemicals in host finding by parasitic insects. *Behav. Biol.*, **16**, 267–289.

Manly, B. F. J., Miller, P. and Cook, L. M. (1972) Analysis of a selective predation experiment. *Amer. Nat.*, **106**, 719–736.

Marshall, A. G. (1981) *The Ecology of Ectoparasitic Insects*, Academic Press, London.

May, R. M., Hassell, M. P., Anderson, R. M. and Tonkyn, D. M. (1981) Density dependence in host parasitoid models. *J. Anim. Ecol.*, **50**, 855–865.

Monteith, L. G. (1964) Influence of the health of the food plant of the host on host-finding by tachinid parasites. *Can. Entomol.*, **96**, 1477–1482.

Monteith, L. G. (1966) Influence of new growth on the food plant of the host on host finding by *Drino bohemica* Mesnil (Diptera: Tachinidae). *Can. Entomol.*, **98**, 1205–1207.

Morrison, G. (1986) 'Searching time aggregation' and density dependent parasitism in a laboratory host–parasitoid interaction. *Oecologia*, **68**, 298–303.

Mueller, T. F. (1983) The effect of plants on the host relations of a specialist parasitoid of *Heliothis* larvae. *Entomol. Exp. Appl.*, **34**, 78–84.

Nealis, V. G. (1986) Responses to host kairomones and foraging behavior of the insect parasite *Cotesia rubecula* (Hymenoptera: Braconidae). *Can. J. Zool.*, **64**, 2393–2398.

Nechols, J. R. and Kikuchi, R. S. (1985) Host selection of the spherical mealybug (Homoptera: Pseudocoidae) by *Anagyrus indicus* (Hymenoptera: Encyrtidae): influence of host stage on parasitoid oviposition, development, sex ratio and survival. *Environ. Entomol.*, **14**, 32–37.

Nordlund, D. A., Strand, M. R., Lewis, W. J. and Vinson, S. B. (1987) Role of kairomones from host accessory gland secretion in host recognition by *Telenomus remus* and *Trichogramma pretiosum* with partial characterization. *Entomol. Exp. Appl.*, **44**, 37–43.

Odell, T. M. and Godwin, P. A. (1984) Host selection by *Blepharipa pratensis* (Meigen), a tachinid parasite of the Gypsy Moth, *Lymantria dispar* L. *J. Chem. Ecol.*, **10**, 311–320.

Orzack, S. H. and Parker, E. D., Jr. (1986) Sex ratio control in a parasitic wasp, *Nasonia vitripennis*. 1. Genetic variation in facultative sex ratio adjustment. *Evolution*, **40**, 331–340.

Osbrink, W. L. A. and Rust, M. K. (1985) Cat flea (Siphonaptera: Pulicidae): factors influencing host-finding behavior in the laboratory. *Ann. Entomol. Soc. Amer.*, **78**, 29–34.

Parker, G. A. and Courtney, S. P. (1984) Models of clutch size in insect oviposition. *Theor. Popul. Biol.*, **26**, 27–48.

Price, P. W. (1980) *Evolutionary Biology of Parasites*, Princeton University Press, Princeton, NJ.

Prokopy, R. J. and Webster, R. P. (1978) Oviposition deterring pheromone of *Rhagoletis pomonella*: a kairomone for its parasitoid *Opius lectus*. *J. Chem. Ecol.*, **4**, 481–494.

Sandlan, K. (1980) Host location by *Coccygominus turnionella* (Hymenoptera: Ichneumonidae). *Entomol. Exp. Appl.*, **27**, 233–245.

Schmidt, J. M. and Smith, J. J. B. (1985) The mechanism by which the parasitoid wasp *Trichogramma minutum* responds to host clusters. *Entomol. Exp. Appl.*, **39**, 287–294.

Schmidt, J. M. and Smith, J. J. B. (1987) The effect of host spacing on clutch size and parasitization rate of *Trichogramma minutum*. *Entomol. Exp. Appl.*, **43**, 125–131.

Sokolowski, M. B. and Turlings, T. C. J. (1987) *Drosophila* parasitoid–host interactions: vibrotaxis and ovipositor searching from the host's perspective. *Can. J. Zool.*, **65**, 461–464.

Sower, L. L. and Torgersen, T. R. (1979) Field application of synthetic Douglis-fir tussock moth pheromone did not reduce egg parasitism by two Hymenoptera. *Can. Entomol.*, **111**, 751–752.

Spradbery, J. P. (1970) Host finding by *Rhyssa persuasoria* (L.), an ichneumonid parasite of siricid woodwasps. *Anim. Behav.*, **18**, 103–114.

Spradbery, J. P. (1974) The responses of *Ibalia* species (Hymenoptera: Ibaliidae) to the fungal symbionts of siricid woodwasp hosts. *J. Entomol. (A)*, **48**, 217–222.

Strand, M. R. and Vinson, S. B. (1983) Factors affecting host recognition and acceptance in the egg parasitoid *Telenomus heliothidis* (Hymenoptera: Scelionidae). *Environ. Entomol.*, **12**, 1114–1119.

Sugimoto, T., Uenishi, M. and Machida, F. (1986) Foraging for patchily distributed leaf miners by the parasitoid, *Dapsilarthra rufiventris* (Hymenoptera:

Braconidae). 1. Discrimination of previously searched leaflets. *Appl. Entomol Zool.*, **21**, 500–508.

Sugimoto, T., Murakami, H. and Yamazaki, R. (1987) Foraging for patchily distributed leaf miners by the parasitoid, *Dapsilarthra rufiventris* (Hymenoptera: Braconidae). 11. Stopping rule for host search. *J. Ethol.*, **5**, 95–103.

Suzuki, Y. and Iwasa, Y. (1980) A sex ratio theory of gregarious parasitoids. *Res. Popul. Ecol.*, **22**, 366–382.

Takabayashi, J. and Takahashi, S. (1985) Host selection behaviour of *Anicetus benificus* Ishii et Yasumatsu (Hymenoptera: Encyrtidae). 3. Presence of ovipositional stimulants in the scale wax of the genus *Ceroplastes*. *Appl. Entomol. Zool.*, **20**, 173–178.

Takabayashi, J. and Takahashi, S. (1988) Effect of the kairomone on the parasitization rates of *Apanteles kariyai* to *Pseudaletia seperata*. *J. Pestic. Sci.*, **13**, 283–286.

Takagi, M. (1986) The reproductive strategy of the gregarious parasitoid, *Pteromalus puparum* (Hymenoptera: Pteromalidae). 2. Host size discrimination and regulation of the number and sex ratio of progeny in a single host. *Oecologia*, **70**, 321–325.

van Veen, J. C. and van Wijk, M. L. E. (1987) Parasitization strategy in the non-paralyzing ectoparasitoid *Colpoclypeus flavus* (Hym., Eulophidae) towards its common summer host *Adoxophyes ovaria* (Lep. Tortricidae). 1. Host finding, acceptance and utilization. *J. Appl. Entomol.*, **104**, 402–417.

Vinson, S. B. (1975) Source of material in the tobacco budworm which initiates host-searching by the egg-larval parasitoid, *Chelonus texanus*. *Ann. Entomol. Soc. Amer.*, **68**, 381–384.

Vinson, S. B. (1976) Host selection by insect parasitoids. *Ann. Rev. Entomol.*, **21**, 109–133.

Vinson, S. B. (1981) Habitat location. In *Semiochemicals, their Role in Pest Control* (eds D. A. Nordlund, R. L. Jones and W. J. Lewis), Wiley, New York, 51–77.

Vinson, S. B. and Guillot, F. S. (1972) Host marking-source of a substance that results in host discrimination in insect parasitoids. *Entomophaga*, **17**, 241–245.

Vinson, S. B. and Iwantsch, G. F. (1980) Host suitability for insect parasitoids. *Ann. Rev. Entomol.*, **25**, 397–419.

Waage, J. K. (1979) Foraging for patchily-distributed hosts by the parasitoid, *Nemeritis canescens*. *J. Anim. Ecol.*, **48**, 353–371.

Waage, J. K. (1986) Family planning in parasitoids: adaptive patterns of progeny and sex allocation. In *Insect Parasitoids* (eds J. K. Waage and D. J. Greathead), Academic Press, London, 63–95.

Waage, J. K. and Godfray, H. C. J. (1985) Reproductive strategies and population ecology of insect parasitoids. In *Behavioural Ecology: Ecological Consequences of Adaptive Behaviour* (eds R. M. Sibly and R. H. Smith), Blackwell Scientific, Oxford, 449–470.

Walter, G. H. (1988) Activity patterns and egg production in *Coccophagus bartletti*, an aphelinid parasitoid of scale insects. *Ecol. Entomol.*, **13**, 95–105.

Ward, S. A. (1987) Optimal habitat selection in time-limited dispersers. *Amer. Nat.* **129**, 568–579.

Wellings, P. W. (1987) Spatial distribution and interspecific competition. *Ecol. Entomol.*, **12**, 359–362.

Wellings, P. W., Morton, R. and Hart, P. J. (1986) Primary sex-ratio and differential progeny survivorship in solitary haplo-diploid parasitoids. *Ecol. Entomol.*, **11**, 341–348.

Werren, J. H. (1980) Sex ratio adaptations to local mate competition in a parasitic wasp. *Science*, **208**, 1157–1159.

Weseloh, R. M. (1976) Discrimination between parasitized and non parasitized hosts by the gypsy moth parasitoid *Apanteles melanoscelus* (Hymenoptera: Braconidae). *Can. Entomol.*, **108**, 395–400.

Weseloh, R. M. (1981) Host location by parasitoids. In *Semiochemicals, their Role in Pest Control* (eds D. A. Nordlund, R. L. Jones and W. J. Lewis), Wiley, New York, 79–95.

Wilson, D. D., Ridgway, R. L. and Vinson, S. B. (1974) Host acceptance and oviposition behavior of the parasitoid *Campoletis sonorensis* (Hymenoptera: Ichneumonidae). *Ann. Entomol. Soc. Amer.*, **67**, 271–274.

Wright, E. J., Muller, P. and Kerr, J. D. (1989) Agents for biological control of novel hosts: assessing an aleocharine parasitoid of dung-breeding flies. *J. Appl. Ecol.*, **26**, 453–461.

Wylie, H. G. (1983) Delayed development of *Microctonus vittatae* (Hymenoptera: Braconidae) in superparasitized adults of *Phylotreta cruciferae* (Coleoptera: Chrysomelidae). *Can. Entomol.*, **115**, 441–442.

Zaki, F. N. (1985) Reactions of the egg parasitoid *Trichogramma evanescens* Westw. to certain insect sex pheromones. *Z. Angew. Entomol.*, **99**, 448–453.

5

Host location and oviposition on plants

Rhondda E. Jones

5.1 INTRODUCTION

Phytophagous insects as a group are not noted for the extent of their parental care: in most cases, all that the female does is to find appropriate foodplants for her offspring and abandon her eggs on them. The spectacular diversity and success of phytophagous insects since the Carboniferous testifies that this amount of care is quite enough. Because the acts of plant choice and oviposition by insects are in general so rapid and apparently straightforward, it is easy to overlook both their crucial importance and the considerable technical problems which insects must overcome in order to achieve them.

Higher plants in particular are not an easy resource to exploit. Of the 29 living insect orders, only nine include species that feed on the living tissue of higher plants (Strong *et al.*, 1984), even though plants are by far the most abundant and accessible source of food in terrestrial communities. Those nine orders, however, have each undergone an extraordinary radiation, and now comprise the majority of insect species. Southwood (1973) has argued that life on higher plants represents a major evolutionary hurdle that most groups of insects have failed to overcome, despite the rewards for doing so. The problems of exploiting higher plants include avoiding desiccation on exposed plant surfaces; remaining attached to the often smooth or hairy surfaces that plants present; avoiding their physical defences against herbivory such as hooks and trichomes; penetrating the often very tough external surface; and most importantly, finding ways to cope with an inferior nutritional resource. Plant tissue, relative to insect tissue, is low in usable nitrogen and energy content and often contains noxious phytochemicals that are unpalatable, toxic, or interfere with digestive assimilation: 'most parts of most plants are poisonous'.

No phytophagous insect can cope with all the problems presented by the whole variety of plants and plant tissues with which it might possibly come in contact. There are many organisms for which we can legitimately say that they come close to being truly 'general' predators—that is, they are prepared to eat any recognizable animal in roughly the right size range which does not eat them first. In this sense, there are no truly 'general' phytophagous insects. Although the degree of specialism varies, all are to some degree specialist, both in the plant species and in the particular tissues within the plant which they are able to exploit.

The host location and choice problems presented by this necessary specialization must, in general, be solved by the ovipositing female, since newly-hatched insects are normally (although not invariably) ill-equipped to seek new hosts for themselves. Moreover, on hatching, they are probably at the stage where they are most sensitive to variations in host quality: when larvae are reared on a range of hosts of decreasing nutritional quality, young larvae may show reduced survival and/or reduced growth rates in response to a degree of deterioration in their hosts which has no measurable effect on the growth or survival of older larvae (Slansky and Feeny, 1977; Jones and Ives, 1979).

As Thorsteinson (1960) points out, and recent workers have emphasized (Singer, 1982, 1986; Miller and Strickler, 1984; Rausher, 1985; Courtney *et al.*, 1989), host selection in insects consists essentially of a series of 'take-it-or-leave-it' situations in which the insect accepts or rejects the plant or habitat it is in and behaves accordingly; insects generally do not appear to 'comparison-shop', although there are some limited exceptions, discussed later. The sequential process involved in host exploitation, adopted from studies of host–parasitoid relationships, has traditionally been subdivided into habitat-finding, host-finding, host recognition, host acceptance, and host suitability. This chapter concentrates on the mechanisms and adaptive consequences of behaviour related to the middle three of these processes—that is, host-finding, host recognition, and host acceptance—although for phytophages the processes of finding a recognizing hosts are often inextricably intertwined and are consequently treated together.

5.2 FINDING AND RECOGNIZING HOSTS

To find a host plant, an insect must move, encounter cues from the host, and then respond to the cues appropriately.

5.2.1 Cues for host-finding

As noted earlier, only a few of the many plants and plant parts that may be present in an area will be suitable hosts for any particular insect species. For a moving insect, therefore, a diverse plant community is a complicated mosaic from which it must somehow pick out cues from plants which are potential hosts from among a regrettably large number that are not. Some insects may find plants simply by blundering into them, after which contact chemoreception allows them to distinguish between potential hosts and non-hosts. But most appear to be able to perceive and use cues at a distance from plants (albeit often a very small distance) and hence to bias their plant encounters toward potential hosts. Miller and Strickler (1984) have reviewed the history of opinions concerning the kinds of cues used by insects in finding hosts. Thirty-eight years ago, Fraenkel (1953) argued that the nutritional composition of all plants was fundamentally similar, and that allelochemicals (*sensu* Whittaker and Feeny, 1971) provided 'token stimuli' which were the major determinants of foraging behaviour and host acceptance. By 1960, it was well recognized that from an insect's point of view, all plants were definitely not nutritionally identical and that nutritional quality also played a major role (Kennedy, 1958; Thorsteinson, 1960). But it was still argued that host-finding and recognition must be almost entirely chemically based, and dependent on a few characteristic chemical stimuli for each species, (a) because there was not sufficient variation (of a kind perceivable by insects) between plant species in other than chemical attributes, and (b) because early work in insect sensory physiology suggested that chemoreceptors were often very specialized. Work over the last three decades, however, has demonstrated conclusively that insects are capable of much more sophisticated sensory and information-processing feats than was at first apparent, and even that it may be the norm for insects to integrate information from several sources to provide cues for host-finding. The use of visual cues is commonplace (Prokopy, 1983), and mechanoreceptor involvement in host recognition has also been demonstrated. There are now a number of very clear examples to illustrate this point: sometimes the different sources of information are clearly used sequentially to trigger successive steps in a chain of behaviour which leads to host-finding; sometimes (especially in the case of multiple chemical cues) different sources of information may be used simultaneously (see e.g. Den Otter *et al.*, 1980).

Butterfly species from at least three different families use a combination of visual cues before alighting and chemotactic cues after alighting. In the species so far studied, the specific visual cue is generally leaf shape. This has been elucidated in most detail for the papilionid *Battus philenor* (Rausher, 1978; Rausher and Papaj, 1983; Papaj, 1986a,b; Papaj and Rausher, 1987), of which more later, but has also been demonstrated in several pierid species. Stanton (1982, 1984) suggested that several *Colias* species also use leaf shape, and that 'landing errors' by searching females on non-hosts tended to be on plants that superficially 'looked like' their legume hosts. A study by Mackay and Jones (1989) of host location in two monophagous tropical pierids (*Eurema* species) came to the same conclusions: grouping all the plant species (except the host species) in their study area on the basis of leaf shape, showed that plants in the leaf-shape category closest to that of the host plant received a disproportionate share of all non-host alightings. Leaf shape is also important to the spruce budworm (*Choristoneura fumiferana*): in tests using artificial substrates, Stadler (1974) showed that attractant compounds painted on artificial Christmas tree needles elicited a greater response from the female moths than the same compounds painted on blotting-paper.

In onionflies (*Delia antiqua*), maximal oviposition requires the synergistic effect of an appropriate chemical (onion-produced alkyl sulphides), a vertical cylindrical shape of the right colour, and moist sand (Harris and Miller, 1982). Neither the chemical without the cylinder, nor the cylinder without the chemical, resulted in substantial oviposition. Tabashnik (1985) has demonstrated that vision, chemoreception, and mechanoreception are all involved in host identification in the diamond-back moth, *Plutella maculipennis*.

Even within a single sensory modality, ovipositing insects may be responding to several different signals. This has been most clearly demonstrated in the case of chemoreception, where insects often respond to a mixture of chemicals in an appropriate balance much more strongly than they do to any one of them alone. For example, the carrotfly *Ptila rosae* exhibits a synergistic response to the complex mixture of chemicals found in the surface wax of carrot leaves (Stadler and Buser, 1984); the cabbage root fly *Eroischia brassicae* will respond to several of the individual chemicals found in swede juice, but to none of them with the enthusiasm with which it responds to swede juice itself (Traynier, 1967). Visser (1986) reviews a range of comparable cases.

In summary, then, we should not expect the cues used by insects to find and recognize hostplants to be either simple or confined to a single sensory modality. Even when it is possible to induce oviposition in the laboratory with single (and especially single chemical) stimuli, the insect in its natural environment is likely to be using a much more complex set of signals.

5.2.2 Plant apparency

Many of the cues discussed above can only be perceived from quite short distances. This is most obvious in the case of contact chemoreception, but even visual cues may require quite close proximity to be useful to the insect. Especially if they are relatively rare, hostplants may still be difficult for searching insects to find.

Moreover, the right plants may not be present continuously, and when they are present, may not be in an appropriate condition to provide satisfactory oviposition sites. Particular plant species or individuals (or plant parts) may be more or less easy to find, depending on their size, conspicuousness, permanence or transience, the complexity and type of their surrounding vegetation, and so on. In recognition of this variation among plants, Feeny (1976) developed the concept of 'apparency' to describe the susceptibility of an individual plant to discovery by its enemies, by whatever means enemies may employ. He originally used this concept as a way of explaining perceived patterns in the kinds of chemical defences employed by plants: although those patterns are now subject to some question, the distinction between 'apparent' and 'un-apparent' plants is proving to be very useful indeed in clarifying and explaining some aspects of the oviposition behaviour of phytophagous insects.

It is a commonplace observation that within a plant population at any one time, some individuals consistently receive many more eggs than do others. Sometimes these differences are clearly adaptive; i.e. result from females choosing hosts which give their offspring the best chance of survival. Differences of this kind are discussed later. But in other cases, differences in apparency provide a more satisfactory explanation for these differences in attack rates than do differences in host quality or other attributes (see e.g. Courtney, 1982).

A general example concerns one of the commonest observations about the egg distributions of many lepidopteran species: that is, the existence of an 'edge effect'. For a number of species, isolated host plants, and plants on the edge of patches of hosts, tend to have more eggs laid on them than do plants in large groups (Cromartie, 1975; Jones, 1977; Thompson and Price, 1977; Shapiro, 1981; Rausher et al., 1981; Mackay and Singer, 1982). Adaptive explanations for this phenomenon have been hypothesized (e.g. avoidance of less mobile specialist parasitoids) but rarely demonstrated, and some studies have demonstrated the opposite: i.e. that juveniles on isolated plants experience lower survival rates (Karban and Courtney, 1987). Jones (1977) used simulation models of the female's movement patterns to show that for the cabbage butterfly *Pieris rapae*, the 'edge effect' could be explained as a statistical consequence of movement patterns that combined some flights out of host plant patches with visual attraction toward a host plant from a short distance. An identical conclusion, although couched in more general terms, was arrived at by Mackay and Singer (1982): i.e. that edge effects result because if a female chooses the nearest hostplant from a random point (her location), this choice process results in the more isolated plants being chosen more often. In other words, the edge effect occurs because isolated and edge plants are more 'apparent'—at least to these lepidop-terans—in the precise technical sense defined by Feeny (1976). Note that as in this case, 'apparent' does not have to mean 'more easily seen' or even 'more easily perceived': it means only 'more likely to be found'.

Many other phytophagous insects show the reverse relationship between colonization probabilities and isolation; indeed Root (1973) developed the 'resource concentration hypothesis' as a result of his study of collard insects to account for this observation. The resource concentration hypothesis states that herbivores are more likely to find and remain on hosts that are growing in dense or nearly pure stands, and that the most specialized species frequently attain higher relative densities in simple environments. There are two predictions involved in this: (a) that herbivores may achieve higher densities per plant in larger patches of host plants, and (b) that herbivores may achieve higher densities per plant in host monocultures than in polycultures that include non-hosts.

Evidence bearing on both these predictions is comprehensively reviewed by Kareiva (1982). The first prediction is of course violated by the lepidopteran species discussed earlier, but is fulfilled by a number of other species (see e.g. Thompson, 1978; van der Meijden, 1979; Lemen, 1981; Kareiva, 1982; MacGarvin, 1982). The second prediction is born out more consistently than the first. For many herbivore species, host monocultures DO result in higher herbivore densities than polycultures. Several hypotheses have been proposed to explain this pattern, including increased natural enemy densities in polycultures, but for at least some species, the pattern seems again to be in part due to 'apparency' effects. That is, phytophages appear to be able to find their hosts more efficiently when there are no other plant species present (Tahvainen and Root, 1972; Root, 1973; Atsatt and O'Dowd, 1976; Bach, 1980, 1984; Perrin, 1980; Rausher, 1981). The mechanism whereby this improved host-finding occurs has been studied in only a few cases, where it resulted partly from a lower probability of long-distance flight after encounter with a hostplant (Bach, 1980, 1984, 1986).

The fact that variations in patch size can have opposite effects on different herbivore species, in both cases at least partly mediated by 'apparency' effects, emphasizes the point that 'apparency' is not simply a property of the plant: it is a property of the perceptual cues provided by the plant to the searching insect, as well as the movement tactics used by the insect in the searching process. Because different insects use different movement tactics and different cues to locate their hosts, a plant which is highly apparent to one species may be effectively invisible to another. There are, however, some plant attributes that are likely to increase apparency at the species level to almost all species of insects searching for them: plant species that are large, widely distributed and abundant, and long-lasting, are likely always to be more apparent than plant species that are small, localized, rare, and transient.

The concept of plant apparency or 'findability' may have the potential to integrate a number of fields in the general area of plant–herbivore relationships and their evolution, and to provide insights into a variety of phenomena related to host location. Strong *et al.* (1984) summarize the arguments used to explain why more widespread and abundant plant species tend to be exploited by more herbivore species: essentially because such plant species will be encountered by chance more often by more species of phytophage, there will be more chances for 'mistaken' ovipositions or feeding attempts on them, and hence more chances for evolutionary adaptation to them. A particularly elegant extension of this general argument is given by Dixon *et al.* (1987) in an hypothesis to explain why aphids decline in diversity toward the tropics. The crucial and unique set of properties possessed by aphids, for the purpose of this argument, are (a) that their telescoped developmental pattern places unusually high demands on nutrient intake, so that they are able to survive for only very short periods away from their host plant, (b) that as a group, they tend (like other phloem feeders) to be highly specialized in their utilization of hostplants, and (c) that because they are such small, weak fliers, who cannot normally control their own flight direction, host finding is a matter of repeated landings and take-offs until they encounter by chance an acceptable host (see Ward, Chapter 8, this volume). The hypothesis presented and formally modelled by Dixon *et al.* is that in these circumstances, the probability that an aphid will successfully find a host on which to settle and reproduce is largely dependent on the proportion of the total vegetational surface area that is covered by the particular host species which the aphid is trying to find. (That is, host apparency, for an aphid, is determined almost entirely by the host's vegetational surface area.) Further, there will be a critical level of host rarity below which so few aphids successfully locate a hostplant that the plant species cannot support a specialist aphid. Because plant diversity increases toward the tropics, the proportion of the total vegetational surface area covered by each individual plant species tends, on average, to decline, and fewer plant species exceed the critical level of abundance that would allow them to support a specialist aphid. Aphid diversity consequently decreases as plant diversity increases. Dixon *et al.* (1987) test this hypothesis by comparing predictions from the model with data on numbers of plant and aphid species in different countries: the fit is remarkably good and suggests that the approach is well worth pursuing.

5.3 HOST ACCEPTANCE AND REJECTION

When a female insect locates a hostplant she may or may not choose to oviposit on it. The process of accepting or rejecting a host is often labelled 'host choice', but, as noted earlier, this is slightly misleading, since the female generally encounters hostplants in sequence and must make separate decisions about the acceptability of each, rather than choosing between simultaneously available hosts. There is a large empirical and theoretical literature on this subject. Much of it overlaps extensively with work on host recognition, since it is seldom possible to distinguish clearly between the case where an insect recognizes a host and rejects it, and the case where it simply fails to recognize the host. Perhaps the best example of the difficulty of making this distinction is provided by the papilionid butterfly *Battus philenor*, studied by Rausher, Papaj, and their coworkers. As noted earlier, although *B. philenor*'s final acceptance or rejection of a host may be based on contact chemoreception, plants are initially located visually using leaf shape as a cue (Rausher, 1978; Rausher and Papaj, 1983). *B. philenor* has two hosts in the area studied, *Aristolochia serpentaria* and *A. reticulata*, which have very different leaf shapes. Butterflies show a strong pre-alighting preference for one or other of these two species, and this preference is conditionable: i.e. females can be trained to land preferentially on one leaf shape or the other. As Rausher and Papaj (1983) point out, it is not possible to distinguish in this case between a conditioning of the perceptual process (i.e. 'search image' formation) and a learned response to equally well-perceived leaf shape cues.

In this section we are concerned with the ways in which ovipositing females discriminate among potential hosts. As the previous paragraph suggests, a discriminatory outcome (in the sense that some sorts of plants receive more eggs than others) may arise as a result of differential host recognition (including the apparency effects discussed earlier) as well as differential acceptance. The two may not always be easily distinguished.

5.3.1 Behavioural components of discriminatory behaviour

Plasticity in any of the sequence of behaviours that culminate in oviposition can be used by the female insect to produce discrimination among potential hosts. This can be illustrated by a study of movement and oviposition in the cabbage butterfly *Pieris rapae* (Jones, 1977). The resulting simulation model of the movement and oviposition process was used to predict how eggs were distributed among hostplants by ovipositing females. In order for the simulation to mimic the oviposition behaviour of real butterflies, the model had to assign probabilities to a set of behaviours exhibited by the butterflies: (a) being attracted toward a host within the perceptual range of the female; (b) landing after approaching the host; (c) ovipositing after landing; (d) resettling on the host after take-off without an intervening flight away. All four of these probabilities varied according to the circumstances, and (b)–(d) in particular were very strongly host-specific. The probabilities did not, however, respond in the same way to changes in particular host attributes. For example, larger, older plants resulted in higher landing probabilities and higher resettle probabilities, but lower oviposition probabilities. That is, females were more likely to alight on larger, older plants, but having alighted, were more likely to oviposit on smaller, younger ones. The net outcome of these conflicting tendencies was that most eggs were laid on plants that were well-grown but not too old. An identical pattern of response to plant age in behavioural components has been observed in tropical pierids (*Eurema* species) (R. E. Jones, unpublished data), and in *Battus philenor* (Papaj and Rausher, 1987, who also noted that females were more likely to land on younger plants than on older ones of the same size). Stanton (1982) compared responses of *Colias eriphyle* to different species of host, and found that here too, different behavioural components varied to some degree quite independently.

Other components of female behaviour may also result in a discriminatory outcome. For example, in insects that lay eggs in batches, the number of eggs per batch may also be plastic (Skinner, 1985). Van Leerdam *et al.* (1984) showed in a pyralid pest of sugarcane that sites preferred for oviposition received larger egg masses, as well as more egg masses.

Because variation in hostplant attributes may affect different components of the oviposition process in different ways, ranking hosts in order of preference based on only one aspect of oviposition behaviour (e.g. on the basis of post-alighting acceptability) may not provide good predictions of the way eggs are distributed among hostplants in the field. Despite this caveat, we probably know more about the kinds of plant attributes that influence ovipositional choices in a range of phytophagous insects than we do about any other aspect of their oviposition behaviour.

5.3.2 Discrimination on the basis of host attributes

There is an extensive literature documenting discrimination by a variety of insect species on the basis of host attributes. Indeed, the development of cultivar varieties resistant to pest attack often involves selection for 'non-preference' (Gould, 1983). Such experiments can reveal remarkably complex variation in preference: Blum (1969) describes experiments on the sorghum shootfly that involved crossing resistant with non-resistant strains of sorghum (the susceptible strain was preferred for oviposition by the fly). The hybrid sorgum was less preferred than the susceptible strain at low fly densities, but preferred at high fly densities.

A variety of techniques have been used to determine whether an insect prefers (is more likely to accept) one host to another: in particular, experimenters may offer the ovipositing insect only one kind of host at a time, or several kinds at once. Several recent workers have suggested that 'no-choice' trials of the first kind are in general to be preferred (Singer, 1982; Courtney and Chen, 1988); partly because insects experience hosts one-at-a-time, and partly because multiple choice tests may confound different behaviours—habitat choice, host location, and host acceptance —which may, as noted earlier, all vary independently. The results of no-choice tests, on the other hand, are more likely to be predominantly due to variation in host acceptance.

The plant attributes responsible for variations in host acceptability, whether in natural or cultivated systems, are various.

As in the examples described in the previous section, insects very commonly discriminate between different species, ages and sizes of their host plants (see e.g. Wiklund, 1974; Ives, 1978; Rodman and Chew, 1980; Singer, 1981; Courtney, 1982; Holdren and Ehrlich, 1982; Williams, 1983; Stanton and Cook, 1984; Mackay, 1985; Robert, 1985; Moore, 1986; Courtney and Chen, 1988; Karban and Courtney, 1987; Price *et al.*, 1987; Robertson, 1987). Insects also commonly show discrimination among different sizes and ages of plant parts within a single plant. Many folivorous insects discriminate strongly among different-aged leaves, frequently but far from universally in favour of younger leaf tissue (Ives, 1978; Raupp and Denno, 1983; Moore, 1986). (That a preference for oviposition on younger leaves is not almost universal is at first sight surprising in view of the fact that when fed younger leaves, the juvenile stages of insect folivores almost always develop and survive better, and moreover often produce larger and more fecund adults (Raupp and Denno, 1983). The adaptive advantages of choosing oviposition sites on the basis of nutritional favourability are considered in more detail later.) *Pemphigus* aphids are much more likely to colonize and reproduce on large leaves of their hostplant than on small ones (Whitham, 1978; Zucker, 1982).

Examples can also be found of discrimination among individual host-plants based on almost any other host attribute one might think of. Perhaps the most universal of the remainder is discrimination on the basis of some component of nutritional quality (see for example Fox and Macauley, 1977; Ives, 1978; Mattson, 1980; Morrow and Fox, 1980; Wolfson, 1980; Rausher, 1981; Moore, 1986). In some cases this is clearly extraordinarily sensitive. Myers (1985) demonstrated that ovipositing cabbage butterflies were able to discriminate between fertilized and unfertilized hosts within 24 h of fertilizer application. Very often, nutrient-based discrimination is correlated with either nitrogen or water content of the hostplants (which are themselves often intercorrelated). White (1978) and Mattson (1980) have suggested that nitrogen is often a limiting nutrient for insect herbivores, and it is therefore not surprising both that discrimination of this kind is so common, and also that it may be so sensitive.

Discrimination based on morphological or physical variability between plants or plant parts has also frequently been documented. Many ovipositing insects discriminate between hairy and non-hairy surfaces; some species are less likely to lay on hairy surfaces (e.g. Schillinger and Gallun, 1968; Kammp and Buttery, 1986; Wilson, 1986); some prefer hairy surfaces (Robinson *et al.*, 1980). Wiklund (1974) showed that the butterfly *Papilio machaon* would oviposit only on leaves strong enough to support the female's weight. Williams (1981) found that ovipositing *Euphydryas gilletti* were choosing leaves for warmth: i.e. that the females were choosing leaves for oviposition which caught the morning sun (probably by laying on the shaded side of sunlit leaves).

A number of investigators have examined the subsequent development and survival of young on different kinds of plants or plant parts between which ovipositing insects discriminate. Although some studies of this kind, as discussed earlier, find no difference in juvenile peformance between preferred and non-preferred plants, others do find that the female was indeed choosing oviposition sites that enhanced the fitness of her offspring, even if the reasons for her choices were not initially obvious. For example, the avoidance of the youngest and most nutritionally favourable foliage for oviposition may serve in some species to minimize competition or egg cannibalism from older larvae (Jones, 1981). Often, of course, the choices made are demonstrably correlated with nutritional value as assayed by higher larval growth rates. For example, Leather (1985) found that females of *Panolis flammea* offered pine foliage of different provenances laid most eggs on the foliage on which larvae achieved the highest growth rates. Similar results have been noted for a variety of other species (see for example Jones and Ives, 1979; Rausher, 1981; Leather, 1986; Whitham, 1978, 1980). In such studies, however, the correlation is rarely exact, suggesting that the females are using cues that allow rapid discrimination but which are imperfectly correlated with host favourability, or that here too, selective factors other than nutritional favourability have influenced the evolution of host acceptance.

One of the most detailed and illuminating analyses of the effects of female discrimination on offspring fitness concerns the choices made by stem mothers of *Pemphigus* aphids (Whitham, 1978, 1980), which, as described earlier, show strong preferences for larger leaves. They also prefer the basal parts of those leaves they colonize. Whitham (1978) showed that these choices more than doubled the fitness expected if leaves were colonized at random.

5.3.3 Discrimination on the basis of prior occupation by conspecifics

Many ovipositing phytophagous insects discriminate against hosts that are, or have previously been, occupied by conspecifics. Discrimination of this kind has been comprehensively documented in many species that attack fruiting structures, and may indeed be the norm for flower, fruit and seed feeders (Averill and Prokopy, 1987). It is relatively rare in foliage-feeding species, but certainly does occur in a number of well-documented cases.

The way in which the ovipositing female detects that the host is already occupied varies considerably between species. Some species use deterrent pheromones: that is, an ovipositing female marks the host with a chemical which deters oviposition by later females. This has, for example, been found in tephritid fruitflies (Prokopy, 1972), and in several species of bean weevils (Oshima *et al.*, 1973; Wasserman and Futuyma, 1981). In the fruitflies, females that encountered a high rate of fruit infestation not only rejected them but were more likely to make long flights away from the area (Roitberg and Prokopy, 1984; Roitberg *et al.*, 1984).

In other cases, the female evidently detects damage produced by prior occupants. In the Australian psyllid *Cardiaspina albitextura*, nymphal feeding kills patches of leaf tissue, and ovipositing females avoid such damaged leaves (Clark, 1963). In this species, the adults stay on particular leaves only briefly, and the damage they produce is not obvious to the human observer. But adult feeding does reduce the favourability of leaves for later juvenile development and survival, and females show measurable discrimination against leaves on which adults have fed for as long as two months afterwards. In a study of the oviposition behaviour of a chalcid species attacking alfalfa seed, Kammp and Buttery (1986) identified a series of volatile compounds that affected the female's behaviour: some acted as attractants; others were deterrents. Two of the latter were compounds found in homopteran alarm pheromones, but others were compounds found in the pods, and it seems likely that high concentrations of these compounds signal pod damage and hence an unsuitable oviposition site. Fitt (1984) showed that two species of tephritid fruitflies in the genus *Dacus* could detect changes in fruit induced by larval feeding and that females would discriminate against these fruit. There are now a number of studies documenting responses to damage or to frass, but it has rarely been possible to evaluate the precise cues that the insect has used to discriminate between damaged and undamaged plants. The variety of sensory modalities that ovipositing insects use to recognize hosts was discussed earlier: the ways in which insects recognize damage may well be equally diverse. Even species that normally are actively attracted toward damaged hostplants, as are many bark beetles and borers, may avoid hosts that are 'mortally wounded' (Dunn *et al.*, 1986).

It is possible that some of the apparent discrimination against damaged plants may actually result from the induction of wound-induced defences by the plants and the perception of these defences by the insect. Studies of induced defences have concentrated primarily on their effects on feeding insects rather than on oviposition (see e.g. Haukioja and Niemela, 1979; Edwards and Wratten, 1983; and Haukioja *et al.*, 1985), and a number of studies have been criticised for design flaws (Fowler and Lawton, 1985). However there is little doubt that induced defences on several time scales do exist, and it would be of considerable interest to examine their effects on oviposition behaviour in more cases; the work of Leather *et al.* (1987) on the pine beauty moth clearly identifies long-term effects on oviposition, probably mediated by changes in the monoterpene profile of the foliage as a result of defoliation of saplings the previous year. It is clear, however, that plant response to damage is a complex phenomenon that may affect the behaviour and fitness of different insects in different ways: Wagner and Evans (1985) showed that damaged Ponderosa pine seedlings did indeed show an increased concentration of allelochemicals, but simultaneously showed an improvement in nutritional quality (i.e. in protein concentrations). Insects able to handle the particular allelochemicals involved might therefore actually prefer damaged plants, even though other species would avoid them.

Species may respond directly to the presence of eggs or juvenile stages. It is likely, however, that some of these will eventually prove to fall into one of the two categories identified above: that is, response to a deterrent pheromonal marker left by the ovipositing female (and perhaps left on the eggs themselves), or response to larval damage. Many butterfly species discriminate against plants that already have eggs on them (Rothschild and Schoonhoven, 1977; Wiklund and Ahrberg, 1978; Rausher, 1979; Courtney, 1981; Shapiro, 1981; Williams and Gilbert, 1981). In the case of the large white butterfly *Pieris brassicae*, the deterrent effect of conspecific eggs is thought to be partly a visual response, but also clearly involves specific deterrent chemicals associated with the eggs (Behan and Schoonhoven, 1978; Den Otter *et al.*, 1980). Others probably result mainly from visual responses: indeed Williams and Gilbert (1981) describe a situation where butterflies are discouraged from laying on potential hosts by projections growing on the leaves that mimic butterfly eggs. Other butterfly species, however, show no discrimination against plants with conspecific eggs (Ives, 1978; Singer and Mandracchia, 1982); this group is likely to be under-represented in the literature since the lack of a response is less likely to be reported.

Similarly, seed beetles discriminate against seeds that already have eggs (Mitchell, 1975; Messina and Renwick, 1985), and will also discriminate between seeds which have fewer or more eggs; Messina and Renwick showed that the beetles generated a nearly uniform egg distribution even after all seeds in their experimental trials had several eggs. Mitchell (1975) examined the effects on fitness of this kind of discriminatory behaviour in his bean weevils: in this case, development and survival was better on larger beans, and on beans without other individuals present. A female could maximize her reproductive fitness if she knew the weights and egg loads of all the beans available to her and chose oviposition sites accordingly. Females actually achieved about 70% of this theoretical maximum, which represented a 20–60% increase in fitness above what they would have achieved if they had distributed eggs at random among beans without reference to either size or prior occupation. Mitchell demonstrated that the observed performance could be accounted for quite well by a model which assumed that females compared the weight and egg load of the current bean with that of the last bean encountered; i.e. used the smallest possible amount of the comparative information which was available to them. This suggests, therefore, (a) that when insects do 'comparison-shop', their comparisons are short-term, and (b) that even such limited comparisons can provide a very significant fitness advantage.

5.3.4 Discrimination based on the presence or absence of other species

There is little evidence to suggest that herbivorous insects discriminate between hosts or host patches on the basis of natural enemy abundance (Chew and Robbins 1984; Jones *et al.*, 1987), but unambiguous data to test the possibility are uncommon. There are, however, well-documented examples of the reverse: that is, of the presence of symbiotic species favouring oviposition. Perhaps the best-documented examples of this are found in lycaenid butterflies, many of which have close relationships with specific ant species (Atsatt, 1981a; Pierce, 1984). The lycaenid larvae provide the ants with nutritive secretions and, in some cases at least, the ants in return provide protection against certain parasites and predators (Atsatt, 1981b; Pierce and Mead, 1981). Many of these lycaenids will normally reject nutritionally suitable hostplants if the right ants are not present (Atsatt, 1981b; Pierce, 1984). Such relationships may impose other forms of selectivity on the lycaenids: Pierce (1984) has suggested that the need to provide ants with nitrogen-rich secretions requires lycaenids to select hostplants which are themselves particularly nitrogen-rich.

5.3.5 'Arbitrary' hostplant rejection

In some species, whether a particular host is accepted or rejected as an oviposition site after the insect has alighted on it appears to be a highly repeatable and (at least in principle) predictable decision. Indeed, Singer (1986) has used this repeatability with the batch-laying butterfly *Euphydryas editha* to devise an elegantly simple technique for assaying the post-alighting acceptability of different hostplant individuals to particular females. In other species, however, 'acceptability' is a much more probabilistic concept. In the cabbage butterfly *Pieris rapae*, which lays its eggs singly, the post-alighting acceptability of a particular host to a female can be measured by the probability that a butterfly that alights on it will lay an egg, but the decision made at any particular alighting does not appear to be predictable in any obvious way (Jones, 1977, 1987). Thompson (1987) has described a similar phenomenon in the seed-parasitic incurvariid moth *Greya subalba*, which lays eggs on umbellifers having a number of seed-pairs per umbellet. The females showed a constant probability of leaving an umbellifer after an egg was laid, regardless of how many unattacked seeds remained in the umbellet, and regardless of umbellet size. Thompson suggested that the reason for this apparent unpredictability may be that the resultant egg distribution provides minimal information to searching parasitoids about where eggs may most profitably be sought. Alternative explanations are also possible. Den Boer (1974) identified 'spreading the risk' to stabilize fluctuations in abundance as one of the benefits of high levels of dispersal. In situations where juvenile survival probabilities vary markedly but at least from the point of view of the ovipositing female unpredictably between hosts or host patches, there may also be some adaptive advantage to the female in including a significant stochastic element in her acceptance or rejection of hosts, and in ensuring that even the most apparently desirable hosts do not receive too large a share of her eggs.

5.4 PHYSIOLOGICAL AND BEHAVIOURAL FACTORS AFFECTING HOST LOCATION AND OVIPOSITION

Previous sections have concentrated on the properties of host plants which may influence their location and acceptance by ovipositing insects. But within an insect species, there may also be a substantial amount of variation between individuals in oviposition behaviour. This section examines the kinds of variation that may result from differences between individuals in age, prior experiences, or current physiological state: that is, variation between individuals that is not genetically based.

5.4.1 Variations in 'motivation'

Motivational variation is reviewed in detail by Courtney *et al.* (1989), who use it as the basis of a detailed verbal model (the 'hierarchy-threshold model') of insect host selection. This model assumes that hosts are ranked hierarchically by ovipositing insects, and that an individual which will accept a lower-ranking host will also accept all hosts above it in the rank order. The rankings should be invariant throughout an individual's life, but acceptance thresholds may vary according to the insect's motivation. That is, a highly-motivated female should accept a wider range of potential hosts than a less-motivated female, but all hosts accepted by the less-motivated female should be included in the acceptable range of the more-motivated female. The basis of intraspecific motivational variation is assumed most often to be egg load: that is, females with many eggs ready to lay should be more highly motivated to lay them than females with few eggs. Egg load will vary considerably over the lifetime of the female and with factors such as host density because at low host densities searching females may not be able to find all the hosts they need), temperature (because oogenesis will proceed more quickly at high temperatures than at low), and weather (in many butterflies, for example, flight and oviposition behaviour are inhibited by overcast or windy weather, so that egg production accumulates within the female until the weather improves (Gossard and Jones, 1977)). Consequently, females may exhibit substantial variation in motivation levels, and hence in host-seeking and host-acceptance behaviour, throughout their lives.

The model is certainly not of universal applicability, and is probably best applied to the post-alighting acceptance or rejection of hosts. (Host location and recognition, as we have previously seen, are very strongly influenced by a variety of factors unconnected with host favourability, and plant attributes which enhance location and recognition may actually decrease post-alighting acceptance probabilities (Jones, 1977).) It also cannot in its present form cope with situations (discussed later) in which host acceptance is modified by learned responses. With these caveats, however, the model succeeds in explaining a diverse array of experimental and field observations, and in making a series of non-trivial predictions that can be (and in some cases have been) experimentally tested. For example, variation in host acceptance thresholds has indeed been shown in a number of cases to vary with age and host deprivation in the ways predicted (Courtney and Courtney, 1982; Singer, 1983, 1986; Fitt, 1986; Courtney and Forsberg, 1988). The model clearly warrants further development and testing.

Courtney *et al.* (1989) suggest that their model also explains the phenomenon of asymmetrical cross-conditioning: that is, where encounter with a low-ranking host may increase the probability that a higher-ranking host is accepted, but encounter with a high-ranking host may decrease (or leave unaffected) the probability that a lower-ranking host is accepted. This phenomenon has been observed in several studies, although not with complete consistency (Mark, 1982; Jaenike, 1983; Papaj and Prokopy, 1988). The model's applicability is rather less apparent in this case: the mechanism for the effect is not specified by Courtney *et al.* (1989), but clearly cannot involve motivational changes induced by variation in egg load. To apply the model in a straightforward way, it would be necessary further to assume that an encounter with a poor host increases motivation (i.e. lowers average acceptance thresholds), while an encounter with a good host decreases motivation. This is possible, but an alternative—and somewhat simpler—model for this phenomenon may be analogous to that discussed earlier and developed by Mitchell (1975) to explain how bruchid beetles manage to distribute their eggs uniformly among seeds: that is, that insects which behave in this way are capable of short-term 'comparison shopping', and are more likely to accept a newly-encountered host if it is better than the one they last encountered, but are less likely to accept it if it is worse. These alternative models are in any case testable, albeit not without difficulty.

Although application of the specifics of the hierarchy-threshold model should probably, as noted earlier, be restricted to post-alighting acceptance or rejection, there is no doubt that motivational effects mediated by the female's reproductive condition do influence other aspects of host location and oviposition behaviour. Cabbage butterfly females with high egg loads are more likely to approach and land on potential hosts than those with low egg loads (Jones, 1977), and Hawkes and Coaker (1979) used wind tunnel experiments to show that only mated, gravid females of the cabbage root fly would exhibit an upwind flight response to hostplant odours.

5.4.2 Variations in experience

Some examples of effects of prior experience on oviposition behaviour have already been discussed: notably learned responses to leaf shape in butterflies, and the cross-conditioning effects noted in the previous section.

Learned responses to hosts may well be very common in phytophagous insects. That an encounter with a host by an adult insect may result in subsequent behavioural modifications is very well documented in several cases, including those mentioned earlier. Among the most striking experimental studies, however, is the induction of preference changes in drosophilids. Jaenike (1983) examined the effects of including various repellent compounds in *Drosophila* culture media and feeding the media to adult flies, on the subsequent acceptance of the media as oviposition sites. In several cases, media which were unacceptable to flies which had not previously fed on them could be made acceptable by allowing the flies to feed on them first: in one case, feeding resulted in a strong preference for the modified medium relative to the control.

Associative learning has also been demonstrated in tephritid fruitflies (Prokopy *et al.*, 1982): flies were allowed to oviposit four times on the same fruit type, and then offered a fifth fruit which might be the same or different. With this experimental protocol, they tended to reject the fifth fruit if it was different from the previous four, but to accept it if it was the same.

Also marginally in the category of modified search behaviour as a result of adult experience are cases where an insect abandons a particular patch of hosts as a result of too high a frequency of encounters with previously attacked hosts. This phenomenon has frequently been documented in parasitoids (see Wellings, Chapter 4, this volume) but may also occur in at least some of those phytophages in which oviposition is deterred by the presence of conspecifics (Roitberg and Prokopy, 1984).

It has also been suggested that ovipositional preferences in adult insects may be conditioned by the hosts on which they fed as larvae (the 'Hopkins host selection principle'). This effect, however, has not been unambiguously demonstrated and probably does not occur. Because conditioning certainly can occur as a result of adult exposure to hosts, to demonstrate an effect of larval conditioning requires that individuals reared on a particular host be removed from any association with the host before they emerge as adults. Most studies where larval conditioning appears to have occurred do not meet this requirement (Jaenike, 1983).

Larval conditioning to particular hosts may not occur, but aspects of a juvenile insect's environment may still exert substantial effects on its reproductive behaviour as an adult. Most obvious are the cases, such as are observed in migratory locusts or in aphids, where environmental conditions experienced by young individuals (or by their mothers in the case of aphids) can invoke a phase shift between sedentary and migratory individuals which includes substantial effects on reproductive and host-seeking behaviour. Similarly, it is likely that, in at least some cases, the occurrence of an adult reproductive diapause may in part be determined by the temperatures or photoperiods experienced by juveniles (Rienks, 1981). Direct effects of rearing temperature on host choice and oviposition behaviour also occur in some butterfly species (R. E. Jones, unpublished data). Effects of this kind have not often been looked for, and it is possible that they are of more widespread occurrence than currently available data suggest.

5.4.3 Variation associated with seasonal forms

Many homopterans—especially aphids—spend the summer on one host species but move to overwinter on a different and often quite unrelated species. In other cases, the hostplant remains the same but different parts of the plant are utilized. Several studies have demonstrated that this alternation of generations is associated with a corresponding shift in reproductive behaviour and host acceptance (see e.g. Claridge and Wilson, 1978; Moran, 1983; Leather, 1986; Butt and Stuart, 1986).

5.5 GENETIC VARIATION BETWEEN AND WITHIN POPULATIONS

The material discussed above makes it clear that there are a variety of sources of environmentally-induced variation in host location and oviposition behaviour in phytophagous insects. Detecting genetically-based variation is correspondingly difficult: there is, nonetheless, ample evidence that substantial genetic variation exists, both within and between populations, and that it takes interesting and suggestive forms.

5.5.1 Variation between populations

Genetic variation between geographically separated populations in behavioural responses to hostplants has been identified in several insect groups. As Mitter and Futuyma (1983) point out, this is not especially surprising, since geographic variation has been found in most features in which it has been sought. Most such studies have looked for variation in preference rankings of different hostplants, and a number of clear-cut cases have been identified (see for example Singer, 1971; Prokopy *et al.*, 1984; Robert, 1985), although other studies which have looked for variation of this kind in situations where it might have been expected have not found it: for example, Tabashnik (1983) compared two populations of the butterfly *Colias philodice eriphyle*, one of which had moved from native legumes to forage alfalfa, and one which had not. He found that larval growth rates of individuals from each population on their own hosts tended to be very slightly better than those not on their own host, but that there was no difference in preference: both preferred alfalfa.

Other aspects of oviposition behaviour can also vary between populations. The behaviour of female cabbage butterflies (*Pieris rapae*) from Canadian (Vancouver), Australian, and English populations differed markedly between national strains (Jones, 1977, 1987; Jones *et al.*, 1980). The Vancouver (and English) females exhibited behavioural traits (high responsiveness to hostplants, high post-alighting acceptance probabilities, high resettling frequencies, very non-directional movement paths) which tended to result in their daily complement of eggs being laid relatively rapidly, at a low cost in flight time, and on relatively few plants. By contrast, the behaviour of Australian females (lower values for responsiveness, host acceptance, and resettling frequencies, and more directional movement) resulted in eggs being less aggregated and spread over a greater number of plants, but at a greater cost in flight time. These differences may have evolved in response to different problems faced by juveniles and adults in each location (Richards, 1940; Dempster, 1967; Jones and Ives, 1979; Jones *et al.*, 1987). In Vancouver and in England, high frequencies of overcast weather (which inhibit flight and oviposition) during the breeding season mean that the major factor limiting population growth is insufficient time for the females to lay all their eggs: despite the aggregated egg distributions, plants are rarely overcrowded with larvae. For the Australian population, on the other hand, the weather is much better, juvenile abundances tend to be much higher, and local juvenile overcrowding resulting in density-dependent mortality occurs frequently. Thus the different behaviour of these geographically separated populations appears to be a response to differences in the factors limiting population growth.

5.5.2 Variation within populations

For evolutionary change to occur, there must also be genetic variability within populations, underlying and influencing the more obvious environmentally-induced behavioural plasticity. Studies which clearly identify genetic variation of this kind are still relatively few and incomplete, but their results are tantalizingly suggestive. Three examples will illustrate this point.

The first concerns a study reported by Ng (1988), on host preference and offspring performance in the butterfly *Euphydryas editha* utilizing the hostplant *Pedicularis semibarbata*. Ng found that the population included 'specialist' and 'generalist' individuals. The 'specialists' consistently accepted some particular hostplant individuals and rejected others; the 'generalists' accepted all the plants. He then grew larvae from each female on accepted/rejected plant pairs. The 'specialists' did consistently better on the accepted plants; the 'generalists' did equally well on all plants. It is not yet clear what (if any) advantage the specialist gains from its restriction to particular plants to compensate for not having all plants available to it, since the generalists appeared to survive just as well as the specialists on the accepted plants, and considerably better than the specialists on the rejected plants. As Ng (1988) points out, the existence of any variation of this kind in pest populations would have significant implications for plant breeders attempting to breed resistant plant strains on the basis of average attack rates or average pest survival rates, since wide usage of the resistant strain may simply result in rapid selection for 'generalist' genotypes in the herbivore population.

The second example concerns the oviposition behaviour of the cabbage butterfly *Pieris rapae* (Jones, 1987). Both Australian and English populations of this species show significant heritable variation in a complex of behavioural traits which might be characterized as 'fussiness', and which results in individual variation both in the rates at which eggs are laid, and degree to which resultant egg distributions are aggregated. These behavioural traits are also correlated with developmental rates, though one generation's selection resulted in a partial decoupling. As in the previous example, we do not yet understand the significance of these relationships.

The third example concerns a mycophagous drosophilid, *Drosophila suboccidentalis* (Courtney and Chen, 1988). This study found substantial heritable variation in both willingness to accept a novel host, and in the number of eggs deposited by accepting females; here again, the two traits were correlated. In this case, Courtney and Chen (1988) suggest that both behavioural traits are driven by heritable variation in egg load and hence in motivation to oviposit.

A number of other studies have also found heritable variation in host choice behaviour (see for example Futuyma and Wasserman, 1981; Jaenike and Grimaldi, 1983; Lofdahl, 1987). These examples demonstrate both that it exists, and that some genetically-influenced traits covary in adaptively significant ways. Identification and elucidation of genetically-influenced behavioural and life history traits, and the ways in which they covary, promises to be one of the most profitable and significant areas of investigation in the field of plant–insect interactions.

5.6 THE SELECTIVE CONTEXT OF HOST LOCATION AND OVIPOSITION BEHAVIOUR

There are innumerable questions which can be asked about the adaptive value of particular kinds of host location and oviposition behaviour: the further development of genetic analyses of the kind just discussed, and of formal modelling studies of the effects on fitness of particular behavioural tactics, some of which were described earlier, are approaches which are likely to have a major impact on our understanding in this area. The adaptive consequences of particular behavioural tactics may involve effects on offspring fitness (see Reavey and Lawton, Chapter 11, this volume), or on the reproductive output and survivorship of the female herself, and any particular behaviour may well have multiple consequences. This chapter concludes with studies which have addressed two apparently simple questions concerning reproductive tactics in phytophagous insects.

5.6.1 Why specialize?

Although all phytophagous insects are to some degree specialist in the hostplants they will accept, some are very much more specialist than others: some accept only one species of host; others may accept hundreds of different host species from several different plant families. Often, ovipositing females accept a much narrower range of hostplants than that on which their offspring can successfully develop and survive (Wiklund,

1974). The most convincing explanation for this difference remains that suggested by Wiklund: that because juveniles which hatch on the 'wrong' host are unlikely to be able to find their way to the right one, it is in their interest to attempt to feed on any plant on which they find themselves: sometimes they will be lucky and survive. Conversely, the female will enhance her own reproductive success by minimizing the frequency of ovipositional mistakes which result in her offspring being on a plant where they cannot survive, so it is in her interest to reduce the risk of such mistakes by rejecting hosts of doubtful favourability, at least to the point where the rate of rejection does not result in a failure to lay her full egg complement. It is at least conceivable that the advantages of avoiding mistakes could be enough on their own to lead to the more extreme forms of host specialization for insects utilizing a common and widespread host species. Not all extreme specialists utilize common hosts, however, and numerous other advantages have been suggested to balance the loss of potential food resources associated with ovipositional and feeding specialization. The most obvious possibility—that host specialization allows larval digestive efficiencies which permit the specialist to develop and survive more successfully on its chosen host than a generalist would —has little empirical evidence in its favour (Smiley, 1978; Scriber and Feeny, 1979; Auerbach and Strong, 1981; Futuyma and Wasserman, 1981; Jones *et al.*, 1987), and several workers have suggested that the advantage of specialism may be that it allows more efficient search behaviour: this hypothesis remains untested. Perhaps the most convincing data bearing on the advantages of specialization are those of Bernays (1988), who demonstrated in an ingenious experiment that specialist phytophagous lepidopteran larvae were in general less susceptible to predator attack than were generalists, presumably because the evolution of cryptic adaptation was easier to achieve in association with a specific plant morphology.

5.6.2 Why lay eggs in batches?

This second question has been given particular attention by workers studying butterflies, in which there are numerous examples of related species utilizing the same or similar host plants, some of which lay eggs in batches while others lay eggs singly with an intervening flight between each oviposition. Suggested answers are almost as numerous as the people suggesting them, but the arguments are complicated by the fact that, as Courtney (1984) points out, once individuals are aggregated in related groups—i.e. once egg clustering already exists—kin selection may then operate to provide a variety of other advantages to living in groups, including aposematism, cooperative feeding, and the enhancement of mutualistic relationships, as described by Stamp (1981) and

Kitching (1981). It is more difficult to conceive of an evolutionary sequence in which the adoption of behaviour and morphology which gives advantages to group-living could precede the existence of groups. If these are therefore excluded as explanations, two further possibilities remain. One is that suggested by Davies and Gilbert (1985) to explain why the large white butterfly *Pieris brassicae* clusters its eggs, while the small white, *Pieris rapae*, which uses the same hostplants, does not. Larval groups of the large white characteristically demolish their host part-way through the larval period and the caterpillars then migrate to adjacent plants. Davies and Gilbert analysed the growth rates of cabbages and the feeding damage inflicted by larvae, and showed that in a large patch of plants the loss of the demolished host was more than compensated for by the higher growth rates of plants without larvae: that is, that the amount of food available to late-stage larvae in the whole patch of plants was greater if larvae were aggregated on a single plant, even though they destroyed it, than would have been the case if larvae were spread more evenly among hosts throughout their lives. They therefore suggested that batch laying was an adaptation to maximize the available food resources in species which characteristically occupied large patches of plants. Implicit in this explanation is the assumption that such species are also characteristically food-limited. This hypothesis has the considerable merit of generating testable predictions, since if it is correct, batch layers should characteristically exhibit behaviours which result in an egg distribution biased toward larger patches of hostplants than do species which lay eggs singly, and should frequently show evidence of food limitation.

An alternative hypothesis is suggested by Courtney (1984): that the evolution of batch laying is a response to a shortage of time available to search for hosts, and will be selected for whenever the cost of laying eggs singly (measured as failure to lay the full egg complement) outweighs the disadvantages of local overcrowding. This explanation is supported by the observation (Jones, 1987) that in the English strain of the cabbage butterfly *Pieris rapae*, which normally lays eggs singly, some individuals began to lay eggs in batches late in the season when weather conditions, which would permit flight and oviposition, became relatively less frequent.

The two questions discussed above provide examples of the kinds of 'strategic' questions about host location and oviposition behaviour to which we may legitimately expect eventually to develop relatively general answers. Our lack of ability to do so convincingly at this point demonstrates that although the study of plant–insect interactions has progressed very remarkably in the last few decades, we still have some way to go, especially in understanding the adaptive significance of many of the behavioural attributes that we observe.

References

Atsatt, P. (1981a) Lycaenid butterflies and ants: selection for enemy-free space. *Amer. Nat.*, **118**, 638–654.

Atsatt, P. (1981b) Ant-dependent food plant selection by the mistletoe butterfly *Ogyris amaryllis. Oecologia*, **48**, 60–63.

Atsatt, P. and O'Dowd, D. (1976) Plant defense guilds. *Science*, **193**, 24–29.

Auerbach, M. J. and Strong, D. R. (1981) Nutritional ecology of *Heliconia* herbivores: experiments with plant fertilization and alternative hosts. *Ecol. Monogr.*, **51**, 63–83.

Averill, A. L. and Prokopy, R. J. (1987) Intraspecific competition in the tephritid fruitfly *Rhagoletis pomonella. Ecology*, **68**, 878–886.

Bach, C. E. (1980) Effect of plant density and diversity on the population dynamics of a specialist insect herbivore, the striped cucumber beetle *Acalymma vittata* (Fab.). *Ecology*, **61**, 1515–1530.

Bach, C. E. (1984) Plant spatial pattern and herbivore population dynamics: factors affecting the movement patterns of a tropical cucurbit specialist (*Acalymma innubum*). *Ecology*, **65**, 175–190.

Bach, C. E. (1986) A comparison of the responses of two tropical specialist herbivores to plant patch size. *Oecologia*, **68**, 580–584.

Behan, M. and Schoonhoven, L. M. (1978) Chemoreception of an oviposition deterrent associated with eggs in *Pieris brassicae. Entomol. Exp. Appl.*, **24**, 163–179.

Bernays, E. (1988) Host specificity in phytophagous insects: selection pressure from generalist predators. *Entomol. Exp. Appl.*, **49**, 139–40.

Blum, A. (1969) Oviposition preference by the sorghum shootfly (*Atherigona varia soccata*) in progenies of susceptible X resistant sorghum crosses. *Crop Sci.*, **9**, 695–696.

Butt, B. A. and Stuart, C. (1986) Oviposition by summer and winter forms of the pear psylla (Homoptera: Psyllidae) on dormant pear budwood. *Environ. Entomol.*, **15**, 1109–1110.

Chew, F. S. and Robbins, R. K. (1984) Egg-laying in butterflies. In (Eds. R. I. Vane-Wright and P. R. Ackery). *The Biology of Butterflies.* Symposium of the Royal Society of London, No. 11, ch. 6, 65–78.

Claridge, M. F. and Wilson, M. R. (1978) Seasonal changes and alternation of food plant preference in some mesophyll-feeding leafhoppers. *Oecologia*, **37**, 247–255.

Clark, L. R. (1963) Factors affecting the attractiveness of foliage for oviposition by *Cardiaspina albitextura* (Psyllidae). *Austr. J. Zool.*, **11**, 20–34.

Courtney, S. P. (1981) Coevolution of pierid butterflies and their cruciferous foodplants. III. *Anthocharis cardamines* survival, development and oviposition on different hostplants. *Oecologia*, **51**, 91–96.

Courtney, S. P. (1982) Coevolution of pierid butterflies and their cruciferous foodplants. IV. Crucifer apparency and *Anthocharis cardamines* (L.) oviposition. *Oecologia*, **52**, 258–265.

Courtney, S. P. (1984) The evolution of egg clustering by butterflies and other insects. *Amer. Nat.*, **123**, 276–281.

Courtney, S. P. (1986) The ecology of pierid butterflies: dynamics and interactions. *Adv. Ecol. Res.*, **15**, 51–131.

Courtney, S. P. and Chen, G. K. (1988) Genetic and environmental variation in oviposition behaviour in the mycophagous *Drosophila suboccidentalis Spcr. Funct. Ecol.*, **2**, 521–8.

Courtney, S. P. and Courtney, S. (1982) The 'edge-effect' in butterfly oviposition: causality in *Anthocharis cardamines* and related species. *Ecol. Entomol.*, **7**, 131–137.

Courtney, S. P. and Forsberg, J. (1988) Host use by two pierid butterflies varies with host density. *Functional Biology*, **2**, 67–75.

Courtney, S. P., Chen, G. K. and Gardner, A. (1989) A general model for individual host selection. *Oikos*, **55**, 55–65.

Cromartie, W. J. (1975) The effect of stand size and vegetational background on the colonization of cruciferous food plants by herbivorous insects. *J. Appl. Ecol.*, **12**, 517–533.

Davies, C. R. and Gilbert, N. E. (1985) A comparative study of the egg-laying behaviour and larval development of *Pieris rapae* L. and *P. brassicae* L. on the same host plants. *Oecologia*, **67**, 278–281.

Dempster, J. P. (1967) The control of *Pieris rapae* with DDT. I. The natural mortality of the young stages. *J. Appl. Ecol.*, **4**, 485–500.

Den Boer, P. J. (1974) Spreading the risk and stabilization of animal numbers. *Acta Biotheoret.*, **18**, 165–194.

Den Otter, C. J., Behan, M. and Maes, F. W. (1980) Single-cell responses in female *Pieris brassicae* (Lepidoptera: Pieridae) to plant volatiles and conspecific egg odours. *J. Insect Physiol.*, **26**, 465–472.

Dixon, A. F. G., Kindlmann, P., Leps, J. and Holman, J. (1987) Why are there so few species of aphids, especially in the tropics? *Amer. Nat.*, **129**, 580–592.

Dunn, J. P., Kimmerer, T. W. and Nordin, G. L. (1986) The role of host treee condition in attack of white oaks by the two-lined chestnut-borer *Agrilus bilineatus* (Weber) (Coleoptera: Buprestidae). *Oecologia*, **70**, 596–600.

Edwards, P. J. and Wratten, S. D. (1983) Wound induced defences in plants and their consequences for patterns of insect grazing. *Oecologia*, **59**, 88–93.

Feeny, P. (1976) Plant apparency and chemical defence. In *Biochemical Interactions Between Plants and Insects* (eds J. W. Wallace and R. L. Mansell), *Recent Adv. Phytochem.*, **10**, 1–40.

Fitt, G. P. (1984) Oviposition behaviour of two tephritid fruit flies, *Dacus tryoni* and *Dacus jarvisi*, as influenced by the presence of larvae in the host fruit. *Oecologia*, **62**, 37–46.

Fitt, G. P. (1986) The influence of a shortage of hosts on the specificity of oviposition behaviour in species of *Dacus* (Diptera: Tephritidae). *Physiol. Entomol.*, **11**, 133–143.

Fowler, S. V. and Lawton, J. H. (1985) Rapidly induced defences and talking trees: the devil's advocate position. *Amer. Nat.*, **126**, 181–195.

Fox, L. R. and Macauley, B. J. (1977) Insect grazing on *Eucalyptus* in response to leaf tannins and nitrogen. *Oecologia*, **29**, 145–162.

Fraenkel, G. S. (1953) The nutritional value of green plants for insects. *Trans. 9th Int. Congr. Entomol., Amsterdam*, **2**, 90–100.

Futuyma, D. J. and Wasserman, S. S. (1981) Food plant specialization and feeding efficiency in the tent caterpillars *Malacosoma disstria* and *M. americanum. Entomol. Exp. Appl.*, **30**, 106–110.

Gossard, T. W. and Jones, R. E. (1977) The effects of age and weather on egg-laying in *Pieris rapae. J. Appl. Ecol.*, **14**, 65–71.

Gould, F. (1983) Genetics of plant–herbivore systems: interactions between applied and basic study. In *Variable Plants and Herbivores in Natural and Managed Systems* (eds R. F. Denno and M. S. McClure), Academic Press, New York.

Harris, M. O. and Miller, J. R. (1982) Synergism of visual and chemical stimuli in the oviposition behaviour of *Delia antiqua* (Meigen) (Diptera: Anthomyiidae). *Proc. 5th Int. Symp. Insect–Plant Relationships*, Wageningen, 107–116.

Haukioja, E. and Niemela, P. (1979) Birch leaves as a resource for herbivores: seasonal occurrence of increased resistance in foliage after mechanical damage of adjacent leaves. *Oecologia*, **39**, 151–159.

Haukioja, E., Suomela, J. and Neuvonen, S. (1985) Long-term inducible resistance in birch foliage: triggering cues and efficacy on a defoliator. *Oecologia*, **65**, 363–369.

Hawkes, C. and Coaker, T. H. (1979) Factors affecting the behavioural responses of the adult cabbage root fly *Delia brassicae*, to host plant odour. *Entomol. Exp. Appl.*, **25**, 45–58.

Holdren, C. E. and Ehrlich, P. R. (1982) Ecological determinants of food plant choice in the checkerspot butterfly *Euphydryas editha* in Colorado. *Oecologia*, **52**, 417–423.

Ives, P. M. (1978) How discriminating are cabbage butterflies? *Austr. J. Ecol.*, **3**, 261–276.

Jaenike, J. (1983) Induction of host preference in *Drosophila melanogaster. Oecologia*, **58**, 320–325.

Jaenike, J. (1985) Genetic and environmental determinants of food preference in *Drosophila tripunctata. Evolution*, **39**, 362–369.

Jaenike, J. and Grimaldi, D. (1983) Genetic variation for host preference within and among populations of *Drosophila tripunctata. Evolution*, **37**, 1023–1033.

Jones, R. E. (1977) Movement patterns and egg distribution in cabbage butterflies. *J. Anim. Ecol.*, **46**, 195–212.

Jones, R. E. (1981) The cabbage butterfly *Pieris rapae* (L.): 'a just sense of how not to fly'. In *The Ecology of Pests: Some Australian Case Histories* (eds R. L. Kitching and R. E. Jones), CSIRO Press, Melbourne, 217–228.

Jones, R. E. (1987) Behavioural evolution in the cabbage butterfly. *Oecologia*, **72**, 69–76.

Jones, R. E. and Ives, P. M. (1979) The adaptiveness of searching and host selection behaviour in *Pieris rapae. Austr. J. Ecol.*, **4**, 75–86.

Jones, R. E., Gilbert, N., Guppy, M. and Nealis, V. (1980) Long-distance movement of *Pieris rapae. J. Anim. Ecol.*, **49**, 629–642.

Jones, R. E., Nealis, V. G., Ives, P. M. and Scheermeyer, E. (1987a) Seasonal and spatial variation in survival of the cabbage butterfly *Pieris rapae*: evidence for patchy density-dependence. *J. Anim. Ecol.*, **56**, 723–737.

Jones, R. E., Rienks, J., Wilson, L. *et al.* (1987b) Temperature, development and survival in monophagous and polyphagous tropical pierid butterflies. *Austr. J. Zool.*, **35**, 235–246.

Kammp, J. A. and Buttery, R. G. (1986) Ovipositional behaviour of the alfalfa seed chalcid (Hymenoptera: Eurytomidae) in response to volatile components of alfalfa. *Environ. Entomol.*, **15**, 388–391.

Karban, R. and Courtney, S. P. (1987) Intraspecific host plant choice: lack of consequences for *Strepthanthus tortuosus* (Cruciferae) and *Euchloe hyantis* (Lepidoptera: Pieridae). *Oikos*, **48**, 243–248.

Kareiva, P. (1982) Influence of vegetation texture on herbivore populations: resource concentration and herbivore movement. In *Variable Plants and Herbivores in Natural and Managed Systems* (eds R. F. Denno and M. S. McClure), Academic Press, New York, 259–289.

Kennedy, J. S. (1958) Physiological condition of the host-plant and susceptibility to aphid attack. *Entomol. Exp. Appl.*, **1**, 49–65.

Kitching, R. L. (1981) Egg clustering and the southern hemisphere lycaenids: comments on a paper by N. E. Stamp. *Amer. Nat.*, **118**, 423–425.

Leather, S. R. (1985) Oviposition preferences in relation to larval growth rates and survival in the Pine Beauty Moth *Panolis flammea*. *Ecol. Entomol.*, **10**, 213–217.

Leather, S. R. (1986) Host monitoring by aphid migrants: do gynoparae maximize fitness? *Oecologia*, **68**, 367–369.

Leather, S. R., Watt, A. D. and Forrest, G. I. (1987) Insect-induced chemical changes in young lodgepole pine (*Pinus contorta*): the effect of previous defoliation on oviposition, growth and survival of the Pine Beauty Moth *Panolis flammea*. *Ecol. Entomol.*, **12**, 275–281.

Lemen, C. (1981) Elm trees and elm leaf beetles: patterns of herbivory. *Oikos*, **36**, 65–67.

MacGarvin, M. (1982) Species-area relationships of insects on host plants: herbivores on rosebay willowherb. *J. Anim. Ecol.*, **51**, 207–223.

Mackay, D. A. (1985) Pre-alighting search behaviour and host plant selection by ovipositing *Euphydryas editha* butterflies. *Ecology*, **66**, 142–151.

Mackay, D. A. and Jones, R. E. (1989) Leaf shape and the host-finding behaviour of two ovipositing monophagous butterfly species. *Ecol. Entomol.*, **14**, 423–31.

Mackay, D. A. and Singer, M. C. (1982) The basis of an apparent preference for isolated host plants by ovipositing *Euptychia libye* butterflies. *Ecol. Entomol.*, **7**, 299–303.

Mark, G. A. (1982) Induced oviposition preference, periodic environments, and demographic cycles in the bruchid beetle *Callosobruchus maculatus*. *Entomol. Exp. Appl.*, **32**, 155–160.

Mattson, W. J. (1980) Herbivory in relation to plant nitrogen content. *Ann. Rev. Ecol. Syst.*, **11**, 119–161.

Messina, F. J. and Renwick, J. A. A. (1985) Ability of ovipositing seed weevils to discriminate between seeds with differing egg loads. *Ecol. Entomol.*, **10**, 225–230.

Miller, J. R. and Strickler, K. L. (1984) Finding and accepting host plants. In *Chemical Ecology of Insects* (eds W. J. Bell and R. T. Carde), Chapman and Hall, London, 127–157.

Mitchell, R. (1975) The evolution of oviposition tactics in the bean weevil *Callosobruchus maculatus* (F.). *Ecology*, **56**, 696–702.

Mitter, C. and Futuyma, D. J. (1983) An evolutionary-genetic view of host-plant utilization by insects. In *Variable Plants and Herbivores in Natural and Managed Systems* (eds R. F. Denno and M. S. McClure), Academic Press, New York, 427–459.

Moore, G. J. (1986) Host plant discrimination in tropical satyrine butterflies. *Oecologia*, **70**, 592–595.

Moran, N. A. (1983) Seasonal shifts in host usage in *Uroleucon gravicorne* (Homoptera: Aphididae) and implications for the evolution of host alternation in aphids. *Ecol. Entomol.*, **8**, 371–382.

Morrow, P. A. and Fox, L. R. (1980) Effects of variation in *Eucalyptus* essential oil yield on insect growth and grazing damage. *Oecologia*, **45**, 209–219.

Myers, J. H. (1985) Effect of physiological condition of the host plant on the ovipositional choice of the cabbage white butterfly, *Pieris rapae*. *J. Anim. Ecol.*, **54**, 193–204.

Ng, D. (1988) A novel level of interactions in plant–insect systems. *Nature*, **334**, 611–613.

Oshima, K., Honda, H. and Yamamoto, I. (1973) Isolation of an oviposition marker from Azuki bean weevil. *Agric. Biol. Chem.*, **37**, 2679–2680.

Papaj, D. R. (1986a) Shifts in foraging behaviour by a *Battus philenor* population: field evidence for switching by individual butterflies. *Behav. Ecol. Sociobiol.*, **19**, 31–39.

Papaj, D. R. (1986b) Conditioning of leaf-shape discrimination by chemical cues in the butterfly *Battus philenor*. *Anim. Behav.*, **34**, 1281–1288.

Papaj, D. R. and Prokopy, R. J. (1988) The effect of prior adult experience on components of habitat preference in the apple maggot fly (*Rhagoletis pomonella*). *Oecologia*, **76**, 538–543.

Papaj, D. R. and Rausher, M. D. (1987) Genetic differences and phenotypic plasticity as causes of variation in oviposition preference in *Battus philenor*. *Oecologia*, **74**, 24–30.

Perrin, R. M. (1980) The role of environmental diversity in crop protection. *Prot. Ecol.*, **2**, 77–114.

Pierce, N. E. (1984) Amplified species diversity: a case study of an Australian lycaenid butterfly and its attendant ants. In *The Biology of Butterflies* (eds R. I. Vane-Wright and P. R. Ackery), *Symp. R. Entomol. Soc. Lond.*, **11**, 197–200.

Pierce, N. E. and Mead, P. S. (1981) Parasitoids as selective agents in the symbiosis between butterfly larvae and ants. *Science*, **211**, 1185–1187.

Price, P. W., Roinen, H. and Tahvanainen, J. (1987) Why does the bud-galling sawfly, *Euura mucronata*, attack long shoots? *Oecologia*, **74**, 1–6.

Prokopy, R. J. (1972) Evidence for a marking pheromone deterring repeated oviposition in apple maggot flies. *Environ. Entomol.*, **1**, 326–332.

Prokopy, R. J. (1983) Visual detection of plants by herbivorous insects. *Ann. Rev. Entomol.*, **28**, 337–364.

Prokopy, R. J., Averill, A. L., Cooley, S. S. and Roitberg, C. A. (1982) Associative learning in egglaying site selection by apple maggot flies. *Science*, **218**, 76–77.

Prokopy, R. J., McDonald, P. T. and Wong, T. T. Y. (1984) Inter-population variation among *Ceratitis capitata* flies in host acceptance pattern. *Entomol. Exp. Appl.*, **35**, 65–69.

Raupp, M. J. and Denno, R. F. (1983) Leaf age as a predictor of herbivore distribution and abundance. In *Variable Plants and Herbivores in Natural and*

Managed Systems (eds R. F. Denno and M. S. McClure), Academic Press, New York, 91–124.

Rausher, M. D. (1978) Search image for leaf shape in a butterfly. *Science*, **200**, 1071–1073.

Rausher, M. D. (1979) Egg recognition: its advantage to a butterfly. *Anim. Behav.*, **27**, 1034–1040.

Rausher, M. D. (1981) Host plant selection by *Battus philenor* butterflies: the role of predation, nutrition, and plant chemistry. *Ecol. Monogr.*, **51**, 1–20.

Rausher, M. D. (1985) Variability for host preference in insect populations: mechanistic and evolutionary models. *J. Insect Physiol.*, **31**, 873–889.

Rausher, M. D. and Papaj, D. R. (1983) Host plant selection by *Battus philenor* butterflies: evidence for individual differences in foraging behaviour. *Anim. Behav.*, **31**, 341—347.

Rausher, M. D., Mackay, D. A. and Singer, M. C. (1981) Pre- and post-alighting host discrimination by *Euphydryas editha* butterflies: the behavioural mechanisms causing clumped distributions of egg clusters. *Anim. Behav.*, **29**, 1220–1228.

Richards, O. W. (1940) The biology of the small white butterfly (*Pieris rapae*) with special reference to the factors controlling its abundance. *J. Anim. Ecol.*, **9**, 243–288.

Rienks, J. H. (1981) Morph determination and ecology of *Catopsilia pomona pomona* (Lepidoptera: Pieridae). Honours thesis, James Cook University.

Robert, P. (1985) A comparative study of some aspects of the reproduction of three *Caryedon serratus* strains in the presence of its potential host plants. *Oecologia*, **65**, 425–430.

Robertson, H. G. (1987) Oviposition site selection in *Cactoblastis cactorum* (Lepidoptera): constraints and compromises. *Oecologia*, **73**, 601–608.

Robinson, S. H., Wolfenbarger, D. A. and Dilday, R. H. (1980) Antixenosis of smooth leaf cotton to the ovipositional response of tobacco budworm. *Crop Sci.*, **20**, 646–649.

Rodman, J. F. and Chew, F. S. (1980) Phytochemical correlates of herbivory in a community of native and naturalized cruciferae. *Biochem. Syst. Ecol.*, **8**, 43–50.

Roitberg, B. D. and Prokopy, R. J. (1984) Host visitation sequence as a determinant of search persistence in fruit parasitic tephritid flies. *Oecologia*, **62**, 7–12.

Roitberg, B. D., Cairl, R. S. and Prokopy, R. J. (1984) Oviposition deterring pheromone influences dispersal distance in tephritid fruitflies. *Entomol. Exp. Appl.*, **35**, 217–220.

Root, R. B. (1973) Organization of a plant–arthropod association in simple and diverse habitats: the fauna of collards (*Brassica oleraceae*). *Ecol. Monogr.*, **43**, 95–124.

Rothschild, M. and Schoonhoven, L. M. (1977) Assessment of egg load by *Pieris brassicae*. *Nature*, **266**, 352–355.

Schillinger, J. A. and Gallun, R. L. (1968) Leaf pubescence of wheat as a deterrent to the cereal leaf beetle, *Oulema melanoplus*. *Ann. Entomol. Soc. Amer.*, **61**, 900–903.

Scriber, J. M. and Feeny, P. (1979) Growth of herbivorous caterpillars in relation to feeding specialization and to the growth form of their food plants. *Ecology*, **60**, 829–850.

Shapiro, A. M. (1981) The pierid red-egg syndrome. *Amer. Nat.*, **117**, 194–276.

Singer, M. C. (1971) Evolution of host plant preference in the butterfly *Euphydryas editha*. *Evolution*, **25**, 383–389.

Singer, M. C. (1981) Determinants of multiple host use by a phytophagous insect population. *Evolution*, **37**, 389–403.

Singer, M. C. (1982) Quantification of host preference by manipulation of oviposition behaviour in the butterfly *Euphydryas editha*. *Oecologia*, **52**, 224–229.

Singer, M. C. (1983) Determinants of multiple host use by a phytophagous insect population. *Evolution*, **37**, 389–403.

Singer, M. C. (1986) The definition and measurement of oviposition preference in plant-feeding insects. In *Insect–Plant Interactions* (eds J. R. Miller and T. A. Miller), Springer-Verlag, New York, 65–94.

Singer, M. C. and Mandracchia, J. (1982) On the failure of two butterfly species to respond to the presence of conspecific eggs prior to oviposition. *Ecol. Entomol.*, **7**, 327–330.

Skinner, S. W. (1985) Clutch size as an optimal foraging problem for insects. *Behavioural Ecology and Sociobiology*, **17**, 231–238.

Slansky, F. and Feeny, P. (1977) Stabilization of the rate of nitrogen accumulation by larvae of the cabbage butterfly on wild and cultivated food plants. *Ecol. Monogr.*, **47**, 209–228.

Smiley, J. (1978) Plant chemistry and the evolution of host specificity: new evidence from *Heliconius* and *Passiflora*. *Science*, **201**, 745–747.

Southwood, T. R. E. (1973) The insect–plant relationship—an evolutionary perspective. *Symp. R. Entomol. Soc. Lond.*, **6**, 3–30.

Stadler, E. (1974) Host plant stimuli affecting oviposition behaviour of the eastern spruce budworm. *Entomol. Exp. Appl.*, **17**, 176–244.

Stadler, E. and Buser, H.-R. (1984) Defense chemicals in leaf surface wax synergistically stimulate oviposition by a phytophagous insect. *Experientia*, **40**, 1157–1159.

Stamp, N. E. (1981) Egg deposition patterns in butterflies: why do some species cluster their eggs rather than deposit them singly? *Amer. Nat.*, **115**, 367–380.

Stanton, M. L. (1982) Searching in a patchy environment: foodplant selection by *Colias p. eriphyle* butterflies. *Ecology*, **63**, 839–853.

Stanton, M. L. (1984) Short-term learning and the searching accuracy of egg-laying butterflies. *Anim. Behav.*, **32**, 33–40.

Stanton, M. L. and Cook R. E. (1984) Sources of intraspecific variation in the hostplant seeking behaviour of *Colias* butterflies. *Oecologia*, **61**, 265–270.

Strong, D. R., Lawton, J. H. and Southwood, R. (1984) *Insects on Plants: Community Patterns and Mechanisms*, Blackwell, London.

Tabashnik, B. E. (1983) Host range evolution: the shift from native legume hosts to alfalfa by the butterfly *Colias philodice eriphyle*. *Evolution*, **37**, 150–162

Tabashnik, B. E. (1985) Deterrence of diamondback moth (Lepidoptera: Plutellidae) oviposition by plant compounds. *Environ. Entomol.*, **14**, 575–578.

Thompson, J. N. (1978) Within-patch structure and dynamics in *Pastinaca sativa* and resource availability to a specialized herbivore. *Ecology*, **59**, 443–448.

Thompson, J. N. (1987) Variance in number of eggs per patch: oviposition behaviour and population dispersion in a seed parasitic moth. *Ecol. Entomol.*, **12**, 311–320.

Thompson, J. N. and Price, P. W. (1977) Plant plasticity, phenology and herbivore dispersion: wild parsnip and the parsnip webworm. *Ecology*, **58**, 1112–1119.

Thorsteinson, A. J. (1960) Host selection in phytophagous insects. *Ann. Rev. Entomol.*, **5**, 193–218.

Traynier, R. M. M. (1967) Stimulation of oviposition by the cabbage root fly *Eroischia brassicae. Entomol. Exp. Appl.*, **10**, 401–412.

van der Meijden, E. (1979) Herbivore exploitation of a fugitive plant species: local survival and extinction of the cinnabar moth and ragwort in a heterogeneous environment. *Oecologia*, **42**, 307–323.

Van Leerdam, M. B., Johnson, K. J. R. and Smith, J. W. (1984) Effects of substrate physical characteristics and orientation on oviposition by *Foreuma loftine* (Lepidoptera: Pyralidae). *Environ. Entomol.*, **13**, 800–802.

Visser, J. H. (1986) Host odour perception in phytophagous insects. *Ann. Rev. Entomol.*, **31**, 121–144.

Wagner, M. R. and Evans, P. D. (1985) Defoliation increases nutritional quality and allelochemics of pine seedlings. *Oecologia*, **67**, 235–237.

Wasserman, S. S. and Futuyma, D. J. (1981) Evolution of hostplant utilization in laboratory populations of the southern cowpea weevil *Callosobruchus maculatus* Fabricius (Coleoptera: Bruchidae). *Evolution*, **35**, 605–617.

Whitham, T. G. (1978) Habitat selection by *Pemphigus* aphids in response to resource limitation and competition. *Ecology*, **59**, 1164–1176.

Whitham, T. G. (1980) The theory of habitat selection examined and extended using *Pemphigus* aphids. *Amer. Nat.*, 115, 449–466.

Whittaker, R. H. and Feeny, P. (1971) Allelochemics: chemical interactions between species. *Science*, **171**, 757–770.

Wiklund, C. (1974) Oviposition preferences in *Papilio machaon* in relation to the host plants of the larvae. *Entomol. Exp. Appl.*, **17**, 189–198.

Wiklund, C. and Ahrberg, C. (1978) Hostplants, nectar source plants, and habitat selection of males and females of *Anthocharis cardamines. Oikos*, **31**, 169–183.

Williams, E. H. (1981) Thermal influences on oviposition in the montane butterfly *Euphydryas gilletti. Oecologia*, **50**, 342–346.

Williams, K. S. (1983) The coevolution of *Euphydryas chalcedona* butterflies and their larval host plants. III. Oviposition behaviour and host plant quality. *Oecologia*, **56**, 336–340.

Williams, K. S. and Gilbert, L. E. (1981) Insects as selective agents on plant vegetative morphology: egg mimicry reduces egg laying by butterflies. *Science*, **212**, 467–469.

Wilson, L. J. (1986) Movement and feeding patterns of *Epilachna cucurbitae* Richards (Coleoptera: Coccinellidae) on pumpkin and zucchini plants. *Austr. J. Ecol.*, **11**, 55–62.

Wolfson, J. L. (1980) Oviposition response of *Pieris rapae* to environmentally induced variation in *Brassica nigra. Entomol. Exp. Appl.*, **27**, 223–232.

Zucker, W. V. (1982) How aphids choose leaves: the roles of phenolics in host selection by a galling aphid. *Ecology*, **63**, 972–981.

6

Host location and oviposition in tephritid fruit flies

Brian. S. Fletcher and Ron J. Prokopy

6.1 INTRODUCTION

The true fruit flies of the family Tephritidae are an economically important group of Diptera distributed throughout most of the temperate, sub-tropical and tropical regions of the world. During their larval stages they feed on and in the vegetative parts of plants, and the better known species develop either in the flower heads or fruit, although some species are leafminers or gall formers. Most of the information relating to host selection in fruit flies arises from studies on the major pest species attacking fruit. This chapter looks at pest species belonging to four genera. *Anastrepha* spp., which are restricted to South and Central America as far north as the southern states of USA; *Ceratitis*, which is of African origin but has now spread to many other parts of the world including Southern Europe, South and Central America, Hawaii and Australia; *Dacus* spp. which occur in the tropical and sub-tropical regions of Africa, Asia, Australia and the Pacific; and *Rhagoletis* which have a mainly Palearctic distribution in North America and Europe, with a few species occurring in the higher regions of Central and South America.

General aspects of tephritid biology, behaviour and ecology have been reviewed previously (Christenson and Foote, 1960; Bateman, 1972; Boller and Prokopy, 1976; Prokopy and Roitberg, 1984; Fletcher, 1987). They are also the subject of a number of Symposium Proceedings (Cavalloro, 1983, 1988; Economopoulos, 1987) and a recent major monograph (Robinson and Hooper, 1989).

The important feature that all the frugivorous species have in common is that their larval resources (i.e. fruit) are discrete ephemeral units which are patchily distributed in the habitat, and are suitable for larval development for a limited period. The eggs are deposited directly into the fruit by the ovipositing females (Fig. 6.1), and there is normally no opportunity for the larvae to move between fruits. Reproductive success, therefore, is strongly influenced by the ability of females to locate and oviposit in fruits

Fig. 6.1 Female *Dacus tryoni* ovipositing into an apple (photograph by J. Green CSIRO).

of the appropriate host species that are at a stage that permits larval survival and development.

This requirement makes the tephritids an ideal group in which to study the relationships between individual species and their host plants, both from an evolutionary and behavioural viewpoint. Most of this chapter is concerned with the factors influencing the host foraging and oviposition behaviour of female tephritids, although some broader issues related to host utilization by individuals and populations, including factors limiting host range and the evolution of host races, are also considered.

6.2 HOST RANGE

Based on their degree of host specialization, tephritid fruit flies can be classed as monophagous (= utilizing a single larval host); stenophagous (=utilizing a few, usually closely related, species in a single plant family); oligophagous (= a restricted host range, with species in only one or a few plant families); or polyphagous (= a wide host range including species in many plant families). The majority of species in all four economically important genera are either monophagous or stenophagous, although many of the major pest species, e.g. *Anatrepha ludens* and *A. suspensa*, *Ceratitis capitata* and *C. rosa*, and *Dacus cucurbitae*, *D. dorsalis* and *D. tryoni* are highly polyphagous. *Rhagoletis pomonella*, which was originally restricted to fruits of hawthorn (*Crataegus* spp.), has shown an interesting expansion of its host range over the past 120 years and now attacks a number of Rosaceae, including apple, apricot, sour cherry, pears and plums, plus the hips of *Rosa rugosa*. The introduction of domestic fruit species into parts of the world where they were previously absent has also resulted in a number of the polyphagous species undergoing major (if somewhat less surprising) expansions in host range.

The ability of fruit flies to expand their host range has been considered to depend on the oviposition specificity of the adults and the ability of the larvae or their associated bacteria to overcome the defence chemicals and digest the tissues of novel host fruit. It has been suggested that genetic changes in both the adult and larvae may be necessary to allow a host shift to occur (Bush and Diehl, 1982), although more recent evidence suggests that populations or species of fruit flies may increase their host range without any shift in physiological adaptations for survival on the new host (Fitt 1986a; Prokopy *et al.* 1988). As proposed by Futuyma (1983) for herbivorous insects generally, expansion of a species onto a new host may first involve genetic changes in the behavioural aspects of host-finding and only subsequently involve selection favouring genes affecting survival or development. Studies carried out by Fitt (1986a) on five species of Australian dacines indicated, for example, that the major factor restricting the utilization of a number of commercial fruits by the specialized species is the oviposition preferences of the adult females rather than the ability of the larvae to survive in the fruits. In the laboratory, larvae of all five species tested survived and completed development in most of the commercial varieties artificially infested with eggs. In contrast, the females of the specialist species, *D. cucumis*, *D. musae* and *D. cacuminatus*, strongly preferred their normal hosts and would not oviposit in novel fruit varieties even in the absence of the preferred host. Specificity in larval survival was most evident in cucumber, in which the cucurbit specialist, *D. cucumis*, was the only species to show normal levels of survival and development.

Commercial fruits, however, have been specifically selected to be palatable to man, and generally have few if any defensive chemicals. In many native fruits it seems likely that survival and development depend upon the species' ability to cope with both the nutritional and other properties of the fruit, with secondary defensive chemicals possibly being of paramount importance (Pree, 1977; Seo and Tang, 1982; Greany *et al.*, 1983; Chan and Tam, 1985). Studies on larval performance of *Dacus* spp. in a series of native fruits indicated that there were high levels of specificity, and only larvae of species utilizing a particular fruit as a host were able to survive in that fruit (Fitt, 1986a). Even in the cultivated fruit commonly utilized by polyphagous species there may be major differences in the survival and development rates of larvae, a factor which is sometimes but not always reflected in the oviposition preferences of the adults (Fletcher, 1987).

In addition, some species show a distinct preference for certain hosts when they are available, but will infest other hosts when the preferred hosts are unavailable. A good example of this type of behaviour occurs in *D. jarvisi*. The preferred host of this Australian dacine is *Planchonia careya*, and both in the field and in laboratory experiments it infests this host almost exclusively when available. When it is not available, however, it will infest a variety of other fruits, many of which are taxonomically quite distinct from its preferred host (Fitt, 1986b).

6.3 HOST RACES

The factors that may give rise to or maintain host races in herbivorous insects include: (a) genetically based differences between conspecific populations in host plant preferences; (b) positive assortative mating due to mating being restricted to host plants, and the resultant coupling of assortative mating with habitat preference; (c) fidelity to a particular host due to the effects of larval or adult experience; (d) differences among hosts that may affect survival ability; (e) temporal differences in host availability that may influence the start and end of diapause (Prokopy *et al.*, 1988; Diehl and Bush, 1989). There is evidence indicating that all of these factors may be involved in host race formation in *Rhagoletis* spp., but here we will consider only those that are directly related to host plant preferences of females.

Studies on the host acceptance patterns of female medflies from two distinct populations occurring in different parts of Hawaii, one from the island of Hawaii, which bred almost exclusively in Jerusalem cherries (*Solanum pseudocapsicum*), a small fruit (*circa* 13 mm diameter), and the other from Maui which bred in a variety of different sized hosts, revealed differences between them (Prokopy *et al.*, 1984). Interestingly, the observed differences in host acceptance patterns were based on fruit size rather than taxonomic relationship. Flies from Hawaii were more ready to accept and oviposit in the smallest fruit, mock oranges (*circa* 10 mm diameter), than the Maui flies. Both groups showed the same preferences for Jerusalem cherries (*circa.* 13 mm) and intermediate size fruit tested (*circa* 20–50 mm), but had a decreased preference for grapefruit (*circa* 115 mm). Flies from a colony that had been laboratory reared for over 300 generations, and only permitted to oviposit in 95 mm diameter plastic spheres for the previous 100 generations, oviposited in grapefruit just as readily as in the other fruit types. As the test females had not previously been exposed to fruit, it was concluded that the results were not due to prior adult experience or larval conditioning and probably reflected genetic differences between the different populations in the size range of fruits they would accept.

R. pomonella females from populations breeding in apple or hawthorn also exhibit different preferences in host acceptance that appear to be due to genetic differences in the populations breeding on the two different hosts, a view now supported by a number of electrophoretic studies on sympatric populations from both hosts that demonstrated significant differences between them in allele frequencies (Feder *et al.*, 1988; Mc-Pheron *et al.*, 1988). Laboratory and field assays on the propensity of *R. pomonella* females, originating from naturally infested hawthorn or apples collected in Amherst, Massachusetts, to attempt oviposition in different host fruits (MacIntosh apples, hawthorn and rosehips) indicated that naive flies, irrespective of whether they originated from apples or hawthorn, showed a greater propensity to oviposit in hawthorn than MacIntosh apples. However, naive females originating from apples showed a greater propensity to oviposit in MacIntosh apples and rose hips than did flies originating from hawthorn. Naive flies from a laboratory culture that had been raised for 50 generations on apples showed a greater tendency to accept apples and reject hawthorn than the field collected flies. Both field flies and laboratory flies showed a greater acceptance of fruit types they had previously oviposited in than of other types (Prokopy *et al.*, 1982c).

Similar results to the above were obtained in more recent studies on apple and hawthorn breeding populations from Amherst, and East Lansing, Michigan (Prokopy *et al.*, 1988). They indicated that males also, regardless of origin, showed a preference for hawthorn fruit compared with apple as sites at which to acquire potential mates, although males of apple origin tended to reside significantly longer on apples than males of hawthorn origin.

Prokopy *et al.* (1982c, 1988) concluded from these results that genetic differences between flies originating from hawthorn and flies originating from apple may have been responsible for the observed differences in fruit acceptance patterns of naive flies. They also concluded that induction during the larval stage had no demonstrable effect on host acceptance patterns, but that host acceptance could be modified by a female's previous ovipositional experience (a phenomenon that is discussed in more detail later in the chapter). They further pointed out that possible genetically based differences in host preference patterns that exist between different *R. pomonella* populations may be one of the processes that can lead to host race formation and eventually sympatric speciation, a theory first proposed by Bush (1974, 1975) and the subject of much recent debate (McPheron *et al.*, 1988; Smith, 1988; Bush *et al.*, 1989).

Papaj and Prokopy (1988) also showed in *R. pomonella* that prior adult experience can influence at least two components of habitat preference, ovipositional and habitat fidelity. Females exposed to either apple or hawthorn in a field cage oviposited at a higher rate in test fruit of the same species than did inexperienced flies or flies exposed to the other fruit species. Females exposed to a particular host fruit species also remained longer in test trees harbouring fruit of that species than did inexperienced females or females exposed to the other fruit species. As indicated earlier, fidelity in habitat selection is considered to be an important factor in promoting host race formation.

Further evidence in support of the sympatric speciation hypothesis is provided by studies on the host acceptance patterns of *R. pomonella*, and what is currently regarded as a closely related sibling species, *R. mendax*, which breeds in blueberry and huckleberry (Diehl and Prokopy, 1986; Bierbaum and Bush, 1987, 1988). In laboratory experiments, *R. pomonella* females bred from field apples collected in Massachusetts showed a greater readiness to attempt oviposition in apple and hawthorn fruits than *R. mendax* reared from lowbush blueberries, collected in New Brunswick; the latter showed a greater propensity to attempt oviposition in blueberries and huckleberries. Similar patterns of propensity to oviposit occurred when the host fruit waxes and surface chemicals of apples and blueberries were transferred to the surface of parafilm-wrapped artificial fruit (Diehl and Prokopy, 1986).

In field cage experiments, oviposition was restricted exclusively to the normal host fruit of each species. In *R. mendax*, differences were also observed in the length of time females of lowbush origin and females collected from highbush blueberries in Michigan spent foraging in test lowbush blueberry plants. Lowbush collected flies spent over four times as long foraging in these plants as highbush origin flies. In tests with two different sized sticky red spheres, males and females of both species were captured most frequently on the larger spheres (7.5 cm diameter) than the smaller spheres (3.4 cm diameter). However relatively more *R. mendax* than *R. pomonella* were captured on the smaller spheres. Field cage tests with Michigan flies carried out by Bierbaum and Bush (1988) gave fairly similar results in relation to host acceptance, except that *R. pomonella* laid eggs into the blueberries. They also observed differences in the amount of time spent on the leaves of hosts and non-hosts in the case of *R. mendax* but not with *R. pomonella*.

Diehl and Prokopy (1986) concluded from their results that both 'species' had the flexibility to adapt to each other's hosts as well as new hosts. They also considered that the data did not preclude the possibility that gene flow between them may occur during the establishment of new populations and it was not possible therefore, to decide if they were in fact distinct species. Bierbaum and Bush (1988), however, concluded that the evidence from electrophoretic studies (Berlocher and Bush, 1982), the limited success of mating crosses between the two species, and their behavioural studies, indicated that the formation of interspecific hybrids in nature was highly unlikely.

6.4 HOST LOCATION

After emergence, the adults of all species of frugivorous fruit flies have a pre-oviposition period during which their activities are frequently devoted to dispersal and food foraging rather than oviposition. This is particularly evident in the case of the polyphagous tropical and subtropical species such as *Dacus tryoni*, *D. dorsalis*, *Anastrepna ludens and A. suspensa*, which are multivoltine and normally emerge from beneath trees that have ceased fruiting. They typically have an initial post-teneral, pre-reproductive dispersal phase that appears to be non-appetitive with regard to hosts and results in flies moving considerable distances from their emergence sites (Fletcher, 1987). Foraging for hosts does not commence until near the end of the pre-maturation period, which may be several days after emergence. In contrast, the temperate *Rhagoletis* species, which are predominantly univoltine, frequently emerge from overwintering pupal diapause in the ground beneath host trees bearing fruit at a suitable stage for infestation. In these circumstances the search for new host patches is normally delayed until fruit in the vicinity of their emergence site has been fully exploited for oviposition (Boller and Prokopy, 1976; Katsoyannos *et al.*, 1986a).

When a female commences host foraging she has to make a series of

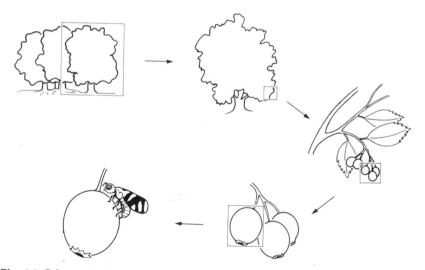

Fig. 6.2 Schematic representation of the hierarchical levels at which *R. pomonella* females forage for oviposition sites (from Roitberg, 1988).

hierarchial decisions in order to locate a host fruit. This may first involve locating a host patch and then a host plant with potentially exploitable fruit. Once she arrives on a host plant she has to locate a fruit suitable for oviposition (Fig. 6.2). After each oviposition she then has to decide whether to continue foraging within that plant or move to another plant if she is not immediately able to find a fruit suitable for oviposition on the first host plant.

The outcome of this is that flies are observed to aggregate in host trees with fruit at a suitable stage of ripeness for oviposition and leave trees when the fruit becomes heavily infested or overripe. The factors influencing this process are relatively complex and not yet fully understood, although it is becoming increasingly clear that the individual flies experience an environment rich in cues. These cues together with the flies' internal state modulates their responses and determines their overall behaviour pattern. In the remainder of this chapter we shall examine the factors that have been shown to be important in females locating and ovipositing in host fruit and the patterns of foraging behaviour they use in exploiting host resources effectively.

6.4.1 Detection of host trees

Except for dispersal flights, almost all activities of fruit flies, including feeding, mating, oviposition and resting take place on or in close proximity to vegetation. The extent to which feeding, mating and resting are restricted to host plants varies considerably from group to group. During the early part of the season adults of *R. cerasi* are observed most frequently on host trees with fruit, but later in the season they are often observed on non-fruiting hosts or non-hosts (Katsoyannos *et al.*, 1986a). In *R. pomonella*, food foraging is predominantly on non-hosts early in the season and later may occur on hosts or non-hosts depending on the relative abundance of food resources (Hendrichs and Prokopy, in press). In exceptional circumstances, adults of *R. pomonella* have also been observed to visit and oviposit in non-host fruit hung in non-host trees (Prokopy, 1977a). Typically, however, mating occurs in fruiting host trees, so that flies may move to these prior to the start of oviposition-site foraging. In the *Dacus* species not only a considerable amount of feeding, but also mating may take place away from host trees. In fact, in many of the curcurbit infesting species, visits to hosts for oviposition take place at fairly well defined times of day, and feeding, mating and resting typically occur on non-hosts at other times (Fletcher, 1987).

The major stimuli that could be used by tephritids in locating plant-associated resources have been reviewed by Prokopy (1977a, 1983) and include volatiles emanating from plant foliage and exudates, odours given off by adult food sources on plant foliage, pheromones of sexually active males either on hosts or non-hosts, volatiles given off by fruit and the visual properties of host plants and fruit. Repellant compounds emanating from the foliage of certain non-host plants and non-host fruits may also be utilized by adults in determining the location of favourable resource sites.

It is reasonable to conclude from a consideration of these possibilities that the arrival of females on host plants may sometimes result from activities not specifically directed to host fruit foraging, such as responses to food foraging stimuli (food odours) or potential mates (sex pheromones and sounds). Indeed, it has been suggested by Drew and Lloyd (1987) that the build-up of volatiles resulting from fruit fly-type bacteria on host fruit surfaces that occurs after flies start arriving on fruiting host plants, may play an important role in attracting other foraging flies into these trees. Here, however, we shall consider those stimuli that appear to be related specifically to adult foraging for larval resources (i.e. host fruit) rather than adult food, mates or shelter.

Although fruit flies have been shown to respond to a number of plant volatiles (as distinct from fruit volatiles), there is very little evidence that they use specific olfactory cues from foliage in detecting and orienting to host trees. Long distance orientation to vegetation appears to be predominantly visual and non-specific to host plants. Olfactory stimuli emanating from host fruit, however, play an important role in the orientation of flies to host plants. They may be the main cues used in the location of fruiting hosts from a distance, especially in habitats such as woodland and rainforests where hosts and non-host plants are in close juxtaposition and form part of a fairly continuous canopy.

(a) Visual stimuli

As is the case with other herbivorous insects (Prokopy and Owens, 1983), fruit flies are sensitive to wavelengths between near ultra-violet and red (360–650 nm), but respond most strongly to colours reflecting most of their energy between 500 and 600 nm and reflecting relatively little energy below 500 nm. The spectral sensitivities of medfly (*Ceratitis capitata*), olive fly (*D. oleae*), cherry fly (*R. cerasi*) and apple maggot (*R. pomonella*) are basically similar with a broad major peak at 485–500 nm (yellow–green) and a secondary peak at 365 nm (ultra–violet) (Remund *et al.*, 1981; Agee *et al.*, 1982; Owens, 1982).

Although green leaves reflect maximum energy in the infra-red region (750–1350 nm), this is outside the known limits of the tephritid visual spectrum. Tephritids are however particularly sensitive to objects reflecting maximal energy between 500 and 600 nm as indicated by their strong attraction to flat fluorescent yellow surfaces that reflect maximally in this range. It appears, therefore, that plant-seeking tephritids detect and respond to foliage due to its high reflectance in the 500–600 nm range, and that this may be further accentuated by the repulsion of non-foliar reflected or transmitted light with a substantial amount of energy below 500 nm, as is the case of sky-light. There is no evidence, however, that tephritids distinguish specific hosts by the reflectance composition of the foliage, even in species like *D. oleae* that utilize a host having leaves that differ in colour from typical foliage (Prokopy and Haniotakis, 1975).

Experiments with tree models suggest that the visual silhouettes of trees elicit positive orientation responses of tephritids and that maximum response is elicited by models that most closely resemble trees in shape, colour, size and physical characteristics. Experiments with *R. pomonella* (Moericke *et al.*, 1975) using large sticky boards of different shapes and colour indicated that more adults responded to flat yellow boards with a general tree like shape than to other types. Experiments with *D. tryoni* (Meats, 1983) indicated that flies were more attracted to large boards painted with an alternating chequer-board pattern of small yellow and green, or yellow and red, squares than similar size boards painted a single colour. Other studies (Fletcher, unpublished data) indicated that when given a choice of three-dimensional tree models with different combinations of artificial and real foliage and fruit, most flies were attracted to the models that more closely resembled real trees. Trees with real foliage attracted more flies than trees with artificial foliage and trees with real fruit attracted more flies than trees with artificial fruit. Trees with real foliage and real fruit were the most attractive combination.

(b) Olfactory stimuli

Both female and male fruit flies have been shown to respond to a wide range of plant chemicals including some components of green-leaf odour, particular the monoenic C_6 alcohols and aldehydes, but none of the leaf-odours appear to be specific to host plants (Keiser *et al.*, 1975; Guerin *et al.*, 1983; Light and Jang, 1987; Light *et al.*, 1988).

Host fruit volatiles, however, seem to be important olfactory cues used by fruit flies in locating hosts. More adults of *R. pomonella* (Prokopy *et al.*, 1973) and *D. tryoni* (Fletcher, unpublished data) were trapped in both host and non-host in trees containing fruit obscured from view in bags, than in similar trees containing bags without fruit. The odour of host fruit has also been shown to attract females of these two species in wind tunnel and olfactometer studies (Fein *et al.*, 1982; Eisemann and Rice, 1989).

The active components of host fruit extracts attractive to *R. pomonella* were found by electroanntenograms and behavioural studies to be a mixture of seven esters, hexyl acetate, (E)-2-hexen-1-yl acetate, butyl 2-methylbutanoate, propyl hexanoate, hexyl propanoate, butyl hexanoate and hexyl butanoate in a $35:2:8:12:5:28:10$ ratio. Conversely, extracts from ripe apples which contained higher ratios of butyl 2-methylbutanoate, propyl hexanoate and hexyl propanoate, compared to the other compounds, reduced the number of landings at the source in wind tunnel tests, and extracts from rotting apples sometimes caused downwind movement away from the source, suggesting a repellent effect (Fein *et al.*, 1982).

In subsequent studies on *R. pomonella* (Carle *et al.*, 1987; Averill *et al.*, 1988), greatest overall activity was obtained from butyl hexanoate, pentyl hexanoate and propyl octanoate. Of these, butyl hexanoate was the only ester found in significant concentrations in head space collections from apple and hawthorn fruit and is therefore considered to be the most important volatile involved in host fruit detection by *R. pomonella*. Studies with *D. tryoni* indicated that a-farnesene, butyric acid and a number of esters and ketones containing 4-6 carbon atoms induced upwind orientation behaviour and probing with the ovipositor at the source (Eisemann and Rice, 1985).

Studies on inter-tree movements of *R. pomonella* using green or white model trees with or without vials containing the aforementioned seven-component ester blend (Prokopy *et al.*, 1987a) indicated that flies tend to move upwind at moderate wind speeds (i.e. 0.5–2.0 m/s), even in the absence of host fruit odour. In the presence of odour, flies showed a greater tendency toward upwind movement and moved between trees at a faster rate, thereby enhancing the chances of encountering a host tree with fruit. The flies appeared to respond positively to tree models containing odour sources up to at least 2.5 m from the point of fly release, but no effects of fruit odour were detected when trees were 4.5 m apart. In more recent studies (Green and Prokopy, unpublished data) using an assay based on a fly's propensity to depart a paper platform in the presence of a fruit-odour plume, the results suggest that the flies are able to detect fruit by odour alone at a distance of at least 20 m.

(c) Other factors

Studies in which individual females of *R. pomonella* (Diehl *et al.*, 1986) and *C. capitata* (Prokopy *et al.*, 1986b) were released onto potted plants in field cages have indicated that the vegetative characteristics of plants can influence both the amount of time flies remain within the tree and their ability to locate host fruits. *R. pomonella* females were found to spend less time on two non-host plants (pine and tomato) than on their normal hosts (apple and hawthorn). They also discovered fewer host fruit hung on non-host plants compared to host plants. Flies that remained on tomato plants for longer than 30s actually started to exhibit behavioural problems, suggesting neurotoxicity. No differences were observed, however, in 'rate' of fruit discovery, or total time spent by R. *pomonella* on non-host birch trees compared with host apple trees.

Studies on C. *capitata* revealed that females likewise spent less time on pine trees or tomato plants, a minor host of this species, than on orange which is a favoured host. However, flies spent an equal amount of time on another non-host, eldorado, as on orange. They also appeared to find fruit just as readily on non-host as on host plants. However, 3-days prior experience of either citrus or tomato plants did not produce any detectable change in the length of time that flies remained on the foliage of either plant species when subsequently tested, compared to inexperienced flies (Prokopy *et al.*, 1990): a marked contrast to what happens after females have prior experience with fruit (see Section 6.5.4).

It seems, therefore, that some species of plants are less favourable, and in some cases actually repellant to females, because of their physical (e.g. leaf size or shape) or chemical characteristics. The residency times of females in non-fruiting host trees are not necessarily any different from those of flies in certain non-host trees.

6.4.2 Detection of host fruit

Once a female arrives on or approaches a host plant, both visual and olfactory cues may be used in locating and selecting individual host fruit on which to alight. Factors influencing the dominant cues may include the proximity and position of the fruit within the tree, its height above the ground, fruit density and maturity plus the architecture of the tree and foliage density, as well as climatic conditions.

Although olfactory cues from ripening fruit appear to be important in the initial location of fruit-bearing host trees, once fruit are in the visual range of the fly their shape, size and colour, and/or contrast against the background foliage and sky are often the predominant factors in determining the fly's response. Studies on some species suggest, however, that fruit volatiles may have an effect on pre-alighting behaviour within plants acting either as attractants or repellants depending upon circumstances.

(a) Shape and size

Studies with fruit-mimicking three-dimensional and two-dimensional objects of different shapes and colours have been carried out with a number of species including R. *pomonella* (Prokopy, 1968, 1973, 1975; Prokopy and Owens, 1978), R. *cerasi* (Boller, 1969; Prokopy, 1969), R. *completa* (Cirio, 1972), R. *fausta*, R. *cingulata* and R. *indiferrens* (Prokopy, 1975, 1977b; Reissig, 1976), *Ceratitis capitata* (Nagakawa *et al.*, 1978), *Dacus oleae* (Prokopy and Haniotakis, 1976; Prokopy *et al.*, 1975) and *D. tryoni* (Hill and Hooper, 1984).

In general these revealed that both sexes of all species were more strongly attracted to spherical shapes of an appropriate colour than to cylinders, cones or cubes. Similarly, flat discs were more attractive than rectangles of the same surface area, indicating that general outline was important. Curved spherical surfaces were preferred to other shapes, undoubtedly due to their closer resemblance to fruit.

The interaction of shape with colour on the host foraging response of females was first indicated by studies on *R. pomonella* (Prokopy, 1968). Both sexes showed a preference for dark coloured spheres (7.5 cm in diameter) over light coloured spheres of the same size. However, as the diameter of the spheres increased there was a marked reduction in attractiveness of dark-coloured spheres and an increase in attractiveness of those painted yellow, suggesting that very large diameter spheres ceased to be regarded as fruit models and instead were perceived as foliage stimuli.

Studies with appropriately coloured spherical fruit models of different diameters, carried out on *R. pomonella*, *R. cingulata*, *R. fausta* (Reissig, 1976) and *D. oleae* (Prokopy and Haniotakis, 1976; Katsoyannos, 1989) indicated that alighting responses of females in a host-foraging mode were greatest on fruit models considerably larger than their normal hosts, a factor which has been attributed to the super-normal stimuli provided by such models. However, there is an upper limit to a stimulating fruit size.

(b) Fruit colour

The significance of colour in determining response to fruit modes has been most extensively studied in *R. pomonella*. As discussed by Owens and Prokopy (1986), colour is determined by surface reflectance, which has three components: hue (= dominant wavelength), intensity or brightness (= total energy) and saturation (= spectral purity, i.e. percentage of dominant wavelength). The perceived colour of a surface is determined by its reflectance in relation to the ambient radiance and the ability of the visual system to detect available reflectance. The contrast necessary for visually distinguishing an object from its background results from perceived differences in any or all of these reflectance components.

To assess the extent to which the hue component of fruit surface reflectance influenced detection of fruit in *R. pomonella*, Owens and Prokopy (1986) compared the spectral reflectance properties of natural host fruit, hawthorn (*Crataegus* sp.) and apple (*Malus sylvestris*), with that of background foliage and sky. They also determined the influence that spectral reflectance had on landings of adults on artificially coloured natural fruit and plastic fruit mimics under field conditions. The unripe green fruit reflected maximally at 550 nm, paralleling that of foliage, and ripe red fruit reflected maximally above 660 nm, with little reflectance at wavelengths below 560 nm.

Captures of both sexes on both artificially coloured natural fruit and plastic spheres were highest on those with the least intensely reflecting pigments, regardless of hues, red hues (maximum reflectance >580 nm) being no more attractive than black (lacking any hue). Attraction of fruit mimics increased with a decrease in total reflectance between 350 and 580 nm. These results suggest that red hues are not used by *R. pomonella* in detecting fruit. Thus, unlike birds, which detect ripe fruit on the basis of hue discrimination between red fruit and green leaves, *R. pomonella* appear to rely more on intensity contrasts associated with spherical shapes. A preference for darker coloured over light coloured spheres was observed regardless of the colour of natural fruit on the host trees in which the experiments were conducted.

The visual mechanism of *R. pomonella* accentuates these contrasts by having peak physiological sensitivity in the spectral region of maximum energy of foliage reflection and transmission and of skylight background, and minimum sensitivity to energy reflected from ripe fruit. In addition, the flies have developed behavioural mechanisms for maximizing contrast during fruit foraging so that fruit are silhouetted against a more intense background (Roitberg, 1985). Studies on *C. capitata* (Nakagawa *et al.*, 1978; Cytrynowicz *et al.*, 1982), *R. cerasi* (Levinson and Haisch, 1984), and *D. tryoni* (Hill and Hooper, 1984) likewise indicated that in these species intensity contrasts acted as important visual cues during fruit foraging with dark spherical objects against light backgrounds being most attractive.

There is some evidence from studies with *R. completa* (Riedl and Hislop, 1985), *Anastrepha fraterculus* (Cytrynowicz *et al.*, 1982), *A. suspensa* (Greany *et al.*, 1977), *C. capitata* (Katsoyannos *et al.*, 1986b; Katsoyannos, 1987) and *D. oleae* (Katsoyannos *et al.*, 1985) that response to hue plays a more important role in fruit selection than is the case for *R. pomonella*.

In studies on *C. capitata* utilizing 7.0 cm diameter spheres of seven different colours (black, blue, red, green, yellow, orange and white) suspended in host trees in Chios, Greece, it was found that yellow was the most attractive colour, followed by red, orange, black and green, with blue and white being the least attractive (Katsoyannos, 1987). The observed colour preferences depended primarily on their hue or specific wavelength, while brightness and contrast against background appeared to have little effect. It was concluded, however, that the strong response to the coloured spheres, particularly yellow, was a feeding response to a super-normal ripe fruit-type stimulus rather than an oviposition response.

In laboratory studies (Katsoyannos *et al.*, 1986b), wild *C. capitata* females from Chios showed a different profile of preferences when provided with 18 mm diameter ceresin domes in which they could oviposit. In this situation, black, blue and red domes were the most preferred and yellow and white domes were the least preferred sites, both in relation to the number of visits and also number of eggs oviposited in them. The observed preference appeared to be dependent on both hue and brightness with contrast to the background having little effect.

Although there seems little doubt from these experiments that *C. capitata* uses hue discrimination in responding to fruit-type stimuli, the actual colour preferences exhibited by flies can vary depending upon their origin and rearing conditions. *C. capitata* reared for 1–3 generations on an artificial diet showed a different colour preference than their wild ancestors from Chios. However, when subsequently reared for one generation on pears, their colour preferences reverted to those of the original wild stock. Females bred from field-collected pupae of Costa Rican origin preferred orange to black domes for oviposition, whereas their progeny reared as larvae on an artificial diet preferred black to orange (reported in Katsoyannos *et al.*, 1986).

(c) Fruit volatiles

Although in the species studied so far, vision plays by far the greater role in locating host fruits once a fly arrives at a host plant with a high density of visually apparent fruits, olfactory cues may still play a supplementary role in host finding. Studies on both *R. pomonella* (Reissig *et al.*, 1982) and *D. tryoni* (Fletcher and Meats, unpublished data) have shown that sticky red spheres baited with fruit volatiles in polyethylene vials catch more adults than unbaited red spheres. Observations on specialist dacine species in laboratory cages have also indicated that flies can recognize and avoid landing on non-host fruits, probably using olfactory cues (Fitt, 1986a). In *D. oleae*, females have been observed in some situations to land more frequently on real fruits than visually similar fruit models (Haniotakis and Voyadjoglou, 1978).

Detailed studies of intra-tree foraging behaviour of *R. pomonella* (Aluja *et al.*, 1989), during which movements of individual flies were tracked in three dimensions, indicated that in a no-choice situation the same percentage of flies visited artificial red fruit, with or without vials containing fruit esters, as visited real red fruit. However, flies tended to travel less distance and spend less time in the tree before locating and landing on an artificial red fruit with fruit odour than a similar one without fruit odour. At low fruit densities, therefore, odour may play an important role in host finding, possibly by facilitating orientation to individual fruit. In contrast, under conditions of abundant fruit the females locate individual fruits during short lateral or upward flights, apparently using visual cues alone. It is likely, however, that fruit odours stimulate the overall rate of fly movement in these circumstances.

6.5 ACCEPTANCE OF FRUIT FOR OVIPOSITION

When a female first lands on a potential host fruit, she normally explores it by walking around its surface, assessing its suitability as an oviposition site. In addition to its visual receptors, olfactory and tactile receptors on the antennae, mouthparts, tarsi and ovipositor are utilized in determining fruit chemical and physical characteristics (Light and Jang, 1987; Crnjar et al., 1989a, 1989b). The stimuli that influence acceptance or rejection of a potential host fruit after landing are relatively complex. They include shape, size, colour, surface texture, odour and chemical composition, plus additional chemical cues that may be present due to previous utilization of the fruit as an oviposition site by intra- or interspecific female fruit flies. Whether a female accepts or rejects the fruit may also be influenced by the genetic characteristics and physiological state of the fly, plus her previous experience with specific fruit types.

6.5.1 Shape, size, colour and other physical characteristics

Laboratory experiments with a variety of different sized and shaped ceresin wax fruit models have indicated that domes and other convex surfaces elicit a greater ovipositional response than flat surfaces (Katsoyannos, 1989). Females of most species studied also exhibit a greater ovipositional preference for models approximately the size of host fruits than markedly smaller or larger models. This is rather different from the situation determining approach and landing, where larger-than-normal models are often selected.

R. cerasi, for example, which usually oviposits in cherries and honeysuckle berries that have diameters ranging from 5–30 mm, showed an oviposition preference for wax domes 10–22 mm in diameter (Prokopy and Boller, 1971; Katsoyannos, 1979). Similarly, Prokopy and Bush (1973), working with four closely related species of North American *Rhagoletis* (*R. cornivora*, *R. mendax*, *R. pomonella* and *R. zephyria*), observed a strong correlation between the size of artificial fruit each species selected for oviposition and the size of their native host fruits. *D. oleae* females have also been shown to select between different sized fruit models, laying more eggs in 35 mm diameter models, which correspond to the approximate size of large ripe olives, than small (5 mm) or large (100 mm) diameter models (Katsoyannos and Pittara, 1983). Fruit size has also been reported to have an effect on the number of eggs laid per clutch in *C. capitata* (McDonald and McInnis, 1975) and some *Dacus* spp. (Fletcher, 1987) that normally lay several eggs at a time.

Colour may also play a role in the post-landing phase of fruit selection, although most experiments involving different coloured models have not distinguished between the increased number of ovipositions due to more individuals landing on models of a preferred colour, and any increased tendency for females that land on a preferred colour to oviposit, compared with what happens on a non-preferred colour.

Initial studies with *R. pomonella* and *R. cerasi* (Prokopy, 1968; Prokopy and Boller, 1970) indicated that females preferred dark coloured to light coloured oviposition devices for egg laying, a similar result to that observed when arrival of adults on different coloured spheres was measured in the field. In *R. cerasi*, however, females have been observed to choose red and orange coloured domes in preference to black for egg-laying (Prokopy and Boller, 1970; Katsoyannos, 1979). In contrast, *R. completa* prefer green or yellow rather than black oviposition devices, which are similar to the colour preferences shown for spheres in field studies (Riedl and Hislop, 1985; Cirio, 1972). Both *C. capitata* and *D. oleae* have also been reported to show preferences for particular colours when allowed to lay eggs in wax domes of different colours in choice experiments (Katsoyannos *et al.*, 1985; Katsoyannos *et al.*, 1986b).

In olive flies, fruit hardness was found to have no effect on the propensity of females to attempt oviposition into several varieties at different stages of ripeness (Neuenschwander *et al.*, 1985) or in natural olives compared to wooden models (Haniotakis and Voyadjoglou, 1978). However, more ovipositions were attempted in smooth-surfaced than rough-surfaced wooden models.

6.5.2 Fruit chemistry

Volatile and surface chemicals associated with different types of fruit can have a significant effect on acceptance or rejection of fruit for oviposition, with attractant chemicals increasing the propensity to oviposit and repellent chemicals deterring oviposition.

Inclusion of slices of host fruit or fruit juice into egging devices has been shown to increase rates of oviposition in a number of species. Guava juice applied to sponges inside black or red plastic spheres for example, resulted in about twice as many eggs being laid as when the sponges were moisted with water only, in tests with *C. capitata* (McInnis, 1989). Treatment of wax models with fruit volatiles, fruit extracts, or surface chemicals has also been shown to influence oviposition attempts or egg laying. In *R. cerasi* for example, fruit-surface chemicals derived from the two main hosts, cherries and honeysuckle berries, increased oviposition rates by

25–30% when applied to wax fruit-models. The fruit-surface chemicals derived from three non-hosts (*Viburnum lantana*, *Cornus mas* and *C. sanguinea*) reduced oviposition by 30–50%, while fruit-surface chemicals from another non-host *Ligustrum vulgare* had no effect (Levinson and Haisch, 1984). Similarly, extracts from the surface of olive fruits have also been shown to act as ovipositional stimulants for *D. oleae* (Girolami *et al.*, 1983; Levinson and Levinson, 1984).

Once a female has accepted a fruit and inserted her ovipositor, other fruit characteristics detected by sensillae on the ovipositor may determine whether egg-laying actually occurs (Stoffolano, 1988; Crnjar *et al.*, 1989a). Sugars, particularly glucose and fructose have been shown to promote egg laying in *R. pomonella* and *R. completa* (Tsiropoulos and Hagen, 1979; Girolami *et al.*, 1986). Fructose was also found to stimulate egg-laying in *D. tryoni* but only when it was accessible to the tarsi and/or labellar sensillae (Eisemann and Rice, 1985). Sodium chloride and malic acid have been shown to stimulate egg-laying by *R. pomonella*. (Girolami *et al.*, 1986). In *D. tryoni* (Eisemann and Rice, 1985), calcium chloride, and in *A. suspensa* (Szentesi *et al.*, 1979) sodium and magnesium chloride, were found to have an inhibitory effect on egg laying. In *C. capitata*, decomposing fruit pulp of hosts and methyl and ethyl alcohol, the final fermentation products of sugars, were shown to have a repellent effect and deter oviposition (Vita *et al.*, 1986). Benzyl isothiocyanate, a compound which is produced naturally by the tissues of green papaya fruit, *Carica papaya*, when damaged during oviposition, also acts as an oviposition deterrent to *C. capitata*, *D. dorsalis* and *D. cucurbitae*, as well as killing larvae (Seo and Tang, 1982).

6.5.3 Host-marking pheromones and other oviposition deterrents

Among frugivorous fruit flies, studies to date suggest that host marking pheromones occur widely in the genus *Rhagoletis* (12 confirmed species), in the two species of *Anastrepha* so far investigated (*A. suspensa* and *A. fraterculus*) and *C. capitata* (Roitberg and Prokopy, 1987). They do not occur, however, in any of the *Dacus* species studied, although *D. oleae* females mark fruit after egg-laying with oviposition deterring compounds derived from olive fruit juice that the fly takes up and disperses on the fruit with its mouth parts (Cirio, 1971).

In *Rhagoletis* species, the pheromone is deposited on the fruit immediately after egg-laying via the ovipositor which is dragged along the fruit surface. The amount of pheromone deposited may be positively correlated with fruit size (Averill and Prokopy, 1987a). In dry conditions the pheromone of *R. pomonella* has a half-life of about 11 days, but it is water soluble and thus is depleted by heavy rain (Averill and Prokopy,

1987b). The chemical composition of *R. pomonella* host-marking phero-mone is not known but the active component of *R. cerasi* pheromone has now been identified (Hurter *et al.*, 1987). In *R. pomonella*, following apparent production in the mid-gut tissue, it is secreted into the gut lining and passed to the hindgut where it accumulates along with other gut contents prior to release (Prokopy *et al.*, 1982a).

After oviposition, females of *D. oleae* suck up the juice that exudes from the oviposition puncture and spread it on the surface of the fruit with their mouthparts. This acts as an oviposition deterrent to other females. The use of host marking compounds has not been observed in any other *Dacus* species, even in those species such as *D. opiliae* and *D. cacuminatus* which only utilize hosts with relatively small fruit (Fitt, 1981). Females of several species readily oviposit in already existing oviposition punctures contain-ing eggs laid by other females. The presence of larvae in fruit does, however, generally deter other females from ovipositing in the same fruit as reported in *D. cucurbitae*, *D. dorsalis* and *D. tryoni* (Prokopy and Koyama, 1982; Fitt, 1984; Prokopy *et al.*, 1989a). Studies on *D. tryoni* and *D. jarvisi* indicated that the deterrent effect of larvae was detectable shortly after the eggs hatched and was interspecific (Fitt, 1984). In *R. pomonella* (Averill and Prokopy, 1987b), *R. completa* (Cirio, 1972), *D. oleae* (Girolami *et al.*, 1981) and *Anastrepha suspensa* (Sivinski, 1987), all species that utilize host marking compounds, females also discriminate against fruit containing larvae, but the effect is not pronounced until the larvae reached the second or third instar.

In general, females tend to reject fruit marked with host marking pheromones, or containing larvae, although willingness to oviposit in such fruit can be influenced by a number of factors including the quantity, quality and distribution of fruit, length of previous host deprivation and the frequency of encountering already infested fruit (Roitberg and Prokopy, 1983; Averill and Prokopy, 1989; Mangel and Roitberg, 1989).

Rhagoletis females do not respond to the presence of marking phero-mones, and female *D. tryoni* do not respond to the presence of young larvae, until they have actually alighted on the fruit (Prokopy, 1981, Fletcher and Prokopy, unpublished data). In the case of *R. pomonella* and *R. cerasi* the principal pheromone receptors have been identified as long D-type hairs situated on the tarsi, although other receptors occur on the mouthparts (Crnjar *et al.*, 1978; Bowdan, 1984; Stadler *et al.*, 1987).

From an evolutionary point of view, the use of host marking com-pounds that deter oviposition when deposited on a fruit is assumed to increase the overall fitness of individuals within a population by increas-ing the likelihood of offspring survival when the resources for larval development are limited, as they are in small fruit. Older larvae usually win in competition with younger larvae. Based on a model utilizing the main features of classical foraging models and incorporating clumping of hosts and aggregation of foragers, Roitberg and Prokopy (1987) concluded that host marking may have evolved primarily as a means whereby individuals avoid laying further eggs in hosts into which they themselves had already oviposited, and that additional benefits in terms of increased offspring survivorship accrued when this also resulted in avoidance of hosts marked by other individuals. A similar conclusion was reached by Roitberg and Mangel (1988) who developed a theory for the evolutionary ecology of marking pheromones by considering the components of fitness associated with marking pheromones in three different ways. Their evolutionary model indicated that a behaviour such as pheromone-marking that benefits other members of the population, will spread through the population if it benefits the perpetrator more than it benefits the others.

They also developed a simple model that suggests that *Rhagoletis* marking pheromones evolved as relatively short lived water soluble compounds because the main advantage of host marking is derived by the females that do the marking, as it avoids wastage of eggs and time, and most of the advantage is gained in the first two days. Water soluble marks are, therefore, almost as effective as more energy costly lipid-based or other longer lasting marks.

6.5.4 The role of learning in the detection and acceptance of host fruit

Studies on *Rhagoletis pomonella*, *Ceratitis capitata* and *Dacus tryoni* have shown that prior ovipositional experience with a particular host fruit type influences the extent to which adults accept or reject that or other fruit types for egg-laying in future encounters (Prokopy *et al.*, 1982c; Cooley *et al.*, 1986; Papaj and Prokopy 1986; Prokopy and Fletcher, 1987). Such a reversible change in behaviour with experience has been considered to indicate learning. The ecological and evolutionary aspects of learning in herbivorous insects are reviewed in Papaj and Prokopy (1989).

Studies in Hawaii using wild *C. capitata*, indicated that when females were exposed to mock orange (*Murraya paniculata*) or sweet orange (*Citrus sinensis*) fruit hung from branches of potted trees in field cages for three day periods, a higher proportion subsequently land on the type of fruit with which they were familiar, when released individually onto trees containing one or other (or a mixture) of the two fruit types (Prokopy *et al.*, 1989b). Subsequent experiments, in which females were exposed to natural or coloured-wax-covered mock oranges or sweet oranges and then their response to coloured-wax-covered natural or artificial fruit tested, indicated that fruit size was the principal character learned and subsequently used in finding fruit with fruit colour and odour being of little or no importance.

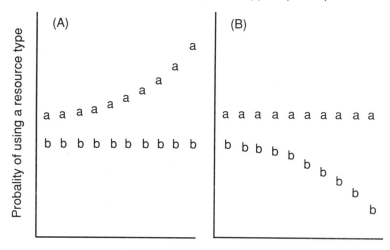

Fig. 6.3 Two possible types of learning. (A) Learning to accept the familiar resource type a so that the probability of using it increases with the increasing numbers of encounters, while the probability of using a novel resource type b remains constant. (B) Learning to reject the novel resource type, so that rate of acceptance of a familiar resource type a remains constant whereas the rate of acceptance of a novel host type b decreases (from Prokopy *et al.*, 1986a).

In studies with *R. pomonella* and *C. capitata*, it was shown that the learning process did not, as was usually assumed, involve only more ready acceptance of a familiar fruit type but also involved rejection of novel fruit types (Fig. 6.3). Female *R. pomonella* that had previously been allowed to oviposit only in hawthorn fruit had no greater propensity to attempt oviposition in test hawthorn fruit than females which had previously been allowed to oviposit alternatively in hawthorn and apple fruit. They were, however, less prone to attempt to oviposit in apple fruit than the untrained females or naive females. Similarly females exposed only to apples were less prone than females allowed to lay alternatively in both fruit, or than naive females, to oviposit in a test hawthorn fruit (Prokopy *et al.*, 1986a). More recent laboratory and field studies have indicated that *R. pomonella* can also learn to discriminate between certain varieties of apple fruits (Early Mackintosh, Red Delicious and Golden Delicious), and show a greater propensity to accept for oviposition the variety with which they are most familiar (Prokopy and Papaj, 1988).

In other studies with *R. pomonella*, the extent to which fruit surface chemistry and fruit size were involved in learning were examined (Papaj and Prokopy, 1986). These indicated that females previously exposed only to apples, attempted to bore into apple and hawthorn models covered in surface chemicals (applied to them by parafilm that had previously been wrapped around apple or hawthorn fruit), at the same level as naive flies. Females previously exposed only to hawthorn fruit were less willing to bore into apple sized models wrapped with parafilm from hawthorn fruit, or either sized model wrapped in parafilm from apple fruit, than naive flies or females previously exposed to apple fruit. These results confirm that the alteration of a female's ovipositional response to both physical (i.e. size) and chemical host fruit cues can result from learning involving rejection of novel stimuli.

Another facet of learning that has recently been examined in *C. capitata* is cross-induction, the tendency for experience with one host fruit species to alter differentially behavioural responses to alternate host fruit species (Papaj *et al.*, 1989). It was found that females exposed to small fruit (mock oranges) accepted five other fruit species with which they were subsequently tested less often as the size of the fruits increased, whereas females exposed to large fruit (sweet oranges) accepted the other fruit species more often as the size of the fruits increased. In contrast, females exposed to medium sized fruit (kumquats) were not cross-induced; females accepted medium sized fruits more frequently and rejected the other fruits to an equal degree regardless of size.

The possible selective advantages of learning in host fruit acceptance by fruit flies in nature have been discussed by Prokopy *et al.* (1986a) and Cooley *et al.* (1986). They considered that females may benefit from rejecting rare hosts if by doing so they enhance their ability to sample and discriminate among abundant conspecific host fruits. It has also been pointed out by Papaj and Prokopy (1989), that learning could promote sympatric divergence, at least in *R. pomonella* populations, by increasing the flies' acceptance of the fruit for which they are genotypically specialized and by increasing the fidelity of flies to trees bearing that fruit.

6.6 INTRA-TREE FORAGING

Quantitative intra-tree host foraging behaviour, involving the detailed tracking of individual females over time in trees harbouring a range of fruit densities and fruit qualities, has been carried out most extensively with *R. pomonella* (Roitberg *et al.*, 1982; Prokopy and Roitberg, 1984; Roitberg and Prokopy, 1984; Aluja *et al.*, 1989), although similar studies have also been carried out with *C. capitata* (Prokopy *et al.*, 1987b) and *D. tryoni* (Fletcher and Prokopy, unpublished data). All three species exhibited similar foraging characteristics that may be summarized as follows.

1. females emigrate within a few minutes after release on small trees devoid of fruit;

2. females visit more fruit, oviposit more often and remain longer on host trees with higher densities of non-infested host fruit than lower densities of non-infested fruit or higher densities of pheromone marked fruit (apple maggot and medfly) or fruit containing larvae;

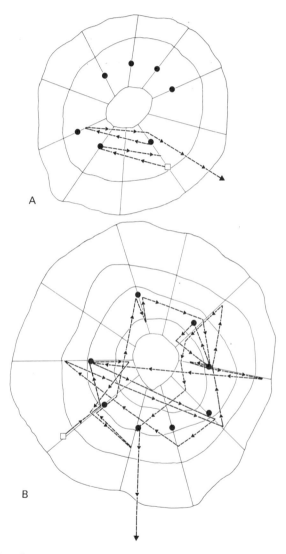

Fig. 6.4 (A) Search path of a female *R. pomonella* in a host tree with eight clusters of host fruit marked with oviposition deterring pheromone. (B) Search path of a female *R. pomonella* in a tree containing eight clusters of unmarked, uninfested fruit (from Roitberg *et al.*, 1982).

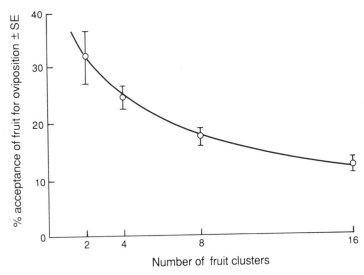

Fig. 6.5 The relationship between level of acceptance of uninfected host fruit and number of different fruit clusters available in the tree, with *R. pomonella* (from Roitberg *et al.*, 1982).

3. females accept a proportionately smaller number of fruit visited as the density of non-infested fruit increases;
4. females exhibit a decreasing giving-up-time with increasing densities of non-infested fruit, or successive contact with pheromone marked fruit or fruit containing larvae.

There was an observed tendency for all three species to invest more time and energy in intra-tree foraging with the increasing richness (quantity and/or quality) of resource items present (Fig. 6.4). This was accompanied by a proportionately reduced acceptance of individual resource items as their abundance increased (Fig. 6.5). This greater selectivity among resource items at increased levels of abundance has also been observed with other insects including butterflies (Rausher, 1983) and several parasitoids (e.g. van Alphen and Galis, 1983; Cloutier, 1984) and is consistent with the theoretical predictions regarding foraging behaviour (Hassel and Southwood, 1978; Pyke, 1984).

The decreased giving-up-time with increased resource density exhibited by these three tephritids could be due to a number of factors, including reduced egg-load. An interesting feature of intra-tree foraging also observed in all three species was that the individual foraging flights tended to increase in duration (from around 5 s to 15 s) towards the end of the foraging period.

6.7 CONCLUSIONS

Although there are many aspects of host foraging in fruit flies that are still not fully understood, it appears that the females locate fruiting host trees from a distance by a combination of relatively non-specific visual stimuli and relatively specific olfactory stimuli. Within host trees, visual stimuli, in particular the spherical shape and contrast of the fruit against the foliage and sky are paramount in locating individual fruit. Prior experience in finding fruit may also enhance future fruit finding ability. Once a fly lands on a fruit, the physical and chemical properties of the fruit determine whether it is accepted or rejected for oviposition. However, this decision can be further influenced by the presence of pheromones or other marking compounds deposited by previous females on the fruit after oviposition or olfactory cues associated with the presence of larvae in the fruit. The physiological state of the fly and learning based on prior experience with particular fruit can also influence a fly's decision to accept or reject a fruit. Indeed, recent attempts to model host acceptance in *R. pomonella* have revealed that the high degree of variability that is observed is not consistent with current theories of host acceptance. Rather, it is best described by models that consider the maximization of life time fitness in relation to the fly's current physiological and informational constraints (Mangel and Roitberg, 1989).

Decisions on the length of time and effort a female puts in to foraging within a particular tree also appear to be based on a relatively complex integration of information relating to the quantity and quality of resources, obtained from visual and olfactory cues and previous encounters with fruit, and the current physiological and informational state of the fly.

References

Agee, H. R., Boller, E., Remund, U., Davis, J. C. *et al.* (1982) Spectral sensitivities and visual attractant studies in the Mediterranean fruit fly, *Ceratitis capitata* (Wiedemann), olive fly, *Dacus oleae* (Gmelin), and the European cherry fruit fly, *Rhagoletis cerasi* (L.) (Diptera, Tephritidae). *Z. Angew. Entomol.*, **93**, 403–412.

Aluja, M., Prokopy, R. J., Elkinton, J. S. and Laurence, F. (1989) A novel approach for tracking and quantifying the movement patterns of insects in three dimensions under semi-natural conditions. *Environ. Entomol.*, **18**, 1–8.

Averill, A. L. and Prokopy, R. J. (1987a) Intraspecific competion in the tephritid fruitfly *Rhagoletis pomonella*. *Ecology*, **68**, 878–886.

Averill, A. L. and Prokopy, R. J. (1987b) Residual activity of oviposition deterring pheromone in *Rhagoletis pomonella* and female response to infested fruit. *J. Chem. Ecol.*, **13**, 167–177.

Averill, A. L. and Prokopy, R. L. (1989) Distribution patterns of *Rhagoletis pomonella* (Diptera: Tephritidae) eggs in hawthorn. *Ann. Entomol. Soc. Amer.*, **82**, 38–44.

Averill, A. L., Reissig, W. H. and Roelofs, W. L. (1988) Specificity of olfactory responses in the tephritid fruit fly, *Rhagoletis pomonella*. *Entomol. Exp. Appl.*, **47**, 211–222.

Bateman, M. A. (1972) The ecology of fruit flies. *Ann. Rev. Entomol.*, **17**, 493–518.

Berlocher, S. H. and Bush, G. L. (1982) An electrophorectic analysis of *Rhagoletis* (Diptera: Tephritidae) phylogeny. *Syst. Zool.*, **31**, 136–155.

Bierbaum, T. J. and Bush, G. L. (1987) A comparative study of host plant acceptance behaviors in *Rhagoletis* fruit flies. In *Proceedings 6th International Symposium of Insect-Plant Relationships*, Labeyrie V., Fabres, G., Lachaise D. (Eds) W. Junk Publishers, Dordrecht, 374–375.

Bierbaum, T. J. and Bush, G. L. (1988) Divergence in key host examining and acceptance behaviors of the sibling species *Rhagoletis mendax* and *R. pomonella* (Diptera: Tephritidae). In *Ecology and Management of Economically Important Fruit Flies*, (ed. M. T. AliNiazee). Oregon State University Agric Exp Station. Special Report 830, 26–55.

Boller, E. F. (1968) An artificial oviposition device for the European cherry fruit fly, *Rhagoletis cerasi*. *J. Econ. Entomol.*, **61**, 850–851.

Boller, E. F. (1969) Neues über die Kirschenfliege: Freilandversuche im Jahre 1969. *Schweizerische Zeitschrift für Obst und Weinbrau*, **105**, 566–572.

Boller, E. F. and Prokopy, R. J. (1976) Bionomics and management of *Rhagoletis*. *Ann. Rev. Entomol.*, **21**, 223–246.

Bowdan, E. (1984) Electrophysiological responses of tarsal contact chemo-receptors of the apple maggot fly *Rhagoletis pomonella* to salt, sucrose and oviposition-deterring pheromone. *J. Comp. Physiol.*, **A154**, 143–152.

Bush, G. L. (1974) The mechanism of sympatric host race formation in the true fruit flies. In White, M. J. D. (ed.) *Genetic Mechanisms of Speciation in Insects*, Australian and New Zealand Book Co., Sydney, 3–23.

Bush, G. L. (1975) Modes of animal speciation. *Ann. Rev. Ecol. Syst.*, **6**, 339–364.

Bush, G. L. and Diehl, S. R. (1982) Host shifts, genetic models of sympatric speciation and the origin of parasitic insect species. In Visser, J. H., Minks, A. K. (eds) *Proceedings 5th International Symposium on Insect-Plant Relationships*, Pudoc, Wageningen, 297–305.

Bush, G. L., Feder, J. L., Berlocher, S. H., McPheron B. A. *et al.* (1989) Sympatric origins of *R. pomonella*. *Nature*, **339**, 346.

Carle, S. A., Averill, A. L., Rule, G. S., Reissig, W. H. *et al.* (1987) Variation in host fruit volatiles attractive to apple maggot fly, *Rhagoletis pomonella*. *J. Chem. Ecol.*, **13**, 795.

Cavalloro, R. (ed.) (1983) *Fruit Flies of Economic Importance*, A. A. Balkema, Rotterdam, 642 pp.

Cavalloro, R. (ed.) (1988) *Fruit Flies of Economic Importance*, A. A. Balkema, Rotterdam/Boston, 87 pp.

Chan, H. T. and Tam, S. Y. T. (1985) Toxicity of a-tomatine to larvae of the Mediterranean fruit fly (Diptera: Tephritidae). *J. Econ. Entomol.*, **78**, 305–307.

Christensen, L. D. and Foote, R. H. (1960) Biology of fruit flies. *Ann. Rev. Entomol.*, **5**, 171–192.

Cirio, U. (1971) Reperti sul meccanismo stimolorisposta nell'ovideposizione del *Dacus oleae* Gmelin (Diptera, Trypetidae). *Redia*, **52**, 577–600.

Cirio, U. (1972) Osservazioni sul comportamento di ovideposizione della *Rhagoletis completa* Cresson (Diptera, Trypetidae) in laboratorio. In: *Atti del IX Congresso Nazionale Italiano di Entomolgia*, Sienna 21–25 June 1972, 99–117.

Cloutier, C. (1984) The effect of host density on egg distribution by the solitary parasitoid *Aphidius nigripes*. *Can. Entomol.*, **116**, 805–811.

Cooley, S. S., Prokopy, P. L., McDonald, P. T. and Wong, T. T. Y. (1986) Learning in oviposition site selection by *Ceratitis capitata*. *Entomol. Exp. Appl.*, **40**, 47–51.

Crnjar, R. M., Prokopy, R. J. and Dethier, V. G. (1978) Electrophysiological identification of oviposition-deterring pheromone receptors in *Rhagoletis pomonella*. *J. New York Entomol. Soc.*, **86**, 283–284.

Crnjar, R. M., Angioy, A., Pietra, P., Stoffolano, J. G. *et al.* (1989a) Electrophysiological studies of gustatory and olfactory responses of the sensilla of the ovipositor of the apple maggot fly, *Rhagoletis pomonella* Walsh. *Boll. Zool.*, **56**, 41–46.

Crnjar, R., Scalera, G., Liscia, A., Angioy, A. M. *et al.* (1989b) Morphology and EAG mapping of the antennal olfactory receptors in *Dacus oleae*. *Entomol. Exp. Appl.*, **51**, 77–85.

Cytrynowicz, M., Morgante, J. S. and De Souza, H. M. L. (1982) Visual responses of South American fruit flies, *Anastrepha fraterculus*, and Mediterranean fruit flies, *Ceratitis capitata*, to colored rectangles and spheres. *Environ. Entomol.*, **11**, 1202–1210.

Diehl, S. R. and Bush, G. L. (1989) The role of habitat preference in adaptation and speciation. In *Speciation and Its Consequences*. D. Otte, J. Endler (eds), Sinauer Assoc. pp. 345–65.

Diehl, S. R. and Prokopy, R. J. (1986) Host selection behaviour differences between the fruit fly sibling species *Rhagoletis pomonella* and *R. mendax* (Diptera: Tephritidae). *Ann. Entomol. Soc. Amer.*, **79**, 266–271.

Diehl, S. R., Prokopy, R. J. and Henderson, S. (1986) The role of stimuli associated with branches and foliage in host selection by *Rhagoletis pomenella*. In *Cavallorlo R. (ed.) Fruit Flies of Economic Importance*, A. A. Balkema, Rotterdam, 191–196.

Drew, R. A. I. and Lloyd, C. A. (1987) The relationship of fruit flies (Diptera: Tephritidae) and their bacteria to host plants. *Ann. Entomol. Soc. Amer.*, **80**, 629–636.

Economopoulos, A. P. (ed.) (1987) Fruit flies, *Proceedings of the Second International Symposium*, 590 pp. Elsevier, Amsterdam (Distributors)

Eisemann, C. H. and Rice, M. J. (1985) Oviposition behaviour of *Dacus tryoni*: The effects of some sugars and salts. *Entomol. Exp. Appl.*, **39**, 61–71.

Feder, J. L., Chilcote, C. A. and Bush, G. L. (1988) Genetic differentiation between sympatric host races of the apple maggot fly *Rhagoletis pomonella*. *Nature*, **336**, 61–62.

Fein, B. L., Reissig, Wh. and Roelofs, W. L. (1982) Identification of apple volatiles attractive to the apple maggot, *Rhagoletis pomonella*. *J. Chem. Ecol.*, **8**, 1473–1478.

Fitt, G. P. (1981) Ecology of northern Australia Dacinae 1. Host phenology and utilization of *Opilia amentacea* Roxb (Opiliaceae) by *Dacus* (Bactrocera) *opiliae*, with notes on some other species. *Aust. J. Zool.*, **29**, 691–705.

Fitt, G. P. (1984) Oviposition behaviour of two tephritid fruit flies, *Dacus tryoni* and *Dacus jarvisi*, as influenced by the presence of larvae in the host fruit. *Oecologia*, **62**, 37–46.

Fitt, G. P. (1986a) The roles of adult and larval specialisations in limiting the occurrence of five species of *Dacus* (Diptera: Tephritidae) in cultivated fruits. *Oecologia*, **69**, 101–109.

Fitt, G. P. (1986b) The influence of a shortage of hosts on the specificity of oviposition behaviour in species of *Dacus* (Diptera: Tephritidae). *Physiol. Entomol.*, **11**, 133–143.

Fletcher, B. S. (1987) The biology of dacine fruit flies. *Ann. Rev. Entomol.*, **32**, 115–144

Futuyma, D. J. (1983) Selective factors in the evolution of host choice by phytophagous insects. In: Ahmed, S. A. (ed.) *Herbivorous Insects: Host Seeking Behaviour and Mechanisms*, Academic Press, New York, 227–244.

Girolami, V., Vianello, A., Strapazzon, A., Raggazzi, E. *et al.* (1981) Ovipositional deterrents in *Dacus oleae*. *Entomol. Exp. Appl.*, **29**, 177–188.

Girolami, V., Stapazzon, A. and De Gerloni, P. F. (1983) Insect-plant relationship in olive flies: general aspects and new findings In: Cavalloro, R. (ed.) *Fruit Flies of Economic Importance*. A. A. Balkema Rotterdam, 258–267.

Girolami, V., Strapazzon, A., Crnjar, R., Angioy, A. M. *et al.* (1986) Behaviour and sensory physiology of *Rhagoletis pomonella* in relation to oviposition stimulants and deterrents in fruit. In R. Cavalloro (ed.) *Fruit Flies of Economic Importance* 84, A. A. Balkema Rotterdam, 183–190.

Greany, P. D. and Szentesi, A. (1979) Oviposition behaviour of laboratory-reared and wild Caribbean fruit flies (*Anastrepha suspensa*: Diptera: Tephritidae): II. Selected physical influences. *Entomol. Exp. Appl.*, **26**, 239–244.

Greany, P. D., Agee, H. R., Burditt, A. K. and Chambers, D. L. (1977) Field studies on colour preferences of the Caribbean fruit fly, *Anastrepha suspensa* (Diptera: Tephritidae). *Entomol. Exp. Appl.*, **21**, 63–70.

Greany, P. D., Styer, S. C., Davis, Ph., Shaw, P. E. *et al.* (1983) Biochemical resistance of citrus to fruit flies. Demonstration and elucidation of resistance to the Caribbean fruit fly, *Anastrepha suspensa*. *Entomol. Exp. Appl.*, **34**, 40–50.

Guerin, P. M., Katsoyannos, Bl., Delrio, G., Remund, U. *et al.* (1983) Fruit fly electroantennogram and behaviour responses to some generally occurring fruit volatiles. In: Cavalloro, R. (ed.) *Fruit Flies of Economic Importance*, A. A. Balkema, Rotterdam, 248–251.

Haniotakis, G. E. and Voyadjoglou, A. (1978) Oviposition regulation in *Dacus oleae* by various olive fruit characters. *Entomol. Exp. Appl.*, **24**, 187–192.

Hassel, M. P. and Southwood, T. R. E. (1978) Foraging strategies of insects. *Ann. Rev. Ecol. Syst.*, **9**, 75–89.

Hendrichs, J. and Prokopy, R. Where do apple maggot flies find food in nature? *Massachusetts Fruit Notes* (in press).

Hill, A. R. and Hooper, G. H. S. (1984) Attractiveness of various colours to Australian tephritid fruit flies in the field. *Entomol. Exp. Appl.*, **356**, 119–128.

Hurter, J., Boller, E. F., Stadler, E., Blattman, B. *et al.* (1987) Oviposition deterring pheromone in *Rhagoletis cerasi*: purification and determination of the chemical composition. *Experimentia*, **43**, 157–164.

Katsoyannos, B. I. (1979) Zum Reproducktions—und Wirtswahlverhal—ten der Kirschenfliege, *Rhagoletis cerasi* L. (Diptera: Tephritidae). Dissertation Nr 6409 ETH Zurich, 180 pp.

Katsoyannos, B. I. (1987) Effects of colour properties of spheres on their attractiveness for *Ceratitis capitata* (Wiedmann) flies in the field. *J. Appl. Entomol.*, **104**, 79–85.

Katsoyannos, B. I. (1989) Response to shape, size and color. In: Robinson, A., Hooper, G. H. (eds) *Fruits Flies Their Biology, Natural Enemies and Control*, Elsevier, Amsterdam, pp. 307–24.

Katsoyannos, B. I. and Pittara, I. S. (1983) Effect of size of artificial oviposition substrates and presence of natural host fruits on the selection of oviposition site by *Dacus oleae*. *Entomol. Exp. Appl.*, **34**, 326–332.

Katsoyannos, B. I., Patsouras, G. and Vrekoussi, M., (1985) Effect of color hue and brightness of artificial oviposition substrates on the selection of oviposition site by *Dacus oleae*. *Ent. Exp. Appl.*, **38**, 205–214.

Katsoyannos, B. I., Boller, E. F. and Benz, G. (1986a) Das Verhalten der Kirschenfliege, *Rhagoletis cerasi* L., bei der Auswahl der Wirtspflanzen und ihre Dispersion. *Mitt. Schweiz. Entomol. Gesell.*, **59**, 315–335.

Katsoyannos, B. I., Panagiotidou, K. and Kechagia, I. (1986b) Effect of color properties on the selection of oviposition site by *Ceratitis capitata*. *Entomol. Exp. Appl.*, **42**, 187–193.

Keiser, I., Harris, E. J., Miyashita, D. H., Jacobson, M. *et al.* (1975) Attraction of ethyl ether extracts of 232 botanicals to oriental fruit flies, melon flies and Mediterranean fruit flies. *Lloydia*, **38**, 141–152.

Levinson, H. Z. and Haisch, A. (1984) Optical and chemosensory stimuli involved in host recognition and oviposition of the cherry fruit fly *Rhagoletis cerasi* L. *Z. Angew. Entomol.*, **97**, 85–91.

Levinson, H. Z. and Levinson, A. R. (1984) Botanical and chemical aspects of the olive tree with regards to fruit acceptance by *Dacus oleae* (Gmelin) and other frugivorous animals. *Z. Angew. Entomol.*, **98**, 136–149.

Light, D. M. and Jang, E. B. (1987) Electroantennogram responses of oriental fruit fly, *Dacus dorsalis*, to a spectrum of alcohol and aldehyde plant volatiles. *Entomol. Exp. Appl.*, **45**, 55–64.

Light, D. M., Jang, E. B. and Dickens, J. C. (1988) Electroantennogram responses of the Mediterranean fruit fly, *Ceratitis capitata*, to a spectrum of plant volatiles. *J. Chem. Ecol.*, **14**, 159.

Mangel, M. and Roitberg, B. D. (1989) Dynamic information and host acceptance by a tephritid fruit fly. *Ecol. Entomol.*, **14**, 181–189.

McDonald, P. T. and McInnis, D. O. (1985) *Ceratitis capitata*: Effect of host size on the number of eggs per clutch. *Entomol Exp. Appl.*, **37**, 207–211.

McInnis, D. O. (1989) Artificial oviposition sphere for Mediterranean fruit flies (Diptera: Tephritidae) in field cages. *J. Econ. Entomol.*, **82**, 1382–1385.

McPheron, B. A., Smith, D. C. and Berlocher, S. H. (1988) Genetic differences between host races of *Rhagoletis pomonella*. *Nature*, **336**, 64–65.

Meats, A. (1983) The response of the Queensland fruit fly, *Dacus tryoni* to tree models. In: R. Cavalloro (ed.), *Fruit Flies of Economic Importance*. A. A. Balkema, Rotterdam, 285–289.

Moericke, V., Prokopy, R. J., Berlocher, S. and Bush, G. L. (1975). Visual stimuli eliciting attraction of *Rhagoletis pomonella* (Diptera: Tephritidae) flies to trees. *Entomol. Exp. Appl.*, **18**, 497–507.

Nakagawa, S., Prokopy, R. J., Wong, T. T. Y., Ziegler, J. R. *et al.* (1978) Visual orientation of *Ceratitis capitata* flies to fruit models. *Entomol. Exp. Appl.*, **24**, 193–198.

Neuenschwander, P., Michelakis, S., Holloway, P. and Berchtold, W. (1985) Factors effecting the fruits of different olive varieties to attack by *Dacus oleae* (Gmel.) (Dipt., Tephritidae). *Z. Angew. Entomol.*, **100**, 174–188.

Owens, E. D. (1982) The effects of hue, intensity and saturation on foliage and fruit finding in the apple maggot. PhD Thesis, University of Massachusetts.

Owens, E. D. and Prokopy, R. J. (1986) Relationship between reflectance spectra of host plant surfaces and visual detection of host fruit by *Rhagoletis pomonella*. *Physiol. Entomol.*, **11**, 297–307.

Papaj, D. R. and Prokopy, R. J. (1986) Phytochemical basis of learning in *Rhagoletis pomonella* and other herbivorous insects. *J. Chem. Ecol.*, **12**, 1125–1143.

Papaj, D. R. and Prokopy, R. J. (1989) Ecological and evolutionary aspects of learning in phytophagous insects. *Ann. Rev. Entomol.*, **34**, 315–350.

Papaj, D. R. and Prokopy, R. J. (1988) The effect of prior adult experience on components of habitat preference in the apple maggot fly (*Rhagoletis pomonella*). *Oecologia*, **76**, 538–543.

Papaj, D. R., Opp., S. B., Prokopy, R. J. and Wong, T. T. Y. (1989) Cross-induction of fruit acceptance by the medfly *Ceratitis capitata*: the role of fruit size and chemistry. *J. Insect. Behavior.*, **2**, 241–253.

Pree, D. J. (1977) Resistance to development of larvae of the apple maggot in crab apples. *J. Econ. Entomol.*, **70**, 611–614.

Prokopy, R. J. (1968) Visual responses of apple maggot flies, *Rhagoletis pomonella* (Diptera: Tephritidae): Orchard studies. *Entomol. Exp. Appl.*, **11**, 403–422.

Prokopy, R. J. (1969) Visual responses of European cherry fruit flies—*Rhagoletis cerasi* L. (Diptera, Trypetidae). *Bull. Entomol. Pologne*, **39**, 539–566.

Prokopy, R. J. (1973) Dark enamel spheres capture as many apple maggot flies as fluorescent spheres. *Environ. Entomol.*, **2**, 953–954.

Prokopy, R. J. (1975) Selective new trap for *Rhagoletis cingulata* and *R. pomonella* flies. *Environ. Entomol.*, **4**, 420–424.

Prokopy, R. J. (1977) Attraction of *Rhagoletis* flies (Diptera: Tephritidae) to red spheres of different sizes. *Can. Entomol.*, **109**, 593–596.

Prokopy, R. J. (1981) Oviposition deterring pheromone system of apple maggot flies. In Mitchell E. R. (ed.) *Management of Insect Pests with Semiochemicals: Concepts and Practice.* Plenum Publishing Corporation, New York, 477–494.

Prokopy, R. J. (1983) Tephritid relationships with plants. In R. Cavalloro (ed.), *Fruit Flies of Economic Importance*, A. A. Bakema, Rotterdam, 230–239.

Prokopy, R. J. (1986a) Stimuli influencing trophic relations in Tephritidae. *Colloques Internationaux du C.N.R.S. Comportment des Insectes et Milieu Trophique*, **265**, 305–336.

Prokopy, R. J. and Boller, E. F. (1970) Artificial egging system for the European cherry fruit fly. *J. Econ. Entomol.*, **63**, 1413–1417.

Prokopy, R. J. and Boller, E. F. (1971) Stimuli eliciting oviposition of European cherry fruit flies, *Rhagoletis cerasi*, (Diptera: Tephritidae), into inanimate objects. *Entomol. Exp. Appl.*, **14**, 1–14.

Prokopy, R. J. and Bush, G. L. (1973) Ovipositional responses to different sizes of artificial fruit by flies of *Rhagoletis pomonella* species group. *Ann. Entomol. Soc. Amer.*, **66**, 927–929.

Prokopy, R. J. and Fletcher, B. S. (1987) The role of adult learning in the acceptance of host fruit for egglaying by the Queensland fruit fly, *Dacus tryoni*. *Entomol. Exp. Appl.*, **45**, 259–263.

Prokopy, R. J. and Haniotakis, G. E. (1975) Responses of wild and laboratory cultured *Dacus oleae* to host plant colour. *Ann. Entomol. Soc. Amer.*, **68**, 73–77.

Prokopy, R. J. and Haniotakis, G. E. (1976) Host detection by wild and lab-cultured olive flies. In *The Host-Plant in Relation to Insect Behaviour and Reproduction. Symp. Biol.* (ed. T. Jermy), Hungary, no. 16, 209–214.

Prokopy, R. J. and Koyama, J. (1982) Oviposition site partitioning in *Dacus cucurbitae*. *Entomol. Exp. Appl.*, **31**, 425–432.

Prokopy, R. J. and Owens, E. D. (1978) Visual generalist—visual specialist phytophagous insects: Host selection behaviour and application to management. *Entomol. Exp. Appl.*, **24**, 409–420.

Prokopy, R. J. and Owens, E. D. (1983) Visual detection of plants and herbivorous insects. *Ann. Rev. Entomol.*, **28**, 337–364.

Prokopy, R. J. and Papaj, D. R. (1988) Learning of apple type biotypes by apple maggot flies. *J. Insect Behav.*, **1**, 67–74.

Prokopy, R. J. and Roitberg, B. D. (1984) Foraging of true fruit flies. *Amer. Sci.*, **72**, 41–49.

Prokopy, R. J., Moericke, V. and Bush, G. L. (1973) Attraction of apple maggot flies to odour of apples. *Environ. Entomol.*, **2**, 743–749.

Prokopy, R. J., Averill, A. L., Bardinelli, C. M. and Bowdan, E. S. *et al.* (1982a) Site of production of an oviposition-deterring pheromone component in *Rhagoletis pomonella* flies. *J. Insect Physiol.*, **28**, 1–10.

Prokopy, R. J., Averill, A. L., Cooley, S. S., Roitberg, C. A. (1982b) Associative learning in egglaying site selection by apple maggot flies. *Science*, **218**, 76–77.

Prokopy, R. J., Averill, A. L., Cooley, S. S., Roitberg, C. A. *et al.* (1982c) Variation in host acceptance pattern in apple maggot flies. In Visser J. H., Minks A. K. (eds), *Proceedings 5th Int. Symposium on Insect-Plant Relationships*, Pudoc Wageningen, 123–129.

Prokopy, R. J., McDonald, P. T. and Wong, T. T. Y. (1984) Inter-population variation among *Ceratitis capitata* flies in host acceptance pattern. *Entomol. Exp. Appl.*, **35**, 65–69.

Prokopy, R. J., Papaj, D. R., Cooley, S. S. and Kallet, C. (1986a) On the nature of learning in oviposition site acceptance by apple maggot flies. *Anim. Behav.*, **34**, 98–107.

Prokopy, R. J., Papaj, D. R. and Wong, T. T. Y. (1986b) Fruit foraging behaviour of Mediterranean fruit fly females on host and non-host plants. *Florida Entomol.*, **69**, 651–657.

Prokopy, R. J., Aluja, M. and Green, T. A. (1987a) Dynamics of host odor and visual stimulus interaction in host finding behaviour of apple maggot flies. In *Proceedings 6th International Symposium of Insect-Plant Relationships*, (eds V. Labeyrie, G. Fabres, and D. Lachaise), W. Junk Publishers, Dordrecht, 161–166.

Prokopy, R. J., Papaj, D. R., Opp, S. B. and Wong, T. T. Y. (1987b) Intra-tree foraging behaviour of *Ceratitis capitata* flies in relation to host fruit density and quality. *Entomol. Exp. Appl.*, **45**, 251–258.

Prokopy, R. J., Diehl, S. R. and Cooley, S. S. (1988) Behavioral evidence for host races in *Rhagoletis pomonella* flies. *Oecologia*, **76**, 138–147.

Prokopy, R. J., Green, T. A., Olson, W. A., Vargas, R. I. *et al.* (1989) Discrimination by *Dacus dorsalis* females (Diptera: Tephritidae) against larval-infested fruit. *Florida Entomol.*, **72**, 319–323.

Prokopy, R. J., Aluja, M., Papaj, D. R., Roitberg, B. D. *et al.* (1990a) Influence of previous experience with host plant foliage on behaviour of Mediterranean fruit fly females. *Proc. Hawaiian Entomol. Soc.*, **29**, (in press).

Prokopy, R. J., Green, T. A. and Wong, T. T. Y. (1989b) Learning to find fruit in *Ceratitis capitata* flies. *Entomol. Exp. Appl.* **53**, 65–72.

Pyke, J. H. (1984) Optimal foraging theory: a critical review. *Ann. Rev. Ecol. Syst.*, **15**, 523–575.

Rausher, M. D. (1983) Alteration of oviposition behaviour by *Battus philenor* butterflies in response to variation in host plant density. *Ecology*, **64**, 1028–1034.

Riedl, H. and Hislop, R. (1985) Visual attraction of the walnut husk fly (Diptera: Tephritidae) to color rectangles and spheres. *Environ. Entomol.*, 810–814.

Reissig, W. H. (1976) Comparison of traps and lures for *Rhagoletis fausta* and *R. cingulata*. *J. Econ. Entomol.*, **69**, 639–643.

Reissig, W. H., Fein, B. L. and Roelofs, W. L. (1982) Field tests of synthetic volatiles as apple maggot (Diptera: Tephritidae) attractants. *Environ. Entomol.*, **11**, 1294–1298.

Remund, U., Economopoulos, A. P., Boller, E. F., Agee H. R. *et al.* (1981) Fruit fly quality monitoring: The spectral sensitivity of field collected and laboratory-reared olive flies, *Dacus oleae* Emel (Dipt., Tephritidae). *Mitt. Schweiz. Entomol. Gesellschaft*, **54**, 221–227.

Robinson, A. and Hooper, G. H. (eds) (1989) *Fruit Flies, Their Biology, Natural Enemies and Control*, Elsevier Amsterdam, Vols. 3A and 3B.

Roitberg, B. D. (1985) Search dynamics in fruit parasitic insects. *J. Insect Physiol.*, **31**, 865–872.

Roitberg, B. D. and Mangel, M. (1988) On the evolutionary ecology of marking pheromones. *Evol. Ecol.*, **2**, 289–315.

Roitberg, B. D. and Prokopy, R. J. (1982) Influence of intertree distance on foraging behaviour of *Rhagoletis pomenella* in the field. *Ecol. Entomol.*, **7**, 437–442.

Roitberg, B. D. and Prokopy, R. J. (1983) Host deprivation influence on response of *Rhagoletis pomonella* to its oviposition deterring pheromone. *Physiol. Entomol.*, **8**, 69–72.

Roitberg, B. D. and Prokopy, R. J. (1984) Host visitation sequence as a determinant of search persistence in fruit parasitic tephritid flies. *Oecologia*, **67**, 7–12.

Roitberg, B. D. and Prokopy, R. J. (1987) Insects that mark host plants. *Bioscience*, **37**, 400–406.

Roitberg, B. D., Van Lenteren, J. C., Van Alphen, J. J. M., Galis, F. *et al.* (1982) Foraging behaviour of *Rhagoletis pomonella*, a parasite of hawthorn (*Crataegus viridis*), in nature. *J. Anim. Ecol.*, **51**, 307–325.

Seo, S. T. and Tang, C. (1982) Hawaiian fruit flies (Diptera: Tephritidae): Toxicity of benzyl isothiocyanate against eggs or 1st instars of three species. *J. Econ. Entomol.*, **75**, 1132–1135.

Sivinski, J. (1987) Acoustical oviposition cues in the Caribbean fruit fly, *Anastrepha suspensa* (Diptera: Tephritidae). *Florida Entomol.*, **70**, 171–172.

Stadler, E., Schoni, R. and Kozlowski, M. W. (1987) Relative air humidity influences the function of the tarsal chemoreceptor cells of the cherry fruit fly (*Rhagoletis cerasi*). *Physiol. Entomol.*, **12**, 339–346.

Stoffolano, J. G. (1988) Structure and function of the ovipositor of the tephritids. In Cavalloro, R. (ed.) *Fruit Flies of Economic Importance*, 87. A. A. Balkema, Rotterdam/Boston, 141–146.

Szentesi, A., Greany, P. D. and Chambers, D. L. (1979) Oviposition behaviour of laboratory-reared and wild Carribean fruit flies (*Anastrepha suspensa*: Diptera: Tephritidae) 1. Selected chemical influences. *Entomol. Exp. Appl.*, **26**, 227–238.

Tsiropoulos, G. J. and Hagen, K. S. (1979) Ovipositional response of the walnut husk fly. *Rhagoletis completa* to artificial substrates. *Z. Angew. Entomol.*, **88**, 547–550.

van Alphen, J. J. M. and Galis, F. (1983) Patch time allocation and parasitization efficiency of *Asobara tabida*, a larval parasitoid of *Drosophila*. *J. Anim. Ecol.*, **52**, 937–952.

Vita, G., Correnti, A., Minnelli, F. and Pucci, C. (1986) Studies on the chemical basis of fruit acceptance by the Mediterranean fruit fly *Ceratitis capitata* Wied: Laboratory trials. In: Cavalloro, R. (ed.), *Fruit Flies of Economic Importance*, 84. A. A. Balkema, Rotterdam, 129–132.

7

Host selection in the Heliothinae

Gary P. Fitt

INTRODUCTION

The Heliothinae (Lepidoptera: Noctuidae) include some of the most damaging pests of agricultural crops worldwide, with species of *Helicoverpa* and *Heliothis* being most important. The generic name *Helicoverpa*, first applied to *H. zea*, *H. armigera*, *H. punctigera*, *H. assulta* and other species by Hardwick (1965), has recently been accepted on taxonomic grounds (Matthews, 1987). Of the major pest species only *H. virescens* remains in *Heliothis* (Table 7.1). However, for convenience throughout this review I use the familiar collective term *Heliothis* to refer to members of both genera.

The population ecology of the pest species of Heliothinae is dominated by their polyphagy, mobility, high fecundity and ability to enter a facultative diapause (Fitt, 1989). These attributes enabled them to survive in unstable habitats where host resources vary in space and time, and to exploit successfully diverse agricultural systems. Collectively and individually the major pest species attack a diverse range of cultivated and uncultivated hostplants (Pearson, 1958; Hardwick, 1965; Nuenzig, 1969; Bilapate, 1984; Stadelbacher *et al.*, 1986; Zalucki *et al.*, 1986). Important cultivated hosts for several species include cotton, soybean, tobacco, maize, sorghum and the pulses (chickpea, pigeonpea), while many horticultural crops are also attacked (e.g. tomatoes, cut flowers). This diversity of hosts offers an array of chemical and physical attributes that may act as cues in host location and recognition. That female *Heliothis* will oviposit on a broad range of plants may imply that host selection is based on generalized plant characteristics or that specific cues that are general to many plants are involved. The polyphagous species show preferences for particular species within their broad host ranges (see later), and many of the non-pest species are specialists; *H. subflexa* on *Physalis* sp. (Laster *et al.*, 1982; Mitchell and Heath, 1987) and *H. fletcheri* on *Sorghum* sp. (Hackett and Gatehouse, 1982). Unfortunately host-related behaviours of specialists have received little attention (e.g. Mitchell and Heath, 1987) and as a consequence this review focuses on the host location and oviposition

Table 7.1 Species of Heliothinae mentioned, together with their distributions and principal hostplants. All but *H. subflexa* are considered important agricultural pests. Major pest species are indicated*.

Species	Distribution	Principal crop hosts	Reference
Helicoverpa			
zea*	North and South America	Maize, sorghum, cotton Tomato, sunflower Soybean	Hardwick (1965) Schneider *et al.* (1986) Nuenzig (1969)
armigera*	Africa, southern Europe and Asia Southeast Asia, Australia, Eastern Pacific	Maize, sorghum, sunflower Cotton, tobacco, soybean Pulses, safflower, rapeseed	Pearson (1958), Reed and Kumble (1982), Zalucki *et al.* (1986) Wardhaugh *et al.* (1980)
punctigera*	Australia	Cotton, sunflower, lucerne Soybean, chickpea, safflower	Zalucki *et al.* (1986) Wardhaugh *et al.* (1980)
assulta	Australia, Southeast Asia, India	Tobacco, *Physalis*, other Solanaceae	Reed and Kumble (1982)
Heliothis			
virescens*	North and South America	Tobacco, cotton, tomato Sunflower, soybean	Schneider *et al.* (1986) Nuenzig (1969)
peltigera	Europe, Africa, Asia	Cotton, safflower	Reed and Kumble (1982)
viriplaca	Southwest Asia, southern USSR	Cotton, legumes	Reed and Kumble (1982)
subflexa	Southern USA	*Physalis angulata*	Laster *et al.* (1982), Mitchell and Heath (1987)

behaviour of the major pest species (Table 7.1), largely in the context of agricultural systems since, apart from host records (Stadelbacher *et al.*, 1986; Zalucki *et al.*, 1986), there has been little work on host selection behaviour in relation to native hosts. Different aspects of host selection have been researched in different species, and for no species do we sufficiently understand the processes whereby hosts are located and accepted for oviposition. Thus throughout the discussion of host selection behaviour I assume that all the pest species adopt similar searching behaviour and rely on broadly similar host cues, while drawing for illustration on what is known for particular species.

For most holometabolous, phytophagous insects, including *Heliothis*, the process of host selection by the ovipositing female is crucial since larval establishment and survival is largely dependent on the selection of appropriate host plants. Although partially grown larvae may be quite mobile and hence able to exercise some choice of feeding site, neonates and early instars are relatively restricted in their movement. This is in contrast to some other noctuids in which neonates disperse on silk from the oviposition site (Gatehouse, 1988). Different plants vary widely in their suitability for *Heliothis* larval development (Pearson, 1958; Gross and Young, 1977; Nadgauda and Pitre, 1983; Stadelbacher *et al.*, 1986; Zalucki *et al.*, 1986; Topper, 1987a) and in their acceptability to ovipositing females (Johnson *et al.*, 1975; Schneider and Roush, 1986; Firempong and Zalucki, 1990a). Larvae may survive and develop on many plants not normally accepted by females as oviposition sites (Isely, 1935) and adults may oviposit on plants of limited suitability for larvae (Farrer and Kennedy, 1987; Juvik *et al.*, 1988; Firempong and Zalucki, 1990b). Host suitability is not the only factor involved in defining host range. Host abundance or 'findability' relative to alternatives will also influence the rate of utilization of different hosts and a trade-off may often be struck between the availability of a host and its suitability. *Heliothis* cope with the often patchy and unpredictable distribution of hosts by spreading their reproductive investment among many small batches of eggs, perhaps on different host species, thus effectively spreading the risks of immature mortality.

Host-finding by moths in general has recently been reviewed by Ramaswamy (1988). Despite their obvious pest status, our understanding of the mechanisms of host selection by *Heliothis* and other noctuids is surprisingly poor. Most field and laboratory research has focused on the *patterns* in the distribution of oviposition; the differential distribution of eggs between alternative hostplants, between individual plants and between plant parts which is the end result of host selection, rather than on the underlying behavioural *processes* of host location, recognition and discrimination. Here I deal firstly with these *patterns* in host use before moving on to what is, and what is not, known of the behavioural processes. Host selection is discussed in terms of two broad processes: (1)

the location of host patches and (2) mechanisms leading to the selection of particular plants for oviposition, although the process can easily be further subdivided (Miller and Strickler, 1984). The first process necessarily involves cues perceived some distance from the plant patch, while the second involves stimuli perceived by the insect at a short distance or after alighting, which allow it to distinguish between host and non-host plants, between different host species and individuals, and which elicit oviposition.

7.2 INDIVIDUALS AND POPULATIONS

Polyphagous species may consist of generalist individuals, each with the capacity to accept and utilize a range of hostplants, or a collection of more specialized individuals forming host races which utilize a narrower spectrum of hosts. Collectively these may still appear to be highly polyphagous (Fox and Morrow, 1980; Singer, 1986). This last alternative may result in regional differences in host utilization by populations, or individual variation in host preference within populations (Fox and Morrow, 1980).

Few studies of *Heliothis* have considered individual variability in host-related behaviour, although questions about regional variation in host preference are now receiving attention (see Section 7.4.2). In general laboratory studies have quantified the responses of groups of individuals. Nevertheless, the second alternative that *Heliothis* populations may consist of host-specific individuals or races, seems unlikely given the ephemeral nature of their hostplants. Most hosts, whether cultivated or otherwise, remain suitable for larval development for only one generation before senescing. An exception is cotton, which may support 3–4 successive generations. Spatial variance in sowing dates of other hosts (e.g. early and late sown sorghum) or plant growth within a local area may allow the completion of more than one generation on a 'preferred' host. However, the inherent variability and unpredictability of the host system exploited by these insects makes it unlikely that such preferences would proceed to the stage of strict host, or even family, specificity. The potential for extensive movement in all the major pest species (Farrow and Daly, 1987; Fitt, 1989) suggests a strategy of successive utilization of many hosts linked by adult movement at both local and regional scales.

7.3 OVIPOSITION BEHAVIOUR OF HELIOTHINAE

Heliothis are primarily crepuscular/nocturnal insects. Adults may occasionally feed and oviposit during the day, particularly when overcast, but most oviposition occurs in short bouts during the first half of the night commencing about dusk (Lingren *et al.*, 1982; Ramaswamy, 1988). As far as is known all significant movements occur at night. Nocturnal activity not only limits the potential involvement of some behavioural modalities (e.g. colour vision) in host selection, but also restricts the possibilities for direct observation of behaviour by typically diurnal researchers. However, the recent application of specialized night vision technology has enabled nocturnal behaviour patterns to be observed and quantified (Lingren *et al.*, 1982; R. A. Farrow and G. P. Fitt, unpublished data). These techniques are beginning to provide some insight into the way mobile moths perceive host patches and how their movements may be constrained by them.

Oviposition behaviour after location of a host patch has not been quantified or described in any detail. Ramaswamy (1988) presents an ethogram of oviposition behaviour in caged *H. virescens*. During oviposition bouts females fly slowly among and hover above plants before landing and quickly depositing an egg. In contrast to many other noctuids which lay their eggs in clusters (*Mythimna, Spodoptera*: Gate-house, 1988), *Heliothis* typically lay eggs singly or in small batches of 1–3 at each oviposition bout. Several eggs may be laid on the same plant, although more often females move a few metres before alighting again. How closely these field cage observations of oviposition mimic that in the field is not known, but many features of an oviposition bout (e.g. the distance moved between plants, the time spent on each plant, etc.) will be influenced by the structure of the local vegetation. Behaviour may differ markedly when females are exploiting diverse patches of ephemeral weed hosts compared with uniform patches of cultivated crops.

7.4 PATTERNS IN HOST UTILIZATION BY HELIOTHINAE

7.4.1 Host ranking

Although all the pest species of *Heliothis* are polyphagous, they display differing preferences for particular hosts (Johnson *et al.*, 1975; Roome, 1975; Schneider *et al.*, 1986; Zalucki *et al.*, 1986). All species readily attack legumes, but *H. zea* and *H. armigera* attack maize and grain sorghum preferentially over most other crop hosts (Pearson, 1958; Neunzig, 1969; Roome, 1975; Stinner *et al.*, 1977). By contrast these graminaceous hosts are not attacked by *H. virescens* and *H. punctigera*. In Australia, *H. punctigera* appears restricted to dicotyledonous hosts, while *H. armigera* occurs on both dicots and monocots (Wardhaugh *et al.*, 1980; Zalucki *et al.*, 1986). In southeastern USA, *H. virescens* attacks tobacco whenever available and is most abundant on cotton and other hosts once the tobacco crop becomes unattractive. Cotton, a crop considered particularly susceptible to *Heliothis* damage, is not a preferred host of any species (Johnson *et al.*, 1975; Firempong and Zalucki, 1989a; Waldvogel and Gould, 1990). In many areas cotton is heavily attacked only after alternative cultivated or weed hosts have senesced. The extent to which these patterns of host utilization in the field derive from behavioural preferences for particular plants or are the result of differing seasonal phenologies and spatial arrangements of potential hosts is not clear. It seems most likely that a combination of the two is involved.

A number of studies have examined the relative preferences of adult *Heliothis* in laboratory assays where host rankings are determined by offering females a choice of plants and recording the numbers of eggs deposited on each. Johnson *et al.* (1975) showed that for plants at a similar phenological state the rank order of ovipositional preference by *H. zea* was maize > tobacco > soybean > cotton. The low ranking for cotton is interesting, and has also been shown for Australian *H. armigera* (Firempong and Zalucki, 1990a) and *H. virescens* from various parts of the USA (Waldvogel and Gould, 1990). Australian *H. armigera* from a variety of locations prefer maize to sunflower and tobacco, and these more than soybean, cotton and lucerne, with pigweed, cabbage and linseed being the least preferred plants (Firmepong and Zalucki, 1990a). Martin *et al.* (1976) showed that *H. virescens* strongly preferred tobacco over other crops, while *H. zea* preferred maize and sorghum above other potential hosts in laboratory tests. Even so, both species used a wide range of other hosts in their study area. Waldvogel and Gould (1990) found *H. virescens* ranked tobacco and velvetleaf (*Abutilon theophrasti*, Malvaceae) above four other plants. Cotton and tomato were consistently the least preferred. The relationship between these behavioural preferences of adults and the relative suitability of plants for larvae has received less attention.

In the case of specialized *Heliothis*, there is some evidence that host range is constrained by both adult and larval adaptations. In the field *H. subflexa* is only recorded on species of *Physalis* (Laster *et al.*, 1982), which contain specific oviposition stimulants (Mitchell and Heath, 1987). *H. subflexa* larvae survive well on *Physalis* and corn (a crop never used as a host in the field):, but poorly or not at all on cotton, soybean, tobacco and wild geranium (*Geranium carolinianum*) (Laster *et al.*, 1982). By contrast, *H. virescens* larvae did not survive on *Physalis*. By means of reciprocal crosses Laster *et al.* (1982) were able to show these differences in survival were genetically based. Hybrid *subflexa* × *virescens* larvae were able to survive well on cotton, soybean and wild geranium, though their performance on *Physalis* was reduced. In this case there appear to be behavioural adaptations in adults and physiological adaptations of larvae that result in a close association between adult and larval host ranges.

7.4.2 Geographic and genetic variation in host utilization

Evidence for geographic variation in host preference of *Heliothis* spp. is sparse; the question has not been widely explored. *H. virescens* populations in southeastern USA utilize cotton extensively, whereas populations in the San Joaquin Valley of California and on the US Virgin Islands in the Caribbean rarely attack this host. Schneider and Roush (1986), by using laboratory preference tests and reciprocal crosses of strains from Mississippi and the Virgin Islands, showed that this difference in host utilization was due to genetic variation in oviposition preference between the sub-populations. They conclude that a greater relative preference for cotton in the Mississippi strain probably evolved as a result of the greater availability of cotton in southeastern USA over the last 150 years. Moreover they present a minimal amount of data, suggesting some individual variation in oviposition preference within the Virgin Islands populations. Similar studies with other *Heliothis* species are needed. Waldvogel and Gould (1990) examined the preferences of *H. virescens* populations from Arizona, North Carolina and Mississippi, where they experienced quite different arrays of hosts but found little variation, whereas preferences of a Virgin Islands population were quite distinct. They dismiss long-distance gene flow as an explanation for lack of divergence and suggest continental *H. virescens* populations may lack the genetic variance necessary for change in host preference.

In Australia, Firempong and Zalucki (1990a) examined variation in oviposition preference between geographic populations of *H. armigera*. Despite some differences in acceptance of lowly ranked hosts, the general rank order of host acceptance was similar in populations from widely separated areas. The absence of major genetic variation among sub-populations of both *H. armigera* and *H. punctigera* throughout eastern Australia (Daly and Gregg, 1985) is consistent with this absence of major variation of host preference.

In many areas of the world the abundance of *Heliothis* spp. on different crops has changed over time (Johnson *et al.*, 1986; Zalucki *et al.*, 1986; Topper, 1987a; Mitter and Schneider, 1987; Fitt, 1989). This suggests the possibility of adaptation in behavioural or physiological traits that enables the exploitation of new crops. Many of the examples relate to cotton. In Australia the abundance of *H. armigera* on cotton appears to have increased over the last 10–15 years since the crop was extensively grown in inland areas (Zalucki *et al.*, 1986). However, in this and most other instances, these changes in host utilization coincided with other changes in both the area planted and agronomy of cotton and in the relative abundance and phenology of alternative hosts, which may have increased the synchrony between emerging adult populations and large areas of attractive cotton. In Australia, the question is clouded by early difficulties in distinguishing *H. punctigera* and *H. armigera* (Zalucki *et al.*, 1986). The latter species has probably always been present at low densities in inland cotton areas and is now known to occur much further into arid inland Australia where ephemeral hostplants are seasonally available (P. C. Gregg, G. P. Fitt, P. H. Twine and M. P. Zalucki, unpublished data). In North America *H. virescens* was not regularly recorded on cotton until the early 1930s, even though the crop had been extensively cultivated since 1810 (Hearn and Fitt, 1990). However, Mitter and Schneider (1987) found evidence of *H. virescens* on cotton somewhat earlier, and conclude that its increased abundance on cotton was due to 'changing cultural practice acting on long-standing low-level infestations rather than genetic change . . .'. In the Sudan Gezira the increased abundance of *H. armigera* on cotton is related to the extent and timing of groundnut and sorghum crops. These crops, which mature soon after cotton is sown, sustain large populations of *H. armigera* which move directly onto cotton upon emergence (Topper, 1987a). Similarly in the high plains of Texas, *H. zea* became a regular pest of cotton in the late 1970s, coincident with the introduction of maize in that region, again providing the previously missing early season host for *H. zea* (Rummel *et al.*, 1986).

7.4.3 Patterns of oviposition between and within individual plants

Heliothis also show considerable preferences for individual plants and for certain plant parts. The most consistent pattern shown in host selection by all *Heliothis* is a preference for flowering stages of their hostplants (Parsons, 1940; Cullen, 1969; Roome, 1975; Wardhaugh *et al.*, 1980; Alvarado-Rodriguez *et al.*, 1982; Adjei-Maafo and Wilson, 1983a,b). This preference is apparent in all stages of host location, recognition and the stimulation of oviposition and may have evolved to sychronize larval feeding with the availability of nitrogen-rich fruiting structures (flowerbuds, flowers and bolls of cotton, cobs of corn, sorghum heads, etc.). However, another factor that may have shaped this preference is the avoidance of the numerous allelochemicals that are present in the foliage of most plants, and which are often absent from flowers and fruits. Such a preference may lead to changes in ranking of host species, or of patches of the same species where these differ in their flowering phenology. Oviposition on non-flowering crops, such as seedling cotton and maize (Wardhaugh *et al.*, 1980; Haggis, 1981; Zalucki *et al.*, 1986) may be a consequence of other alternatives being unavailable. Parsons (1940) found that *H. armigera* strongly preferred flowering crops, but when these were not available they appeared to distribute eggs indiscriminantly, sometimes on plants not normally selected as hosts.

The clumped pattern of egg dispersion among plants seen in some crops (e.g. cotton: Wilson and Room, 1983; Mabbett and Nachapong, 1984) suggests that all plants are not equally attractive to ovipositing *Heliothis*. Some of the factors involved in discrimination at this level, such as the presence of nectar or changes in the production of volatiles associated with flowering, are discussed in detail later (Section 7.5.2). Other factors that may be involved include plant height (Mabbet and Nachapong, 1984; Firempong and Zalucki, 1990b), surface texture, the presence of conspecific eggs, larvae or frass, as well as differences in surface chemistry and volatiles operating over a very short range.

Heliothis eggs are deposited on many plant structures (Alvarado-Rodriguez *et al.*, 1982; Wilson *et al.*, 1983; Mabbet and Nachapong, 1984; Farrer and Bradley, 1985a), though these are not necessarily those most appropriate for the initiation of larval feeding. In crops such as cotton most eggs are laid in the top third of the plant canopy (Wilson *et al.*, 1983; Farrer and Bradley, 1985a) and are typically clustered on young veg-

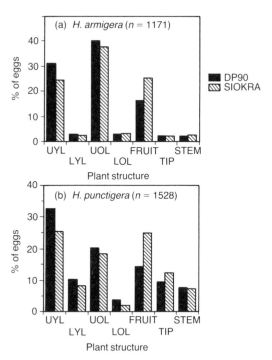

Fig. 7.1 The within-plant distribution of eggs by (a) *H. armigera* and (b) *H. punctigera* on two varieties of cotton, DP90 (a normal leaf variety) and SIOKRA (an okra leaf variety). Plants were categorized into seven structures: UYL, LYL, upper surface and lower surface of young leaves; UOL, LOL, upper and lower surface of old (mature) leaves; TIP, the growing terminal with unexpanded leaves and fruit; FRUIT, squares (flowers buds), flowers and bolls; STEM, mainstems and petioles.

etative growing points, but only rarely are they laid directly on the developing flower buds. *H. zea* and *H. virescens* lay a high proportion of eggs directly on the growing tips of cotton (Farrer and Bradley, 1985a). Farrer and Bradley (1985b) found little effect of the site of oviposition within plants on the probability of establishment and survival of larvae. However, in Australia, although most eggs of *H. armigera* and *H. punctigera* are found in the top third of the canopy, most are deposited on structures other than the growing tip (Fig. 7.1). Many eggs are deposited a considerable distance from suitable sites for larval establishment. For example, up to 40% of *H. armigera* eggs are found on the upper surfaces of mature fully expanded leaves on which larval establishment may be quite poor while only 15–25% of eggs were deposited directly on fruits. On tomato most eggs (>90%) are laid on leaflets (Alvarado-Rodriguez *et al.*, 1982), usually those near to blooms. Hatchling larvae thus have only a short distance to travel to locate a feeding site, but in many cases feeding will be initiated on leaf tissue, so exposing neonates to numerous toxins (Duffey, 1986). In some other crops eggs are laid close to the site of larval feeding (e.g. on maize silks, legume flowers and pods) thus allowing larvae to move directly onto or into the preferred feeding site.

Many insects use pheromones (e.g. oviposition-deterring pheromones) or other chemical stimuli to minimize larval competition (Prokopy, 1981). For *Heliothis* there is no evidence that females discriminate against plants bearing eggs (Fitt, 1987; Waldvogel and Gould, 1990). However, oviposition deterrents associated with larval feeding damage or frass have been observed in *Heliothis* (Fitt, 1987; Firempong, 1987), and some other noctuids; *Hadena* (Brantjes, 1976), *Trichoplusia* (Renwick and Radke, 1980) and *Spodoptera* (Gross, 1984). Renwick and Radke (1980) demonstrated the existence of a powerful oviposition deterrent in the larval frass of *Trichoplusia ni* that dramatically reduced egg-laying on plants with feeding larvae or to which a crude frass extract had been applied. Gross (1984) showed that oviposition by *Spodoptera frugiperda* on maize was significantly reduced by aqueous homogenates of *S. frugiperda* or *H. zea* larvae, which suggested the influence of semiochemicals derived from the larvae or their by-products (frass). Plants treated with extracts of larval frass are also avoided by ovipositing *H. punctigera* (Fitt, 1987).

7.5 THE BEHAVIOURAL PROCESSES

Having examined the patterns in *Heliothis* egg distribution it is important to recognize the influence of the spatial and temporal structure of agro-ecosystems, and the association of *Heliothis* spp. with such systems, on the processes of host location and the expression of preferences for particular hosts. The term 'preference' implies active choice from among a range of options, all of which are perceived simultaneously or in sequence, if learning is invoked (Miller and Strickler, 1984; Singer, 1986; Papaj and Prokopy, 1989). Choice necessarily implies that the insects possess the sensory characteristics needed to differentiate between different plants. While in stands of uncultivated plants host-seeking females may encounter a diverse choice of alternative hostplants, in many agricultural situations they will be faced with large patches of single species. In some cases these patches may be surrounded by native vegetation consisting of hosts and non-hosts, while in other situations there may be a patchwork of different cultivated plants (e.g. Stinner, 1979); again there may be both hosts and non-hosts. This spatial pattern of ecosystems will not only influence the population dynamics of *Heliothis* spp. (Kennedy *et al.*, 1987), but will also have a major influence on host selection by altering a female's perception of the range of hosts available. Whether females simultaneously perceive alternative hosts over long distances (as assumed in some population models: Stinner *et al.*, 1974) depends on whether long-range cues are involved in host location. If not, as seems more probable (see below), females may rarely experience a simultaneous choice between alternatives except perhaps at field boundaries, but must move throughout the local area in order to locate and sample the available hosts. Using cues derived from plants in their current locality they can either remain and deposit eggs or else move elsewhere in search of perhaps more suitable hosts. Distinguishing these two possibilities is complicated by our ignorance of the distances over which moths are able to perceive host patches. This point is discussed more fully below. However, if host patches are perceived only after moths move within a narrow attractive space around them, then differential utilization of hosts will be a reflection of the hierarchy of host *acceptance* (*sensu* Singer, 1986) and subsequent residence times in host patches, rather than of *preferential responses* during host location.

As discussed above (Section 7.4), major differences in egg densities are often observed in the field between different hosts (Wardhaugh *et al.*, 1980; Haggis, 1981; Schneider *et al.*, 1986; Topper, 1987a), or between individuals of the same host at different stages of development (Weisenborn and Trumble, 1988). Laboratory bioassays also demonstrate preferences for particular hosts by different species (Section 7.4.1). However, it is pertinent to consider here whether these patterns in spatial distribution of eggs accurately reflect the behavioural processes (differential host attractiveness or acceptability) discussed later. Some hosts may be highly attractive to females from a distance, but may lack suitable oviposition-inducing stimuli, while other hosts may only be perceived at short range but be highly stimulating for oviposition. These possibilities arise since many crop plants, and the spatial arrangement under which they are often grown (large single-species blocks), would not have been prominent or even present during the evolutionary shaping of *Heliothis* host preferences or selection behaviour. Many of the host associations discussed earlier (Section 7.4.2) may be relatively new and there is some evidence that evolutionary adaptation to host plants at behavioural and physiological levels is ongoing (Section 7.4.3). Perhaps as a result the correspondence between the potential host range of *Heliothis* larvae and the rank order of host acceptability of ovipositing females is not always close. The same is true in many other insects (Wiklund, 1975; Courtney, 1981).

Moreover, because of limitations imposed by temporal or spatial patterns of host availability in the field, the behavioural rankings revealed by choice experiments in the laboratory may not be expressed in the field. The most highly preferred hosts may be available only briefly or in small areas where they remain undiscovered by the majority of moths. Nevertheless, it is encouraging that Johnson *et al.* (1975) were able to explain seasonal patterns of egg abundance on various crops in North Carolina based on their behavioural model.

The possibility that learning of physical or chemical characteristics of hostplants may influence subsequent searching or acceptance behaviour of females (Traynier, 1986; Papaj and Prokopy, 1989) has not been explored for *Heliothis*, but should also be considered. Learning may occur early in adult life when the moth emerges into a particular host environment or later during oviposition encounters. Barton-Browne (1977) defined three mechanisms by which learning might influence subsequent host encounters. First, a female might learn to associate long-distance cues with the levels of stimulation elicited when contact is made, and so come to rely more heavily on distance perception. Second, the level of excitatory input elicited by the most stimulating plant yet visited might be used for comparison with plants visited subsequently; those providing a lower level of excitatory input being rejected at least initially (Singer, 1986). Third, after ovipositing on a plant which provided an adequate level of excitatory input, the insect might then localize its searching.

There is no evidence that adult *Heliothis* preferentially select plants on which they developed as larvae, though feeding preferences of late instar larvae can be influenced by experience during earlier instars (Jermy *et al.*, 1968). In most cases adult *Heliothis* emerging from beneath hostplants on which they fed as larvae will encounter senescent, unattractive plants. It seems unlikely therefore that this initial experience of plants will greatly influence subsequent behaviour. However, there are exceptions. Cotton can support several consecutive generations during the growing season and moths produced early in the season may well emerge into an attractive crop environment. Are these moths more likely to prefer cotton over other crops? While there is no evidence for such a phenomenon, it warrants some investigation.

7.5.1 Mechanisms of host patch location

Defining what constitutes a host patch is difficult for mobile insects. In agricultural systems a patch can conveniently be defined as an area containing a monoculture of a particular crop, though the insect may not perceive the full extent of such patches, which may vary from a few square metres to thousands of hectares. In diverse communities of native vegetation a patch may be an area or habitat where acceptable hostplants are predominant. Discovery of a patch may be defined as the 'moment an insect herbivore crosses some arbitrarily defined host-plant density isocline, travelling from lower to higher abundance' (Stanton, 1983, p. 136). The patch is then the area within that isocline. For polyphagous herbivores such as *Heliothis* an area of vegetation containing a mixture of several hosts may be perceived from a distance merely as a dense stand of hosts; the individual species are distinguished using short-range cues once searching within the patch has begun. As discussed in the next section, the critical question is when does a searching insect perceive that it has entered a 'patch', and perhaps switches from a long-distance searching mode to behaviours involved in selecting and finally accepting individual plants for oviposition. The patch is thus best defined by this change in the behaviour of the insect itself.

(a) Categories of movement involved in host patch location

Heliothis are highly mobile insects. Their ability to undertake extensive local and inter-regional movements is a major factor contributing to their success as pests. At least three categories of movement can be defined —migratory, long-range and short-range (Farrow and Daly, 1987)— though distinctions between them remain unclear.

Long distance migratory movements occur well outside the boundary layer where the insect can control its flight direction and speed, often at high altitudes (200–1000 m: Farrow and Daly, 1987), where the direction of movement, the distance travelled and the point of arrival are controlled by synoptic scale weather patterns (Drake and Farrow, 1985, 1988; Wolf *et al.*, 1986). The presence, or absence of suitable hosts, or nectar sources, is unlikely to influence precisely where migrating adults land after a night's flight, though their presence in that locality could determine whether migratory movement is then terminated. This is an area requiring further research. Short-range movements occur within or just above the plant canopy and involve activities such as feeding, oviposition, mating and sheltering. In relation to host selection these movements occur after location of host patches and involve processes of host recognition and acceptance discussed later.

Long-range movements at low altitudes (up to 15 m above ground), however, are probably most important in the location of host patches. The cues perceived from the ground cover by *Heliothis* moths flying at low altitude which might be involved in host patch location are poorly known. Host-seeking females may fly at random with respect to host patches (e.g. fly steadily downwind) until they perceive hostplants in response to short-range cues. In this case, the moth may have to fly over a crop before recognizing it. Alternatively, moths may perceive plant patches over a relatively long distance by vision or olfaction, then undertake movements (either directed movement—taxes, or adjustments to the pattern of movement—kineses) which tend to increase the probability of contact with the host. In this case the probability of contact with a host depends on the size and density of patches, which influence their visual characteristics, or the size and intensity of odour plumes, and on the olfactory or visual sensitivity of the insect. In each case searching after the initial encounter with a host patch may be biased towards keeping the insect within the patch.

(b) Stimuli involved in host patch location; visual stimuli

The importance of visual stimuli in host selection by phytophagous insects is uncertain. Prokopy and Owens (1983) define three properties of plants that may serve as visual cues to foraging insects: spectral qualities (hue, saturation, intensity), dimensions, and patterns (see also Chapter 6). All of these are more likely to be involved in the host location of diurnal foragers than of nocturnal species such as *Heliothis* (Schneider *et al.*, 1986; Ramaswamy, 1988). Nevertheless, Callahan (1957) suggested that short wavelength light reflected from fields may provide sufficient pattern stimuli to flying moths to attract them to feed or oviposit. These stimuli may not be host specific but might allow flying moths to distinguish patches of vegetation from bare ground (fallow fields, etc.). Callahan (1957) also demonstrated in the laboratory that ovipositing *H. zea* could discriminate between different wavelengths of low, equal-intensity light. Oviposition preference decreased with increasing wavelength from blue to red. The spectral sensitivity of the compound eyes of *H. zea* and *H. virescens* is identical (Agee, 1973) and colour vision may play a minor role in foraging activities at low light intensities around dusk. Some studies of male responses to pheromone traps (Gross *et al.*, 1983) and to hostplants (Firempong and Zalucki, 1990b) also demonstrate the visual acuity of *Heliothis* adults and indicate that general visual characteristics of plants (shape and outline, height in relation to surroundings), which comprise their dimension and pattern, may well play a part in short distance discrimination prior to landing. Nevertheless, Ramaswamy *et al.* (1987) found the eyes were not necessary for short-range host discrimination by

H. virescens and the precise involvement of visual stimuli in host selection by *Heliothis* remains uncertain.

(c) Olfactory stimuli

In some insects orientation to olfactory cues from a distance has been demonstrated (Carde, 1984), though usually over distances of only a few metres (Kennedy, 1986; Finch, 1986). Chemically-mediated, optometer-guided anemotaxis appears to be the principal mechanism of long-distance orientation to hostplant odours by flying insects (Carde, 1984; Finch, 1986), although experimental evidence remains limited.

For *Heliothis* the role of olfactory cues in host patch location is far from clear. Most research in this field has concentrated on responses of male noctuids to point sources of sex pheromone. The difficulty is in determining the distances at which hosts are perceived via olfactory, or other cues. Do *Heliothis* perceive host patches from 1 km, 100 m or 1 m? Early studies suggested that volatiles emanating from rapidly growing cotton and maize plants may act as long-distance attractants or as arrestants tending to concentrate *Heliothis* moths within areas of flowering hosts (Thomas and Dunham, 1931; Barber, 1938; Parsons, 1940). While host volatiles are undoubtedly involved in selection of individual plants and the stimulation of oviposition (Section 7.5.2), there is little evidence of long-distance perception or orientation in response to volatiles (Ramaswamy *et al.*, 1987; Ramaswamy, 1988). In the only work yet to examine olfactory responses of *Heliothis* females to plant volatiles, Ramaswamy *et al.* (1987) could find no evidence of upwind flight in *H. virescens* to individual cotton plants.

However, in agricultural situations plumes of volatiles will emanate from large blocks of synchronously developing crops, sometimes hundreds of hectares in extent, rather than from individual plants as often used in behavioural experiments. Moreover, where crops are grown in virtual monoculture, there will be little interference or masking of volatiles by those of other plants in the vicinity, as may occur in more diverse plant communities (Papaj and Rausher, 1983). The size and complexity of the olfactory space around large host patches will vary according to the density and dimensions of the patch (Stanton, 1983) and the strength and turbulence of airflow (Murlis, 1986), but the olfactory stimulus may be sufficient to allow oriented upwind movement of moths from distances of several tens of metres.

(d) Nocturnal observations of Heliothis behaviour

Observations of *Heliothis* moths using night vision devices and low level radar studies (Wolf *et al.*, 1986; V. A. Drake, unpublished data; J. R. Riley *et al.*, unpublished data) provide evidence of two major classes of low level movements (up to 15 m above ground) often occurring simultaneously within a local population. Some moths appear to fly consistently downwind within and between adjacent crops (Stinner *et al.*, 1982; G. P. Fitt and R. A. Farrow, unpublished data) and do not seem responsive to different crop types or crop boundaries. Both *H. armigera* and *H. punctigera* have been observed moving in a general downwind direction out of apparently attractive cotton crops and over fallow ground supporting no hosts. The age structure of these moths relative to those which remain foraging within the crop is not known; they may for example be newly emerged sexually immature individuals undertaking a pre-reproductive dispersal flight.

The other type of flight activity is upwind and is strongly suggestive of a response to volatile stimuli emanating downwind from highly attractive hosts (e.g. silking maize) to distances of several metres. Perhaps these moths are the reproductively mature host-seeking individuals within the population. Upwind and crosswind flight is also observed late at night in mate-seeking male *Heliothis*. Most nocturnal observations of *Heliothis* flight activity have been done to coincide with the emergence and emigration of large populations from blocks of senescent hosts (e.g. Lingren *et al.*, 1988; J. R. Riley *et al.*, unpublished data). However recent observations of *H. armigera* moths flying over blocks of highly attractive (silking) maize indicate that their movements are largely constrained within the patch (G. P. Fitt and R. A. Farrow, unpublished data). A high proportion (up to 60%) of moths flying across the maize block were seen to turn back into that crop rather than fly out over adjacent, but less attractive cotton crops. This turning behaviour occurred at both downwind and upwind sides of the crop, though moths tended to travel further from the maize edge before turning on the downwind side than on the upwind side. The moths were clearly able to perceive the patch boundaries, whether by olfactory or visual cues, and to respond very quickly. Similar turning behaviour has been observed to a lesser extent at the boundary of cotton and fallow.

If we assume that major differences in egg density between patches of different hosts or of the same host at different stages of development are a reflection of underlying behavioural processes (see earlier discussion in Section 7.5), then these patterns suggest either that moths are efficiently sampling available hosts within a wide area and being arrested in the most acceptable host, or are able to detect those hosts from a distance and are attracted to them. Perhaps a combination of both occurs, with initially randomly searching individuals being attracted to acceptable crops over distances of a few metres and then arrested within the crop, leading over time to the aggregation of ovipositing females into the most acceptable and 'arresting' host crops.

It should be clear from the above discussion that our understanding of host patch location by *Heliothis* is poor and warrants further study. Many questions remain. For example, do females which have located an acceptable host crop on one night leave those crops in search of other patches on subsequent nights? Observations by Topper (1987b) suggest regular nightly redistributions of moths between oviposition, feeding and resting sites in different crops. Analyses of pheromone trap catches (Morton *et al.*, 1981; Fitt *et al.*, 1989) suggest differences between *H. armigera* and *H. punctigera* in the scale of local movements; *H. armigera* tends to be sedentary after colonizing a host crop, while populations of *H. punctigera* appear to be continually redistributed among different blocks. Clearly much more research on local movement in relation to host location is needed.

7.5.2 Mechanisms of selection between individual plants

(a) Sensory mechanisms and possible cues

Once a host patch has been located, a range of chemical, physical or visual attributes may be used as proximate cues in host selection over short distances, even though a dominant stimulus (often a host-specific chemical) may appear to be the major factor leading to oviposition. These proximate cues may be perceived via a number of sensory modalities: contact and olfactory chemoreceptors, mechanoreceptors of various types and the visual system. Olfactory and visual cues may be perceived over short distances and so may be used in pre-alighting discrimination between plants, while contact chemoreceptors and mechanoreceptors are useful only in post-alighting assessment. Coupled with the role of external stimuli, is the influence of the physiological state of the insect which may alter its perception of external cues or thresholds for the performance of particular responses. For example, if acceptable hosts are scarce, host-seeking insects may become more ready to accept less preferred hosts as the length of time since the last successful oviposition increases (Singer, 1986; Fitt, 1986). As a result they may accept plants of low ranking, which may or may not be suitable for larval development, but which are normally unacceptable in the presence of more highly ranked plants. It is the balance between these external and internal excitatory and inhibitory stimuli which ultimately results in the selection and acceptance of a plant for oviposition or feeding (Miller and Strickler, 1984).

The processes of host recognition and oviposition by *Heliothis* are a little better understood than that of host patch location, in that specific attractants or oviposition stimulants have been identified from several plants (Jones *et al.*, 1970; Jackson *et al.*, 1984; Tingle and Mitchell, 1984; Juvik *et al.*, 1988). Unfortunately, the behavioural assays used in many studies of *Heliothis* host selection do not discriminate between different sensory modalities (Ramaswamy *et al.*, 1987; Fitt and Boyan, 1990). For example, assays which simply involve counting the number of eggs laid on substrates impregnated with different compounds do not distinguish short-range olfaction from contact chemoreception (e.g. Tingle and Mitchell, 1984). If our understanding of *Heliothis* host selection is to improve it is essential that assays and observations be designed so as to reveal the stimuli and sensory receptors involved in different aspects of the behavioural process.

All species of *Helicoverpa* and *Heliothis* show distinct preferences for flowering stages of their hostplants, and in some situations females will also accept flowering individuals of species that are unsuitable for larval development (Firempong and Zalucki, 1990b). Whatever the mechanism, the broad host ranges of pest *Heliothis* suggest either that females recognize and respond to a diversity of chemicals (or other stimuli) found in different hostplants, or that a single chemical or group of chemicals associated with flowering, and common to many plant families (analogous to the 'green odour' of Visser and Ave, 1978), is involved in stimulating oviposition. The latter possibility offers an explanation for many observations of egg-laying on flowering plants that are not hosts (M. P. Zalucki, personal communication). Explanations for this preference have concentrated on two factors as the proximate cues involved: (1) fluctuations in the release of volatile secondary plant metabolites related to the active stages of growth and fruiting, and (2) the phenology of extrafloral nectar production or of other sources of adult food.

(b) Olfactory stimuli

Several early studies concluded that volatiles were important in host selection by *Heliothis* (Thomas and Dunham, 1931; Barber, 1938; Parsons, 1940). Thomas and Dunham (1931) suggested that *H. zea* were attracted to an odour associated with the 'succulence' of cotton. Parsons (1940) demonstrated that crude extracts from maize tassels and cotton flowers acted as short-range attractants and possibly stimulated oviposition. Nevertheless, there remains no clear evidence for olfaction in the orientation of *Heliothis* to hostplants or in the distinction at short-range between hosts and non-hosts (Ramaswamy *et al.*, 1987). This is an area requiring much further work.

Analyses of the volatiles released by crop plants usually reveal a diversity of components. Hedin (1976) identified a total of 70 volatiles (mainly aliphatic and aromatic hydrocarbons, alcohols and carbonyl compounds) released by actively growing cotton in the field, although only 15 of them were also present in cotton bud essential oil. Although there was no growth-related change in the yield of extracted essential oils, he found a distinct fluctuation in the production of volatiles with peak production coinciding with the period when the plant had the highest square count and was supporting significant numbers of flowers and young bolls. During fruit set there was a gradual decline in volatile production followed by a rapid decrease as square production ceased. This pattern of volatile production closely resembles the fluctuation in attractiveness of cotton to ovipositing *Heliothis*. Numerous volatiles have been identified from the silks and tassels of maize (Flath *et al.*, 1978; Cantelo and Jacobsen, 1979; Buttery *et al.*, 1980), one of which, phenyl acetaldehyde, was shown to be attractive to both male and female *H. zea* (as well as several other insects found in maize crops: Cantelo and Jacobsen, 1979). Unfortunately, these volatiles have only been assessed by exposing them in traps in the field. This at least demonstrates that they are attractants, but their precise role in host selection remains to be demonstrated.

(c) Contact chemoreception

The role of surface chemistry in host recognition and the stimulation of oviposition is perhaps better understood since at least some active chemicals have been identified for some species. Jones *et al.* (1970) identified triacetin, a constituent of corn silks, and several related compounds as oviposition stimulants for *H. zea*. Jackson *et al.* (1984) demonstrated the importance in leaf surface chemistry in the recognition and acceptance of tobacco cultivars by *H. virescens*. Genotypes receiving fewer eggs in choice tests were found to lack certain alpha and beta diols. Preparations of these active components when sprayed onto the surface of resistant plants caused an increase in oviposition to levels similar to those on susceptible genotypes. Although they did not determine whether the compounds were attractants, arrestants or oviposition stimulants, they suggest that they were active only over very short distances and so could be important only as post-alighting cues. Tingle and Mitchell (1984) also presented evidence of the role played by surface chemistry in post-alighting behaviour by *H. virescens*, and emphasized the importance of oviposition deterrents in site selection by phytophagous insects. Using a substrate with a standard surface texture (paper towelling), they showed in laboratory assays that aqueous extracts from a number of plants (mainly non-hosts) caused reductions in the number of eggs deposited relative to untreated controls. These effects were repeated in field cage and field tests where the extracts were applied to growing tobacco plants. Unfortunately, it was not determined whether the extracts modified the insects' pre- or post-alighting behaviour, though the latter seemed most probable. Mitchell and Heath (1987) demonstrated the presence of an oviposition stimulant for the specialist *H. subflexa* in methanolic extracts of its preferred hostplant, *Physalis angulata*, though they did not attempt to identify the compound.

The attractiveness of various *Lycopersicon* species to *Heliothis* illustrates that adult responses to hostplant stimuli are not necessarily adaptive. Most *Lycopersicon* species are highly attractive to ovipositing *Heliothis*, but all contain allelochemicals in their foliage that kill or restrict the growth of larvae (Duffey, 1986; Farrer and Kennedy, 1987). Juvik *et al.* (1988) identified a group of sesquiterpenes produced by leaf trichomes of wild tomatoes (*Lycopersicon hirsutum*), and to a lesser extent cultivated tomatoes, as oviposition stimulants for *H. zea*. The chemicals had low volatility and did not stimulate oviposition in another noctuid, *Trichoplusia ni*. Whole plants and extracts from the wild tomatoes *L. hirsutum* (Accession LA1777) and *L. hirsutum* f. *glabratum* were most attractive to *H. zea*, but paradoxically foliage from these species was the least suitable for larval growth and survival of the 45 cultivars tested. Indeed several trichome exudates from *Lycopersicon*, including 2-tridecanone, are known to be toxic to *Heliothis* larvae when they are forced to consume leaf tissue (Duffey, 1986; Farrer and Kennedy, 1987). Most *Heliothis* eggs are laid on leaflets of tomato (Alvarado-Rodriguez *et al.*, 1982), particularly those adjacent to blooms, and these chemicals may greatly restrict the establishment of larvae before they can reach the preferred feeding sites in fruiting structures. *L. hirsutum* LA1777 is also highly preferred by *H. armigera* females. Juvik *et al.* (1988) suggest this preference for an unsuitable hostplant may indicate that the oviposition stimulant (possibly α-santalene) is common to other hostplants of *H. armigera*.

Although there are now several examples of attractant or stimulatory compounds or fractions identified from particular crops, individually none of these compounds are as active as whole-plant extracts in stimulating oviposition. This suggests that female *Heliothis* respond to a complex of chemical stimuli during hostplant recognition and acceptance, in addition to any physical stimuli.

The sensory modalities involved in the perception of these allelochemicals by various *Heliothis* species have not been fully determined and deserve much more attention. Ramaswamy *et al.* (1987) used ablation techniques to show that surface chemicals were perceived largely by tarsal chemoreceptors in *H. virescens*. Using only these receptors females were able to discriminate between cotton (a host) and groundcherry (*Physalis*, a non-host). Juvik *et al.* (1988) used simple bioassays (unfortunately lacking airflow) to determine whether olfaction or contact chemoreception was involved in the perception of oviposition stimulants in ethanolic extracts of tomato. They suggest that volatile sesquiterpenes may be involved in olfactory orientation to tomato since in the absence of any direct contact they observed a 1.5-fold increase (relative to hexane-treated controls) in the number of eggs deposited over discs impregnated with the extract from the most attractive cultivar. However, this was much less than the 13.5-fold difference when tarsal and ovipositor contact was possible, suggesting that much of the effect was due to non-volatile chemicals rather than to volatiles as claimed.

(d) Mechanoreception and other physical cues

The substrate surface texture is an important cue in host recognition and acceptance. Many authors have observed that villous or rugous surfaces receive more eggs than smooth surfaces (Cullen, 1969; Lukefahr et al., 1971; Adjei-Maafo and Wilson, 1983b; Porter, 1984; Ramaswamy et al., 1987). Cullen (1969) considered that surface texture, rather than chemistry or the presence of food, was most important in the selection of oviposition sites within the plant by H. punctigera and he was able to demonstrate the importance of tarsal mechanoreceptors in this discrimination. By contrast, Ramaswamy et al. (1987) suggest receptors on the ovipositor may be involved.

The preference for hairy surfaces has been used to some advantage by plant breeders, most notably in cotton (Thomson, 1988). Glabrous genotypes of cotton attract fewer eggs than hirsute varieties (Lukefahr et al., 1971; Fitt, 1987), with egg densities often linearly related to the density of trichomes (Robinson et al., 1980). Fitt (1987) showed that female H. punctigera had a briefer residence time on glabrous cotton plants than on hirsute varieties. One explanation for the preference for hairy surfaces is that eggs are dislodged more easily by rainfall from glabrous plants (Porter, 1984), though a number of other characteristics of hirsute plant surfaces (lower wind surface velocities, thicker boundary layer, less extreme variation in surface temperature) that influence egg and larval survival may have been involved in the evolution of the preference (Ramaswamy, 1988 and references therein).

Other physical features that may be perceived visually include plant outlines, contrast with background and height relative to surrounding plants. Mabbett and Nachapong (1984) observed that cotton plants which were noticeably taller than their neighbours received a disproportionately high number of eggs. Similarly, Firempong and Zalucki (1990b) observed that prostrate plants (cabbage, pigface) of low acceptability became more acceptable to female H. armigera when elevated to the same height as acceptable plants (maize, sunflower). These observations imply a role for visual cues in host recognition.

(e) The role of nectar availability in host selection

Most adult Lepidoptera, including *Heliothis*, require a source of nectar to provide energy for flight and for maximal longevity and fecundity (Adjei-Maafo and Wilson, 1983a; Willers et al., 1987). Adult *Heliothis* seek sweet secretions within minutes of emergence (Lingren *et al.*, 1988) and in some situations mature females may move nightly between sources of nectar and oviposition sites (Topper, 1987a). Topper (1981) and Adjei-Maafo and Wilson (1983a,b) found close correlations between peaks of oviposition in cotton (by *H. armigera* in the Sudan and *H. punctigera* in Australia, respectively) and peaks of nectar production. Both studies found most eggs were deposited near sites on the plant where nectar production was greatest, which in turn suggests a close correspondence between feeding and egg-laying activities. The relationship between nectar production and oviposition probably arises largely from the influence of nectar availability on daily fecundity of moths, but both Topper (1981) and Adjei-Maafo and Wilson (1983a,b) suggest that the presence of nectar may itself be a stimulus eliciting oviposition since nectariless cottons typically receive fewer eggs than nectaried genotypes (Thomson, 1988). The presence of nectar *per se* cannot act as a long-distance cue for host-finding or recognition, and could influence behaviour only after the moths have colonized a patch. However, it also seems unlikely that nectar is an oviposition stimulant. Cullen (1969) and Ramaswamy (1988) observed quite distinct patterns of movement associated with feeding and oviposition behaviours. They found that while bouts of each may be interspersed throughout the active period, they did not occur simultaneously. Wardhaugh *et al.* (1980) also comment that nectar availability may act as an arrestant leading to more sedentary behaviour of colonists or newly emerged moths. Since peaks of nectar production in cotton coincide with peak production of fruiting forms (Adjei-Maafo and Wilson, 1983b), and with the release of volatiles (Hedin, 1976), it is quite possible that these factors are more important in host recognition, though the presence of nectar may influence acceptance. Moreover *Heliothis* adults obtain nectar from a wide range of flowering plants, many of which are not used for oviposition (Callahan, 1958; Hendrix *et al.*, 1987). For example, in southern Texas newly emerged *H. zea* and *H. virescens* feed avidly at night on nectar at flowers of the genus *Gaura* spp. of the evening primrose family, before moving into adjacent crops (J. R. Raulston and P. D. Lingren, personal communication). *Gaura* spp. release volatiles which are attractive to both sexes (P. D. Lingren, J. R. Raulston and T. Shaver, personal communication), but eggs are not deposited on the plants. In cropping situations females may also feed in one crop and oviposit in another (e.g. Topper, 1987a).

Other factors known to influence the attractiveness of plants to ovipositing females include nutrition and water status of the plant (Slosser, 1983), though in both cases the effect may be mediated through the production of volatiles or nectar, which would tend to make crops more apparent over a distance, or hold larger numbers of recruits once they had arrived.

7.6 CONCLUSIONS

The host selection strategy adopted by polyphagous *Heliothis* seems well attuned to the exploitation of a succession of patchily distributed ephemeral host plants (Farrow and McDonald, 1988; Fitt, 1989). Females have a strong preference for flowering plants and respond to a number of hostplant features during host selection (volatiles, leaf surface chemicals, nectar, surface texture, plant height), some of which are host specific, others not. The preference for flowering stages results in oviposition being concentrated on those plants or stages most suitable for larval growth, and within-plants the placement of eggs on or near to fruiting structures is conducive to larval establishment. However, acceptance of plants for oviposition extends beyond the normal host range, resulting in some oviposition on flowering non-hosts. These observations and the work of Ramaswamy *et al.* (1987) suggest that early stages in the catenary process of host selection may be based on generalized cues associated with flowers, and that discrimination between different plants occurs only after close approach or contact.

Considering their great economic importance, our understanding of the details of *Heliothis* host selection is surprisingly poor and research on their sensory physiology remains rudimentary. Not surprisingly most research has focused on cues operating at close range or after the plant has been contacted, i.e. after the insect has located a host patch. For example, a number of specific plant chemicals have been identified that are attractive to females or stimulate oviposition, though often laboratory assays do not allow behavioural modalities to be distinguished. Little is known about the location of host patches and the involvement of long-distance cues remains unclear. A possible scenario is that randomly searching individuals are attracted to acceptable crops over distances of a few metres and then arrested within the crop. The effect is that over time aggregations of ovipositing females appear in the most acceptable and 'arresting' host crops.

The management of *Heliothis* in agro-ecosystems currently relies heavily on pesticides (Fitt, 1989; Hearn and Fitt, 1991). A better understanding of host selection in the pest species of *Heliothis* will be necessary for the successful implementation of alternative management systems that employ hostplant resistance, trap crops, or the use of behaviour-modifying chemicals (deterrents or repellants) (Rice, 1985; Kennedy *et al.*, 1987).

References

Adjei-Maafo, I. K. and Wilson, L. T. (1983a) Association of cotton nectar production with *Heliothis punctigera* (Lepidoptera: Noctuidae) oviposition. *Environ. Entomol.*, **12**, 1166–1170.

Adjei-Maafo, I. K. and Wilson, L. T. (1983b) Factors affecting the relative abundance of arthropods on nectaried and nectariless cotton. *Environ. Entomol.*, **12**, 349–352.

Agee, H. R. (1973) Spectral sensitivity of the compound eyes of field collected adult bollworms and tobacco budworms. *Ann. Entomol. Soc. Amer.*, **66**, 613–615.

Alvarado-Rodriguez, B., Leigh, T. F. and Lange, W. H. (1982) Oviposition site preference by the tomato fruitworm (Lepidoptera: Noctuidae) on tomato, with notes on plant phenology. *J. Econ. Entomol.*, **75**, 895–898.

Barber, G. W. (1938) The concentration of *Heliothis obsoleta* moths at food. *Entomol. News*, **49**, 256–258.

Barton-Browne, L. (1977) Host related responses and their suppression: some behavioral considerations. In *Chemical Control of Insect Behaviour. Theory and Application* (eds H. H. Shorey and J. J. McKelvey), John Wiley & Sons, New York, 117–127.

Bilapate, G. G. (1984) *Heliothis* complex in India—a review. *Agric. Rev.*, **5**, 13–26.

Brantjes, N. (1976) Prevention of super parasitism of *Melandrium* flowers by *Hadena*. *Oecologia*, **24**, 1–6.

Buttery, R. G., Ling, L. C. and Teranishi, R. (1980) Volatiles of corn tassels: possible corn earworm attractants. *J. Agric. Food Chem.*, **28**, 771–774.

Callahan, P. S. (1957) Oviposition response of imago of the corn earworm *Heliothis zea* (Boddie) to various wavelengths of light. *Ann. Entomol. Soc. Amer.*, **50**, 444–452.

Callahan, P. S. (1958) Behaviour of the imago of the corn earworm, *Heliothis zea* (Boddie), with special reference to emergence and reproduction. *Ann. Entomol. Soc. Amer.*, **51**, 271–283.

Cantelo, W. W. and Jacobsen, M. (1979) Corn silk volatiles attract many species of moths. *J. Environ. Sci. Health*, **14**, 695–709.

Carde, R. T. (1984) Chemo-orientation in flying insects. In *Chemical Ecology of Insects* (eds W. S. Bell and R. T. Carde), Chapman and Hall, London, 111–126.

Courtney, S. (1981) Coevolution of pierid butterflies and their cruciferous food plants. III. *Anthocaris cardamines* (L) survival, development and oviposition on different hostplants. *Oecologia*, **51**, 91–96.

Cullen, J. M. (1969) The reproduction and survival of *Heliothis punctigera* in South Australia. Ph.D. Thesis, University of Adelaide, Australia.

Daly, J. C. and Gregg, P. C. (1985) Genetic variation in *Heliothis* in Australia: species identification and gene flow in two pest species *H. armigera* (Hubner) and *H. punctigera* Wallengren (Lepidoptera: Noctuidae). *Bull. Entomol. Res.*, **75**, 169–184.

Drake, V. A. and Farrow, R. A. (1985) A radar and aerial trapping study of an early spring migration of moths (Lepidoptera) in inland New South Wales. *Austr. J. Ecol.*, **11**, 223–235.

Drake, V. A. and Farrow, R. A. (1988) The influence of atmospheric structure and motions on insect migration. *Ann. Rev. Entomol.*, **33**, 183–210.

Duffey, S. S. (1986) Plant glandular trichomes: their partial role in defence against insects. In *Insects and the Plant Surface* (eds B. Juniper and T. R. E. Southwood), Edward Arnold, London, 151–172.

Farrer, R. R. and Bradley, J. R. (1985a) Within-plant distribution of *Heliothis* spp. (Lepidoptera: Noctuidae) eggs and larvae on cotton in North Carolina. *Environ. Entomol.*, **14**, 205–209.

Farrer, R. R. and Bradley, J. R. (1985b) Effects of within-plant distribution of *Heliothis zea* (Boddie) (Lepidoptera: Noctuidae) eggs and larvae on larva development and survival on cotton. *J. Econ. Entomol.*, **78**, 1233–1237.

Farrer, R. R. and Kennedy, G. G. (1987) Growth, food consumption and mortality of *Heliothis zea* larvae on foliage of the wild tomato *Lycopersicon hirsutum* f. *glabratum* and the cultivated tomato, *L. esculentum*. *Entomol. Exp. Appl.*, **44**, 213–219.

Farrow, R. A. and Daly, J. C. (1987) Long-range movements as an adaptive strategy in the genus *Heliothis* (Lepidoptera: Noctuidae): a review of its occurrence and detection in four pest species. *Aust. J. Zool.*, **35**, 1–24.

Farrow, R. A. and McDonald, G. (1988) Migration strategies and outbreaks of noctuid pests in Australia. *Insect Sci. Appl.*, **8**, 531–542.

Finch, S. (1986) Assessing host-plant finding by insects. In *Insect–Plant Interactions* (eds J. R. Miller and T. A. Miller), Springer-Verlag, New York, 23–64.

Firempong, S. K. (1987) Some factors affecting host plant selection by *Heliothis armigera* (Hübner) (Lepidoptera: Noctuidae). Ph.D. Thesis, University of Queensland, Brisbane, Australia.

Firempong, S. K. and Zalucki, M. P. (1990a) Host plant preferences of populations of in *Heliothis armigera* (Hubner) (Lepidoptera: Noctuidae) from different geographic locations. *Austr. J. Zool.*, **37**, 665–73.

Firempong, S. K. and Zalucki, M. P. (1990b) Host plant selection by *Heliothis armigera* (Hübner) (Lepidoptera: Noctuidae); role of certain plant attributes. *Austr. J. Zool.*, **37**, 675–83.

Fitt, G. P. (1986) The influence of a shortage of hosts on the specificity of oviposition behaviour in species of *Dacus* (Diptera, Tephritidae). *Physiol. Entomol.*, **11**, 133–143.

Fitt, G. P. (1987) Ovipositional responses of *Heliothis* spp. to host plant variation in cotton (*Gossypium hirsutum*). In *Insects–Plants, Proceedings of the 6th International Symposium on Insect–Plant Relationships*, Pau, 1986 (eds V. Labeyrie, G. Fabres and D. Lachaise), Dr W. Junk, Dordrecht, 289–294.

Fitt, G. P. (1989) The ecology of *Heliothis* species in relation to agroecosystems. *Ann. Rev. Entomol.*, **34**, 17–52.

Fitt, G. P. and Boyan, G. S. (1990) Methods for studying behaviour. In *Heliothis: Research Methods and Prospects* (M. P. Zalucki ed.), Springer-Verlag, New York (in press).

Fitt, G. P., Zalucki, M. P. and Twine, P. (1989) Temporal and spatial patterns in pheromone trap catches of *Helicoverpa* spp. in cotton growing areas of Australia. *Bull. Entomol. Res.*, **79**, 145–162.

Flath, R. A., Forreg, R. R., John, J. O. and Chan, B. G. (1978) Volatile components of corn silk (*Zea mays*, L.): possible *Heliothis zea* (Boddie) attractants. *J. Agric. Food Chem.*, **26**, 1290–1293.

Fox, L. R. and Morrow, P. A. (1981) Specialisation: species property or local phenomenon. *Science*, **211**, 887–893.

Gatehouse, A. G. (1988) Migration and low population density in armyworm (Lepidoptera: Noctuidae) life histories. *Insect Sci. Appl.*, **8**, 1–9.

Gross, H. R. (1984) *Spodoptera frugiperda* (Lepidoptera: Noctuidae): deterrence of oviposition by aqueous homogenates of fall armyworm and corn earworm larvae applied to whorl-stage corn. *Environ. Entomol.*, **13**, 1498–1501.

Gross, H. R. and Young, J. R. (1977) Comparative development and fecundity of corn earworm reared on selected wild and cultivated early-season hosts common to the Southeastern USA. *Ann. Entomol. Soc. Amer.*, **70**, 63–65.

Gross, H. R., Carpenter, J. E. and Sparks, A. N. (1983) Visual acuity of *Heliothis zea* (Lepidoptera: Noctuidae) males as a factor influencing the efficiency of pheromone traps. *Environ. Entomol.*, **12**, 844–847.

Hackett, D. and Gatehouse, A. G. (1982) Diapause in *Heliothis armigera* (Hübner) and *H. fletcheri* (Hardwick) (Lepidoptera: Noctuidae) in the Sudan Gezira. *Bull. Entomol. Res.*, **72**, 409–422.

Haggis, M. J. (1981) Spatial and temporal changes in the distribution of eggs by *Heliothis armigera* (Hübner) (Lepidoptera: Noctuidae) on cotton in the Sudan Gezira. *Bull. Entomol. Res.*, **71**, 183–193.

Hardwick. D. F. (1965) The corn earworm complex. *Mem. Entomol. Soc. Can.*, **40**, 247 pp.

Hearn, A. G. and Fitt, G. P. (1991) Cotton cropping systems. In *Field-Crop Ecosystems* (ed. C. Pearson), Elsevier, Amsterdam.

Hedin, P. A. (1976) Seasonal variations in the emission of volatiles by cotton plants growing in the field. *Environ. Entomol.*, **5**, 1234–1238.

Hendrix, W. H., Mueller, T. F., Phillips, J. R. and Davis, O. K. (1987) Pollen as an indicator of long-distance movement in the bollworm, *Heliothis zea* (Lepidoptera: Noctuidae). *Environ. Entomol.*, **16**, 1148–1151.

Isely, D. (1935) Relation of hosts to abundance of cotton bollworm. *Arkansas Agric. Exp. Stn. Bull.*, **320**, 30 pp.

Jackson, D. M., Severnson, R. F., Johnson, A. W. *et al.* (1984) Ovipositional response of tobacco budworm moths (Lepidoptera: Noctuidae) to cuticular chemical isolates from green tobacco leaves. *Environ. Entomol.*, **13**, 1023–1039.

Jermy, T., Hanson, F. E. and Dethier, V. G. (1968) Induction of specific food preference in lepidopterous larvae. *Entomol. Exp. Appl.*, **11**, 211–230.

Johnson, M. W., Stinner, R. E. and Rabb, R. L. (1975) Ovipositional response of *Heliothis zea* (Boddie) to its major hosts in North Carolina. *Environ. Entomol.*, **4**, 291–297.

Johnson, S. J., King, E. G. and Bradley, J. R. (eds) (1986) *Theory and Tactics of Heliothis Population Management. 1. Cultural and Biological Control. Southern Co-operative Series Bull.*, **316**. Oklahoma State University, Stillwater, Oklahoma.

Jones, R. L., Burton, R. L., Bowman, M. C. and Beroza, M. (1970) Chemical inducers for oviposition for the corn earworm, *Heliothis zea*. *Science*, **186**, 856–857.

Juvik, J. A., Babka, B. A. and Timmermann, E. A. (1988) Influence of trichome exudates from species of *Lycopersicon* on oviposition behaviour of *Heliothis zea* (Boddie). *J. Chem. Ecol.*, **14**, 1261–1278.

Kennedy, J. S. (1986) Some current issues in orientation to odour sources. In *Mechanisms of Insect Olfaction* (eds T. L. Payne, M. C. Birch and C. E. J. Kennedy), Clarendon Press, London, 11–26.

Kennedy, G. G., Gould, F., Deponti, O. M. and Stinner, R. E. (1987) Ecological, agricultural, genetic, and commercial considerations in the deployment of insect-resistant germplasm. *Environ. Entomol.*, **16**, 327–338.

Laster, M. L., Pair, S. D. and Martin, D. R. (1982) Acceptance and development of *Heliothis subflexa* and *H. virescens* (Lepidoptera: Noctuidae), and their hybrid and backcross progeny on several plant species. *Environ. Entomol.*, **11**, 979–980.

Lingren, P. D., Sparks, A. N. and Raulston, J. R. (1982) The potential contribution of moth behaviour research to *Heliothis* management. In *International Workshop on* Heliothis *Management Proceedings*, Patancheru, India, 1981 (eds W. Reed and V. Kumble), ICRISAT, India, 39–47.

Lingren, P. D., Warner, W. B., Raulston, J. R. *et al.* (1988) Observations on the emergence of adults from natural populations of corn earworm, *Heliothis zea* (Boddie) (Lepidoptera: Noctuidae). *Environ. Entomol.*, **17**, 254–258.

Lukefahr, M. J. Houghtaling, T. E. and Graham, H. M. (1971) Suppression of *Heliothis* populations with glabrous cotton strains. *J. Econ. Entomol.*, **64**, 486–488.

Mabbett, T. H. and Nachapong, M. (1984) Within-plant distribution of *Heliothis armigera* eggs on cotton in Thailand. *Trop. Pest Manage.*, **30**, 367–371.

Martin, P. B., Lingren, P. D. and Greene, G. L. (1976) Relative abundance and host preferences of cabbage looper, soybean looper, tobacco budworm, and corn earworm on crops grown in Northern Florida. *Environ. Entomol.*, **5**, 878–882.

Matthews, M. (1987) The classification of the Heliothinae (Noctuidae). Ph.D. Thesis, British Museum and Kings College, London, 253 pp.

Miller, J. R., and Strickler, K. L. (1984) Finding and accepting host plants. In *Chemical Ecology of Insects* (eds W. J. Bell and R. T. Carde), Chapman and Hall, London, 127–158.

Mitchell, E. R. and Heath, R. R. (1987) *Heliothis subflexa* (GN) (Lepidoptera: Noctuidae): demonstration of oviposition stimulant from ground cherry using novel bioassay. *J. Chem. Ecol.*, **13**, 1849–1858.

Mitter, C. and Schneider, J. C. (1987) Genetic change and insect outbreaks. In *Insect Outbreaks* (eds P. Barbosa and J. C. Schultz), Academic Press, New York, 505–532.

Morton, R., Tuart, L. D. and Wardhaugh, K. G. (1981) The analysis and standardisation of light trap catches of *Heliothis armigera* (Hubner) and *H. punctiger* Wallengren (Lepidoptera: Noctuidae). *Bull. Entomol. Res.*, **71**, 207–225.

Murlis, J. (1986) The structure of odour plumes. In *Mechanisms of Insect Olfaction* (eds T. L. Payne, M. C. Birch and C. E. J. Kennedy), Clarendon Press, London, 27–38.

Nadgauda, D. and Pitre, H. (1983) Development, fecundity, and longevity of the tobacco budworm (Lepidoptera: Noctuidae) fed soybean, cotton, and artificial diet at three temperatures. *Environ. Entomol.*, **12**, 582–86.

Nuenzig, H. H. (1969) The biology of the tobacco budworm and the corn earworm in North Carolina with particular reference to tobacco as a host. *Nth Carolina State Univ. Agric. Exp. Stn Tech. Bull.*, **196**, 76 pp.

Papaj, D. and Prokopy, R. J. (1989) Ecological and evolutionary aspects of learning in phytophagous insects. *Ann. Rev. Entomol.*, **34**, 315–350.

Papaj, D. and Rausher, M. (1983) Individual variation in host location by phyto-phagous insects. In *Herbivorous Insects: Host-seeking Behaviours and Mechanisms* (ed. S. Ahmed), Academic Press, New York, 77–124.

Parsons, F. S. (1940) Investigations on the cotton bollworm, *Heliothis armigera*, Hübn. Part III. Relationships between oviposition and the flowering curves of food-plants. *Bull. Entomol. Res.*, **31**, 147–177.

Pearson, E. O. (1958) *The Insect Pests of Cotton in Tropical Africa*. Commonwealth Institute of Entomology, London.

Porter, R. P. (1984) Role of host plant trichomes and tarsal sensilla in oviposition by *Heliothis virescens*. M. S. Thesis, Mississippi State University, Mississippi, USA.

Prokopy, R. J. (1981) Epideictic pheromones that influence spacing patterns of phytophagous insects. In *Semiochemicals: Their Role in Pest Control* (eds D. A. Nordlund, R. L. Jones and W. J. Lewis), Wiley, New York, 181–223.

Prokopy, R. J. and Owens, E. D. (1983) Visual detection of plants by herbivorous insects. *Ann. Rev. Entomol.*, **28**, 337–364.

Ramaswamy, S. B. (1988) Host finding by moths: sensory modalities and behaviours. *J. Insect Physiol.*, **34**, 235–249.

Ramaswamy, S. B., Ma, W. K. and Baker, G. T. (1987) Sensory cues and receptors for oviposition by *Heliothis virescens*. *Entomol. Exp. Appl.*, **43**, 159–168.

Reed, W. and Kumble, V. (eds) (1982) *Proceedings of the International Workshop on Heliothis Management*, 15–20 November 1981, ICRISAT Center, Patancheru, India.

Renwick, J. A. and Radke, C. D. (1980) An oviposition deterrent associated with frass from feeding larvae of the cabbage looper, *Trichoplusia ni* (Lepidoptera: Noctuidae). *Environ. Entomol.*, **9**, 318–320.

Rice, M. (1985) Semiochemicals and sensory manipulation strategies for be-havioural management of *Heliothis* spp. In Heliothis *Ecology Workshop 1985 Proceedings* (eds M. P. Zalucki and P. H. Twine), Queensland Dept of Primary Industries, Brisbane, 27–46.

Robinson, S. H., Wolfenbarger, D. A. and Dilday, R. H. (1980) Antixenosis of smooth leaf cotton to the ovipositional response of tobacco budworm. *Crop Sci.*, **20**, 646–649.

Roome, R. E. (1975) Activity of adult *Heliothis armigera* (Hb) (Lepidoptera: Noctuidae) with reference to the flowering of sorghum and maize in Botswana. *Bull. Entomol. Res.*, **65**, 523–530.

Rummel, D. R., Leser, J. F., Slosser J. E. *et al.* (1986) Cultural control of *Heliothis* spp. in southwestern US cropping systems. In *Theory and Tactics of* Heliothis *Population Management. 1. Cultural and Biological Control* (eds S. J. Johnson, E. G. King and J. R. Bradley), *Southern Cooperative Series Bull.*, **316**, 38–53. Oklahoma State University, Stillwater, Oklahoma.

Schneider, J. C. and Roush, R. T. (1986) Genetic differences in oviposition preference between two populations of *Heliothis virescens*. In *Evolutionary Genetics of Invertebrate Behaviour: Progress and Prospects* (ed. M. D. Huettel), Plenum Press, New York, 163–171.

Schneider, J. C., Benedict, J. H., Gould, F. *et al.* (1986) Interaction of *Heliothis* with its host plants. In *Theory and Tactics of* Heliothis *Population Management. 1. Cultural and Biological Control* (eds S. J. Johnson, E. G. King and J. R. Bradley), *Southern Cooperative Series Bull.*, **316**, 3–21.

Singer, M. C. (1986) The definition and measurement of oviposition preference. In *Insect–Plant Interactions* (eds J. R. Miller and T. A. Miller), Springer-Verlag, New York, 65–94.

Slosser, J. E. (1983) Potential of *Heliothis* spp. (Lepidoptera: Noctuidae)-resistant cottons in limited irrigation situations. *J. Econ. Entomol.*, **76**, 864–868.

Stadelbacher, E. A., Graham, H. M., Harris, V. E. *et al.* (1986) *Heliothis* populations and wild host plants in the southern US. In *Theory and Tactics of* Heliothis *Population Management. 1. Cultural and Biological Control* (eds S. J. Johnson E. G. King and J. R. Bradley), *Southern Cooperative Series Bull.*, **316**, 54–74.

Stanton, M. L. (1983) Spatial patterns in the plant community and their effects upon insect search. In *Herbivorous Insects. Host-seeking Behaviour and Mechanisms* (ed. S. Ahmed), Academic Press, New York, 125–160.

Stinner, R. E. (1979) Biological monitoring essentials in studying wide-area moth movement. In *Movement of Highly Mobile Insects: Concepts and Methodology* (eds R. L. Rabb, and G. G. Kennedy), University Graphics, Raleigh, NC, 199–211.

Stinner, R. E., Rabb, R. L. and Bradley, J. R. (1974) Population dynamics of *Heliothis zea* (Boddie) and *H. virescens* (F.) in North Carolina: a simulation model. *Environ. Entomol.*, **3**, 163–168.

Stinner, R. E., Rabb, R. L. and Bradley, J. R. (1977) Natural factors operating in the population dynamics of *Heliothis zea* in North Carolina. *Proceedings of the XVth International Congress on Entomology*, Washington, 622–642.

Stinner, R. E., Regniere, J. and Wilson K. (1982) Differential effects of agroecosystem structure on dynamics of three soybean herbivores. *Environ. Entomol.*, **11**, 538–543.

Thomas, F. L. and Dunham, E. W. (1931) Factors influencing infestation in cotton by *Heliothis obsoleta*. *J. Econ. Entomol.*, **24**, 815–821.

Thomson, N. J. (1988) Host plant resistance in cotton. *Austr. J. Agric. Sci.*, **53**, 261–270.

Tingle, F. C. and Mitchell, E. R. (1984) Aqueous extracts from indigenous plants as oviposition deterrents for *Heliothis virescens* (F.). *J. Chem. Ecol.*, **10**, 101–113.

Topper, C. P. (1981) The behaviour and population dynamics of *Heliothis armigera* (Hb) (Lepidoptera: Noctuidae) in the Sudan Gezira. Ph.D. Thesis, Cranfield Institute of Technology, Cranfield, UK.

Topper, C. P. (1987a) The dynamics of the adult population of *Heliothis armigera* (Hübner) (Lepidoptera: Noctuidae) within the Sudan Gezira in relation to cropping pattern and pest control on cotton. *Bull. Entomol. Res.*, **77**, 525–539.

Topper, C. P. (1987b) Nocturnal behaviour of adults of *Heliothis armigera* (Hübner) (Lepidoptera: Noctuidae) in the Sudan Gezira and pest control implications. *Bull. Entomol. Res.*, **77**, 541–554.

Traynier, R. M. M. (1986) Visual learning of sinigrin solution as an oviposition releaser for the cabbage butterfly, *Pieris rapae*. *Entomol. Exp. Appl.*, **40**, 25–33.

Visser, J. H. and Ave, D. A. (1978) General green leaf volatiles in the olfactory orientation of the Colorado beetle, *Leptinotarsa decemlineata*. *Entomol. Exp. Appl.*, **24**, 538–549.

Waldvogel, M. and Gould, F. (1990) Genetic variation in oviposition preference in

Heliothis virescens in relation to macroevolutionary patterns of Heliothine host range. *Evolution*, **44**, 1326–1337.

Wardhaugh, K. G., Room, P. M. and Greenup, L. R. (1980) The incidence of *Heliothis armigera* (Hübner) and *H. punctigera* Wallengren (Lepidoptera: Noctuidae) on cotton and other host plants in the Namoi Valley of New South Wales. *Bull. Entomol. Res.*, **70**, 113–131.

Weisenborn, W. D. and Trumble, J. T. (1988) Optimal oviposition by the corn earworm, (Lepidoptera: Noctuidae) on whorl-stage sweet corn. *Environ. Entomol.*, **17**, 722–726.

Wiklund, C. (1975) The evolutionary relationship between adult oviposition preferences and larval host plant range in *Papilio machaon*. *Oecologia*, **18**, 185–197.

Willers, J. L., Schneider, J. C. and Ramaswamy, S. B. (1987) Fecundity, longevity and caloric patterns in female *Heliothis virescens*: changes with age due to flight and supplemental carbohydrate. *J. Insect Physiol.*, **33**, 803–808.

Wilson, L. T. and Room, P. M. (1983) Clumping patterns of fruit and arthropods in cotton, with implications for binomial sampling. *Environ. Entomol.*, **12**, 50–54.

Wilson, L. T., Gutierrez, A. P. and Leigh, T. F. (1983) Within-plant distribution of immatures of *Heliothis zea* on cotton. *Hilgardia*, **48**, 12–23.

Wolf, W. W., Sparks, A. N., Pair, S. D. and Westbrook, J. K. (1986) Radar observations and collections of insects in the Gulf of Mexico. In *Insect Flight: Dispersal and Migration* (ed. W. Danthanarayana), Springer-Verlag, Heidelberg, 231–234.

Zalucki, M. P., Daglish, G., Firempong, S. and Twine, P. (1986) The biology and ecology of *Heliothis armigera* (Hübner) and *H. punctigera* Wallengren (Lepidoptera: Noctuidae) in Australia: what do we know? *Austr. J. Zool.*, **34**, 779–814.

8

Reproduction and host selection by aphids: the importance of 'rendezvous' hosts

Seamus A. Ward

8.1 INTRODUCTION

Most species of aphids (Homoptera: Aphididae) occur on plants of only a few species, often within a single genus (Eastop, 1973). This extreme specificity is costly, since it means that the vast majority of migrating aphids die before they can find an acceptable host. [Taylor (1977) estimates that perhaps fewer than one migrant in a thousand finds a host.] In seeking to explain how specificity has evolved despite this cost, we might invoke two main processes: adaptation and speciation. Very different approaches are taken by students of these two processes.

In the functional approach to the study of adaptation, it is assumed that the feature of interest (in this case, host range) has been optimized by natural selection, and reached an evolutionary equilibrium. Attention is then focused on the ecological or plant-physiological factors that cause natural selection to maintain the observed level of host specificity.

An important feature of this approach is the simplifying assumption that the evolution of host range depends on the relative fitnesses of aphids on the various hosts, but that these fitnesses are constant, i.e. the alleles determining the adaptations to the host have already been fixed by selection. In fact, however, these characters also evolve. This means that aphid populations may show considerable genetic variation in both host preference and adaptations to the various hosts. If these characters are genetically correlated (i.e. individuals prefer those host species to which they are best adapted) and mating is assortative with respect to host preference, then polyphagous species may split into 'host races', and eventually speciate (Müller, 1985). If this process has been important, then host specificity is a means of avoiding not only unsuitable hosts, but also genetically unsuitable mates.

In this chapter I shall consider the roles of adaptation and speciation in the evolution of host specificity in aphids. I shall first discuss various suggested functional explanations, and attempt to clarify the relations between the aphids' complex life cycles and the selection for specificity. I shall then examine the evidence concerning host races and sympatric speciation.

8.2 FUNCTIONAL EXPLANATIONS OF HOST SPECIFICITY

8.2.1 Optimization: the decision and the currency

Functional explanations of host-specificity include the implicit assumption that the observed behaviour is an optimal strategy. Any such explanation must contain statements concerning:

1. the nature of the decision, i.e. the set of possible behaviours from which the optimal strategy is to be chosen;
2. the 'currency' to be maximized by the 'optimal' strategy; and
3. the constraints under which the decision is to be optimized (Oster and Wilson, 1978; Lewontin, 1987; Stephens and Krebs, 1986).

The host range of a species of aphids is a consequence of the decisions taken by all individual migrants. Since an aphid can encounter only one plant at a time, this decision is best expressed as 'acceptance' (settling to reproduce) or 'rejection' (taking off again to seek another host) rather than as a true preference. In this chapter 'preference' means that one host is more often, or more readily, accepted than another.

In what follows I shall assume that the currency to be maximized is the arithmetic mean fitness of each generation's migrants. If p_i is a migrant's probability of settling on a host of species i, and w_i is the number of migrating descendents produced per colonist on host i, then the mean fitness is $\Sigma p_i w_i$.

8.2.2 Constraints: failure to find a host

Clearly, an ideal aphid would fly straight from its source plant to a host of the species on which its reproductive success is greatest, but real aphids cannot do this. To predict the optimal behaviour we must take into account the constraints on a migrating aphid.

Aphids are small phloem-feeding homopterans. Weak flyers, they have little control over their direction of flight unless the wind speed is less than 1–2 m/s (Dixon, 1985), so once airborne they cannot easily choose on which plant to land until the wind has carried them into its immediate vicinity. Here, they are attracted downward by yellow–green or grey–green colours (Moericke, 1955) rather than by any host-specific cues. Thus, although they alight more often on plants than elsewhere, aphids land at random on hosts and non-hosts alike (Kennedy *et al.*, 1959a, b). Host specificity, therefore, results from selective re-emigration from non-hosts.

The second major constraint is that the time available for a migrant to find a host is limited. Since phloem is a very rich source of sugars, aphids have little need for large energy stores; instead, most of the abdomen is filled with developing embryos. While this results in a high potential fecundity (Leather and Wellings, 1981; Ward *et al.*, 1983a) it also means that aphids can survive for only a short time before exhausting their reserves (Ward and Dixon, 1982, 1984; Ward *et al.*, 1983b). Starvation, and probably also desiccation, thus limit the time available for host selection.

8.2.3 The optimal strategy

I have shown elsewhere (Ward, 1987a) that if preference results from selective re-emigration and the time available to a migrant is limited, then the optimal behaviour involves a series of 'discrimination phases'. Hosts should be ranked according to their suitability; and each host except the ideal one should be rejected if encountered during an initial discrimination phase, but accepted thereafter. Inferior hosts should be readily accepted if the preferred host is rare (see also Levins and MacArthur, 1969; Doyle, 1978; Jaenike, 1978).

Although the model contains terms representing variables that are difficult to measure (e.g. the rate of searching), it can still be used to derive simple criteria for assessing functional explanations of host specificity (S. A. Ward, in press). The most relevant criterion is as follows. An aphid specializing on host HA, where it has fitness w_a, should colonize host HB *immediately* (i.e. show no preference for HA over HB) if w_b, its fitness on HB, satisfies

$$w_b/w_a > P_a(2 - P_a)/[2 - P_a(2 - P_a)] \qquad (1)$$

where P_a is the proportion of migrating specialists that eventually find a host of species HA, and A and B are the densities of hosts HA and HB, respectively. (This approximation is accurate provided not more than a few per cent of migrants find hosts within the time available.) For example, consider a monophagous aphid searching in an area that also contains an equally abundant non-host. If 1% of the migrants find hosts, then natural selection will favour alleles for acceptance of the non-host if $w_b/w_a > 0.01005$ (from Equation (1)): the aphids should readily settle on any plant on which their fitness is more than about 1% of that on their preferred host. (This threshold ratio is insensitive to changes in B, the abundance of the non-preferred host, so unless the two hosts' densities differ by orders of magnitude it is not important to quantify B precisely.)

In fact, 1% is probably an overestimate of the success rate of aphids searching for hosts. Taylor (1977) suggests that perhaps fewer than one in a thousand finds an acceptable host; and field data on *Rhopalosiphum padi* yield an estimate of approximately 0.2% (S. A. Ward, S. R. Leather and L. A. O. Turl, in preparation). Even if we assume a 1% success rate, however, any functional explanations of host specificity must somehow account for a 99% reduction in the fitness of aphids which colonize the 'wrong' host.

8.2.4 Sensory constraints: aphids as botanists

Before asking whether the various suggested selection pressures are strong enough to account for host specificity, we must deal with a third possible constraint: the limits of aphids' abilities to distinguish between suitable and unsuitable hosts. In other words, we must make inferences or assumptions about what the possible host ranges are.

The various cues used by aphids to recognize their hosts have been well reviewed recently (e.g. Dixon, 1985; Risebrow and Dixon, 1987; Klingauf, 1988), so I shall describe them only briefly here.

The decision to alight on a plant depends mainly on fairly unspecific visual cues: the plant's colour (Moericke, 1955), shape or size (Åhman *et al.*, 1985). Several authors have also demonstrated that aphids can respond to olfactory stimuli (e.g. Petterson, 1970), but it is not yet clear that these are important in *attracting* aphids to hosts in the field from any distance.

After alighting, the aphid spends some time walking and briefly probing the cuticle, or, at most, the epidermis. The frequency and duration of these probes depend on the strength of the 'probing stimuli' (Klingauf, 1988). In most cases these are chemical—secondary chemicals such as sinigrin (Wensler, 1962), sparteine (Smith, 1966) and phlorizin (Klingauf, 1971), or surface waxes (Klingauf *et al.*, 1971). In the absence of appropriate probing stimuli the aphid may decide to leave the plant without any further examination.

If the results of the brief probes are satisfactory, the aphid will probe further in search of the phloem. On preferred hosts this may occur after very few exploratory probes; on non-hosts or resistant cultivars the initial phase may be long and involve repeated probing (Traynier and Hinis, 1988). Once the phloem is pierced, the aphid's decision to feed depends on the presence of phagostimulants or the absence of feeding deterrents. Feeding may be influenced by the concentrations of basic nutrients such as sucrose and amino acids, but also by more specific secondary chemicals such as sinigrin (van Emden, 1972).

Thus, a plant is accepted because it possesses a particular combination of physical and chemical attributes, not because it belongs to a particular species. The problem facing anyone seeking a functional explanation of host specificity is that of defining the strategy set: what are the possible host ranges?

Consider *Brevicoryne brassicae*. This aphid feeds on crucifers, which it selects on the basis of their high concentrations of mustard oil glycosides such as sinigrin (van Emden, 1972). Now suppose that selection would favour an extension of its host range, to include a particular leguminous plant. Mutant *B. brassicae* that accept this plant must do so by no longer requiring sinigrin. This, however, might result in a large number of other plants becoming acceptable. The decision is thus not between 'Cruciferae only' and 'Cruciferae plus one legume', but between 'Cruciferae only' and perhaps 'Cruciferae plus all leguminous plants'. Future mutations at other loci might improve the aphids' ability to choose among the newly-captured hosts, but these will not have been selected before the change in the response to sinigrin. To reach the optimal host range, therefore, the aphids must first become *too* polyphagous. [Such constraints on host recognition were first invoked as a cause of monophagy by Levins and MacArthur (1969).] What this means is that the constraints on the mechanisms of host selection may prevent aphids from attaining the optimal host range; that to reach the adaptive peak the aphid population must first cross the adaptive valley of excessive polyphagy.

The importance of such sensory constraints depends on the questions asked. A full explanation of why, for example, the sycamore aphid, *Drepanosiphum platanoidis*, does not colonize a particular species of *Acer*, requires not only that the 'potential host' is inferior to the normal host, but also that this limited expansion of host range does not entail the acceptance of a much wider range of plants. If we try to explain why *D. platanoidis* colonizes fewer species than *B.brassicae* the constraint is still important: crucifers may simply be more difficult to distinguish than species of *Acer*.

In the absence of data on the host ranges that aphids *might* utilize, it is impossible to take such neurological constraints into account. In practice, this rarely matters, but the likelihood that such constraints do exist means that comparisons among aphids that do not share hosts must be viewed with caution.

8.3 SUGGESTED FUNCTIONS OF HOST SPECIFICITY

8.3.1 Introduction

In this section I shall consider various functional explanations of host specificity. I shall discuss here only those explanations which assume that adaptations to the various hosts are at an evolutionary equilibrium, so that the only characters free to evolve are the behavioural responses of aphids to plants on which they have alighted.

The suggested benefits of host specificity are:

1. stabilization of the population;
2. avoidance of chemically or architecturally unsuitable hosts;
3. escape from predation or interspecific competition; and
4. finding mates.

8.3.2 Population stability

One of the explanations discussed by van Emden (1978) can be summarized as follows. Natural selection favours features leading to population stability; stability depends mainly on intraspecific competition; host specificity results in a concentration of the aphid population on a few host species, and thus increases the degree of competition; by contrast, generalists will escape the effects of competition by dispersing over a wide range of hosts. Populations of generalists will thus be unregulated and unstable, and more likely to become extinct than populations of specialists. There are two main difficulties with this explanation. First, it relies on group selection to prevent the spread of genes for polyphagy; individual aphids must pay the costs of specialization (reduced success in finding hosts) for the benefit of the population.

The second problem concerns the relation between monophagy and stability. If aphid populations are regulated by intraspecific competition on the host plant, then an expansion of host range will result only in an increase in the carrying capacity of the region. There is no reason to suppose that the new equilibrium will be any less stable than the old; indeed, polyphagy may lead to a spreading of the risk of extinction (den Boer, 1968) and *increase* the stability of the population (Redfearn and Pimm, 1988). We can thus reject the hypothesis that the function of host specificity is to stabilize populations.

8.3.3 Chemical and physical features of the hostplant

Several chemical, structural and ecological features of the hostplant have been shown to have direct effects on aphids' growth, survival and reproduction, and thus presumably to influence the selection on host range.

(a) Chemical features

Three characteristics may be important here:

1. the nutritional quality of the phloem;

2. the presence of toxins, anti-feedants or repellants; and
3. the ability of aphids to manipulate the host's physiology to their own advantage.

(b) Nutritional quality

Phloem sap is extremely rich in energy, and aphids are apparently unable to utilize all of the sugar they imbibe. What limits their growth is the low concentration of amino acids (Mittler, 1958; Dixon, 1970).

Aphids do tend to prefer the more nutritious of the available hosts (van Emden, 1972; Kieckhefer and Lunden, 1983), but this is not the sole criterion for host selection: for example, *Brevicoryne brassicae* is capable of growing and reproducing on broad bean (which it normally rejects) treated with sinigrin (Wensler, 1962). This species is thus more specific than it need be on purely nutritional grounds.

Nutritional quality clearly cannot be ignored as a factor in the evolution of host range, but its role as the major factor must be questioned. Purely nutritional features of the plant are unlikely to explain differences between the host ranges of related species: e.g. between *Myzus persicae*, which colonizes several hundred host species, and the highly specific *M. certus*.

(c) Toxins, antifeedants and repellants

These three indirect effects of plant chemistry must be distinguished. Toxins affect growth or survival, but do not inhibit feeding; antifeedants may affect only the rate at which the aphid imbibes the phloem sap; and repellants prevent the aphids from settling on the plant. The difference is important since a chemical is an antifeedant or repellant only insofar as the aphids respond to it as such. But it is precisely this response that we are trying to explain.

The distinction is difficult for three reasons: first, the effects of the aphids' behavioural response to an antifeedant are very difficult to distinguish from the direct metabolic effects of a toxin, without detailed measurements of consumption rates and assimilation efficiencies— anything that reduces the consumption rate also reduces growth and reproduction; second, what is initially a toxin is likely to become a repellant as the aphids evolve to avoid it; and finally, if a repellant has no toxic effects aphids should be selected to ignore it (unless, as I shall argue below, other factors favour specificity).

This being so, many plant chemicals may combine toxic, antifeedant and repellant effects. The literature contains numerous examples of chemicals whose concentrations are negatively correlated with aphids' growth, reproduction or survival, and many of these also inhibit feeding or colonization. The compounds implicated include sinigrin (van Emden, 1972), hydrogen cyanide (Dritschilo *et al.*, 1979), phenols (Zucker, 1982), alkaloids (Wink *et al.*, 1982) and hydroxamic acids (Argandoña *et al.*, 1980).

(d) Gall formation

Many aphids induce galls on their hosts; these provide an improved food supply and protection against desiccation. I know of no direct experimental work on the specificity of the chemical signals used, but galling may require a precise match: the aphid must be able to instruct the plant to grow into a protective gall. If the aphids' instructions are obeyed by only a few species of plant, then failure to induce galls may be an important barrier to the capture of new hosts.

(e) Host structure

Two features of the structure of the host influence the performance of aphids settling on them: the depth of the phloem and the presence of hairs or glands on the leaf surface.

If the phloem lies deep below the surface of the plant, then only aphids with long stylets are able to feed (Dixon, 1985). For example, Shearer (1976) showed that the nymphs in the first spring generation of *Periphyllus testudinaceus* (which feed through the bark of sycamore) have relatively long stylets compared with those of their summer descendants, which feed on the minor veins in the leaves. There is also a strong correlation between the depth of the phloem and the length of the stylets in the five species feeding on different parts of the oak (Dixon, 1985).

The presence of hairs on the leaf surface can have a similar effect: *Uroleucon* species feeding on hosts with long, dense trichomes have relatively long rostra (Moran, 1987).

Finally, surface hairs prevent some aphids from feeding on resistant potatoes (Tingey and Sinden 1982) and the wild lime, *Tilia petiolaris* (Carter, 1982, cited in Dixon, 1985).

(f) Can these factors account for host specificity?

Although the presence of toxins or physical defences may be important in preventing aphids from utilizing particular plants, it is not clear that this can explain host specificity. For example, *Lipaphis erysimi* and *B. brassicae* contain an enzyme to detoxify sinigrin, and are thus able to colonize crucifers (MacGibbon and Benzenberg, 1978), but this does not explain why they do not also feed on other hosts. To explain that, we must be able to infer either a benefit from host specificity *per se* (see Section 8.3.6) or a fitness trade-off: adaptation to one host must reduce fitness on others, by at least 99%.

Of the chemical and architectural features of the host, only gall induction seems likely to demand specificity rather than merely ruling out particular plant species as hosts.

8.3.4 Host phenology

The final host attribute of importance in determining the fitness of aphids is phenology. First, clearly it is maladaptive for aphids to oviposit (or larviposit) on a host which is to die or deteriorate before the offspring can reach maturity. It is perhaps for this reason that *Uroleucon nigrotibium* does not colonize the annual *Solidago juncea*, even though this is nutritionally superior to the normal host, *S. nemoralis* (Moran, 1984).

There are also three important features of the phenology of perennial plants. First, in tree-dwelling species such as *Drepanosiphum platanoidis*, nymphs emerging before bud-burst starve, while those hatching too late miss the peak in the concentration of nitrogen, and thus suffer a reduction in fitness (Dixon, 1976). Second, aphids must have some means of coping with any seasonal fluctuations in host quality—e.g. the production of aestivating nymphs by *Periphyllus* spp., or reproductive diapause in *D. platanoidis* (Dixon, 1975). Finally, the production of overwintering eggs should be postponed as long as possible, but not so long that the final generation is produced after leaf-fall (Ward *et al.*, 1984).

Unfortunately, little information is available on the extent to which adaptation to one host's phenology means that other hosts are no longer suitable. Phenological adaptations do seem, however, to provide trade-offs of the sort required for the evolution of specificity: if egg-hatch is synchronized with bud-burst in one host, it may be too late on others (Komazaki, 1986). Such asynchronies may even reduce fitness by the 99% required to prevent host capture (see Section 8.2.3).

8.3.5 Ecological specialization

A number of authors have suggested that niches may be restricted by interspecific competition (MacArthur 1972) or predation (Colwell and Fuentes, 1975; Gilbert and Singer, 1975). Aphids' host ranges, however, seem unlikely to be limited by these interactions, for several reasons.

(a) Interspecific competition

Aphids might adversely affect each others' growth rates in several ways. Direct competition for phloem sap seems unlikely, but there may be competition for galls (Akimoto, 1988) or feeding sites, and effects such as accelerated senescence of the host or inhibition of its growth (reviewed by

Wellings *et al.*, 1989) will influence the performance of all aphids on the plant (Tamaki and Allen, 1969). Crowding by competitors may increase the percentage of alatae, and thus limit the size of the colony (P. W. Wellings, personal communication); and, finally, there may be competition for mutualistic ants (Addicott, 1978). (This last results in competition between aphids on *different* hosts in the same habitat, so cannot lead to selection for specificity.)

The importance of the interactions on the host must depend on the chance that the competing species find themselves on the same individual host. On small herbaceous or annual hosts this may occur only rarely, since in most years only a small proportion of the host population is likely to be infested. On large perennial hosts the chances are greater. Here, however, the competitive interactions may be weaker, for the following reasons: the architectural diversity of a tree means that several different feeding sites are available (Dixon, 1985); phloem sap is not likely to be limiting; and senescence is accelerated leaf by leaf, not simultaneously throughout the tree.

Even if interspecific competition is important in aphids' population dynamics, it almost certainly cannot reduce their fitness by enough to restrict their host range. In fact, since most of the competitive interactions would also result in stronger intraspecific competition, they should result in an expansion of host range (Ward, 1987a).

(b) Predators

Recently Bernays and Graham (1988) argued that natural enemies are the major factor in the evolution of host range in phytophagous insects; that selection favours specialization in 'enemy-free space' (Gilbert and Singer, 1975).

The role of predators and parasitoids in regulating aphid populations is unclear (Frazer, 1988), and in any case likely to vary among aphid species and from year to year or habitat to habitat. For several reasons, however, natural enemies alone seem unlikely to cause selection for specificity.

First, although aphid parasitoids may be hostplant-specific (Read *et al.*, 1970), they cannot form a barrier to the capture of new hostplants—if there are no aphids on a particular host there will be no parasitoids (or aphid-specific predators) either.

Second, although generalist predators may sometimes have a dramatic effect on aphid populations (Chambers *et al.*, 1983), it is not clear that (1) they can reduce colonists' fitness by the amount required, or (2) their effects on aphid populations on different hosts are sufficiently different, to bar aphids from novel hosts.

It is possible, however, that natural enemies and relatively low nutritional quality might interact to reduce aphids' fitness by more than the sum of their effects (e.g. van Emden, 1978). Strong interactions might then prevent colonization of some plants, although again it is not clear that this can lead to specificity.

8.3.6 'Rendezvous' hosts

In this section I consider the possibility that selection favours host specificity *per se* (Rohde, 1979; Ward, 1987b), since hostplants serve not only as habitats and sources of nutrients but also as *rendezvous* for the sexes. If mating occurs on the hostplant, and the sexes find each other by first finding a host, then individuals that settle on uninfested hosts have little chance of mating. This means that selection favours genes for extreme specificity on whichever host is most commonly colonized, even if other available plants are more suitable in other respects (Ward, 1987a).

Before deriving testable predictions from this hypothesis, it is necessary to provide some background information on aphids' life cycles (Fig. 8.1). Aphids are cyclically parthenogenetic: in spring, 'fundatrices' hatching from sexually produced eggs give rise to a series of parthenogenetic generations that culminate in the production of males and sexually reproducing females in autumn. A number of species have lost the sexual generation and become 'anholocyclic'. Of the 'holocyclic' species, most are 'autoecious'; i.e. they colonize a single group of hosts throughout the year. Here, therefore, there is only one system of host selection. About 10% of the holocyclic species, however, are 'heteroecious': these alternate between 'primary' hosts, on which they mate and oviposit, and 'secondary' hosts, on which most of the parthenogenetic generations occur. Since the primary and secondary hosts are usually unrelated, and species sharing primary (or secondary) hosts may colonize different secondary (primary) hosts, these aphids apparently have two independently evolving sets of genes for host preference. Table 8.1 shows the importance of mating success and oviposition sites in the evolution of host range in aphids with the various life cycles.

If failure to mate has been important in restricting host range, then the host ranges of aphids with the various life cycles should differ as follows:

1. Heteroecious winter < Heteroecious summer;
2. in summer: Anholocyclic = Heteroecious > Autoecious.

Table 8.1 The roles of hostplants in the reproduction of aphids with various life cycles, showing whether the primary (usually winter) and secondary hosts are (+) or are not (−) used as rendezvous sites.

Life cycle	Primary hosts	Secondary hosts
Anholocyclic	−	−
Autoecious	+	+
Heteroecious	+	−

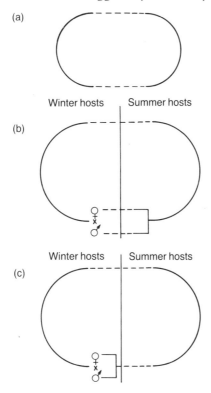

Fig. 8.1 Diagrams of the main types of life cycle in the Aphididae. Solid lines, parthenogenetic reproduction; broken lines, migration. An aphid is autoecious if its summer and winter host ranges are similar; it is heteroecious if they are not. The three life cycles distinguished here are (a) anholocyclic; (b) holocyclic, outbreeding; and (c) holocyclic with self-fertilization—i.e. males may mate with their genetically identical sisters.

Heteroecious species are indeed more specific in winter than in summer (Hille Ris Lambers, 1950, 1980). For example, *Myzus persicae* colonizes only a few species of *Prunus* in winter whereas its summer hosts include monocots and members of many families of dicots (van Emden *et al.*, 1969). Although many of these are annuals, which may not be available to the autumn migrants, others are colonized by other species of aphid throughout the year, so host phenology cannot account for the seasonal change in specificity.

The predicted difference between summer host ranges of autoecious and heteroecious species is shown by the results of Eastop's (1973) survey: of 455 autoecious species only 48 (11%) infest plants of more than

Table 8.2 A comparison of the summer host ranges of autoecious
and heteroecious aphids feeding on grasses in central Europe
(data from Börner, 1952)

Summer host range	Heteroecious	Autoecious	Total
1 genus	1	10	11
> 1 genera	13	11	24
Total	14	21	35

Chi-squared = 4.74, on one degree of freedom, $P < 0.05$.

one genus, whereas 48% (37) of the 77 heteroecious species considered show polyphagy in their secondary host range. But Eastop's survey was not designed as a test of the *rendezvous* hypothesis, and its use as such has two shortcomings. In a number of the species included, the primary host is colonized by 'sexuparae'—females which then produce both apterous males and sexually reproducing females. In these species, therefore, self-fertilization is possible (the males mating with their genetically identical sisters), so aphids on the 'wrong' host are able to mate; the selection against an expansion of host range is considerably weakened. The second shortcoming is that the survey included aphids from gymnosperms and angiosperms, monocots and dicots, trees and forbs; it is difficult to compare aphids with such diverse tastes—is an aphid feeding on ten genera of grasses less or more specialized than one which colonizes five genera of crucifers?

Table 8.2 shows the results of a preliminary attempt to overcome these problems (S. A. Ward, in preparation). Only aphids with alate males are included, and all of the species occur on grasses in central Europe during the summer (Börner, 1952). These data support the hypothesis that failure to mate prevents expansion of the host range of autoecious aphids, but not the summer host range of heteroecious species.

In this context it is also worthy of note that some of the most polyphagous autoecious aphids (e.g. *Aulacorthum solani*, *Macrosiphum euphorbiae*) produce apterous males—the ability to self-fertilize may uncouple selection on host range from that for mate-finding.

Finally, the loss of the sexual generation uncouples host range from mating success. This may be the reason for the inreased host ranges of anholocyclic strains in normally holocyclic species (Müller, 1968), and of anholocyclic species relative to their holocyclic congeneric species (Müller, 1966). Although some of these may result from between-clone variation there is evidence that single anholocyclic clones are more polyphagous than their holocyclic relatives (e.g. Müller, 1970).

In addition to these features of the present host ranges of aphids, the fact that many lineages show great conservatism in their primary hosts

(Hille Ris Lambers, 1980) is also consistent with the existence of a sex-related barrier to host capture.

These results, however, do not conclusively demonstrate that the barrier results solely from the failure to mate. The requirement for oviposition sites and feeding sites for the fundatrices may also be import-ant (Moran, 1988). The fact that the fundatrices of some secondarily autoecious species (i.e. those that have abandoned the original primary host) differ greatly from those of related heteroecious species, suggests that considerable adaptation is necessary before they can efficiently utilize the secondary hosts (Moran, 1988). In some groups, then, the fundatrices may be irrevocably specialized. (It is perhaps important here that it is usually the fundatrices that induce galls.) The arguments against this as a major factor in host range restriction are (1) that many originally heteroecious aphids have been able to omit their autumn migration, to become secondarily autoecious on the secondary hosts; and (2) that fundatrices are sometimes able to feed on plants unrelated to their normal hosts (Tamaki *et al.*, 1980; Müller, 1985).

8.3.7 The functions of host specificity

In conclusion, although nutritional quality cannot be ignored as a factor in host range evolution (especially the evolution of host alternation), it seems inadequate as a cause of specificity. For a functional explanation of specificity we may have to invoke features of the host that require specific adaptations in the aphid—plant phenology or gall formation—and selec-tion for specificity *per se* as a means of finding mates. Presumably, however, several factors interact to prevent host-range expansion. The study of such interactions is difficult, and still in its infancy, but it is an essential area of future research if we are ever fully to understand the functions of specialization.

8.4 HOST RACES AND SYMPATRIC SPECIATION

8.4.1 Introduction

The functional explanations discussed above were based on optimization of host preference, assuming that all other characters remain constant. Host selection, however, has two important consequences: it determines the strength of the selection on characters determining viability on the various hosts; and it influences the choice of a mate—if mating occurs on the host it is assortative with respect to host preference.

If different adaptations are required for aphids to utilize different hosts (e.g. Komazaki, 1986), then selection should result in local adaptation, and genetic differences between aphid populations on different plants.

There are many examples of aphid biotypes able to attack formerly resistant crops (e.g. Cartier *et al.*, 1965; Lammerinck, 1968; Nielson *et al.*, 1971); and local adaptation to particular host species has been shown for *Myzus persicae* (Weber, 1983, 1985b, 1986), *Sitobion avenae* (Weber, 1985a) and *Metopolophium dirhodum* (Weber, 1985c). Eventually, selection may result in negative genetic correlations between the fitnesses on different hosts—aphids adapted to one host perform poorly on the others. Such correlations have been shown to exist within some species (Service, 1984), and between partially isolated biotypes (Mosbacher, 1962; Müller, 1962, 1971, 1980; Markkula and Roukka, 1970; Bournoville, 1977); although Weber (1985a, 1985c) found none in *S. avenae* or *M. dirhodum*.

If there is also genetic variation in host preference, within the aphid species (e.g. Müller, 1968, 1970, 1980), then preference and viability may become genetically correlated: genes for strong preference for a particular host will be associated with those for high fitness on that plant. In this case, a single aphid species may contain a number of host-specific 'biotypes' or host races (Eastop, 1972; Müller, 1985).

Müller (1985, 1986) has argued persuasively that these may ultimately diverge to become separate species: that sympatric speciation is an important process in aphid evolution.

Clearly not all groups of subspecies or sibling species will have arisen by sympatric divergence of host races; the point here, however, is that if *any* sympatric speciation occurs in aphids then one function of host specificity is to avoid genetically unsuitable mates.

8.4.2 Theories of sympatric speciation

Sympatric speciation has been the focus of lively debate in recent years. While Müller and others (Bush, 1975; Zwölfer and Bush, 1984; Feder *et al.*, 1988; McPheron *et al.*, 1988) contend that geographical isolation is not a prerequisite for speciation, particularly in phytophagous insects, several authors argue that the conditions for sympatric speciation are too stringent for the process to be important in nature (e.g. Futuyma and Mayer, 1980; Paterson, 1981).

What then are these conditions? The literature contains a number of models of the various steps in this process (Maynard Smith, 1966; Bush, 1975; Pimm, 1978; Kondrashov, 1983a,b; Rice, 1984; Rausher, 1984). The results of these models suggest that sympatric speciation is possible if mating is assortative with respect to a character showing stable underdominance (i.e. with disruptive selection on a stably polymorphic character).

Stable underdominance requires that the extreme phenotypes (or homozygotes) are selected for in different habitats, in *each* of which density-dependent processes regulate the population (Levene, 1953;

Maynard Smith, 1966, 1970). Mating will be assortative if the character under disruptive selection is correlated with host preference. The original generalist species may then split into two specialists if selection eliminates those individuals whose habitat preference is inappropriate—those that prefer the habitat in which they are least fit, or in which they are likely to produce maladapted hybrids. Thus preference may evolve as a means of avoiding unsuitable mates.

8.4.3 Can aphids speciate sympatrically?

Two difficulties arise when the existing models are applied to aphids. First, their definition of 'preference' contains the implicit assumption that all migrants find habitats (hosts)—the effect of an allele substitution at the 'preference' loci is to change the proportion of migrants choosing one habitat rather than the other, but all genotypes are equally successful in finding habitats. In addition, with preference determined by a single locus, or as a single quantitatively varying character, the models permit no true-breeding generalists.

The existence of polyphagous biotypes (Müller, 1971), however, suggests that host range is free to increase or decline: that in aphids, mutations resulting in the capture of a new host do not inevitably lead to the rejection of the original hosts. Thus, host range must be determined by alleles for 'acceptance' at several loci: there is not only a single preference locus. Now, to avoid genetically unsuitable mates, aphids must reduce their host range, and run the risk of failing to find hosts at all. This means that for such specialization to be favoured by selection, hybrids between host races must be extremely unfit on both hosts; the conditions for speciation are more stringent than is suggested by the existing models (S. A. Ward, in preparation).

The second problem is that of gene flow via matings on a third, fourth, etc., host. Unless the races have non-overlapping host ranges they cannot become reproductively isolated. In fact, however, aphids' host races do share hosts—e.g. *Acyrthosiphon pisum* (Müller, 1971, 1980).

These considerations suggest that sympatric divergence in aphids may not be the result of disruptive selection and assortative mating alone. But the fact of such divergence cannot be denied: Müller's results provide overwhelming evidence that many aphid species do contain sympatric strains differing in preference for, and adaptation to, the various host-plants.

If disruptive selection and assortative mating cannot account for sympatric divergence in aphids, how then have the biotypes arisen? One partial answer to this question may be as follows (S. A. Ward, in preparation). In aphid species with apterous males, particularly if the hosts are small herbaceous plants unlikely to receive more than one migrant, males will usually mate with their genetically identical sisters: self-fertilization is prevalent. Indeed, Yamaguchi (1985) argues that this occurs even on heavily infested trees. Self-fertilization may limit gene flow through the population, and permit the temporary divergence of races whose host preference matches their ability to survive on the various available hosts. Indeed, self-fertilization may result in speciation if mutants arise with abnormal karyotypes; here, if structural heterozygotes are eliminated then new karyotypes can persist only as a result of self-fertilization.

In many of the races and subspecies reported in the literature the males are indeed apterous: e.g. various biotypes of *Acyrthosiphon pisum* (Müller, 1971, 1980; Müller and Steiner, 1985), *A. pelargonii* (Müller, 1983), *Aphis frangulae* (Thomas, 1968) and *Aulacorthum solani* (Müller, 1970, 1976), and the members of the *Uroleucon cichorii* and *U. jaceae* groups (Mosbacher, 1962). Others are secondarily autoecious strains of host-alternating species (the loss of the autumn migration to the primary host must inevitably result in self-fertilization in the new mutants): examples include subspecies of *A. frangulae* (Thomas, 1968), *Brachycaudus prunicola* (Thomas, 1962) and *Myzus cerasi* (Dahl, 1968). Still others have omitted sexual reproduction altogether, e.g. in *A. pelargonii* (Müller, 1983), *A. frangulae* (Thomas, 1968), *A. solani* (Müller, 1970) and the predominantly anholocyclic members of the *Macrosiphum euphorbiae* group (Möller, 1971).

In all of these cases reproductive isolation between host races or subspecies can be explained without invoking assortative mating. However, there still remain several examples of sympatric subspecies or sibling species with alate males, distinguished by their choice of primary hosts [in *B. prunicola*, (Thomas, 1962) and *M. cerasi* (Dahl, 1968)] or—even more difficult to explain—secondary hosts [the *Aphis fabae* complex (Müller, 1982), *Aphis frangulae* (Thomas, 1968) and the sibling species *Ovatus crataegarius* and *O. insitus* (Müller and Hubert-Dahl, 1979)]. It remains to be shown whether these diverged sympatrically or have instead regained sympatry since diverging in allopatry. The theoretical considerations discussed above suggest the latter.

8.5 DISCUSSION AND CONCLUSIONS

8.5.1 Functional explanations of host specificity

The constraints and selection pressures on host range in aphids may be very different from those on host selection in other well-studied groups of phytophagous insects. First, only a very small fraction of aphid migrants find hosts (Taylor, 1977; S. A. Ward *et al.*, in preparation), so polyphagy is at a premium. Second, unlike lepidopterans and other folivores, aphids may not normally encounter many of the toxins and secondary chemicals implicated in plant defence—if these are enclosed in vacuoles in cells in the epidermis or parenchyma, they may be by-passed by aphids feeding on the phloem sap. This is certainly not always the case (Dixon, 1985) but as causes of host specificity, the most important features of the plant may be its phenology and the aphids' ability to manipulate its growth.

In this chapter, however, I have tried to argue that while features of the hostplant are certainly important, the evolution of host range must also be seen in the context of aphids' complex life cycles, in particular the seasonal change from parthenogenetic to sexual reproduction.

The preliminary comparisons of host range discussed above (Section 8.3.6) support the hypothesis that the need to find mates has been an important barrier to the capture of new hosts; morphs and species seeking *rendezvous* hosts are on average more host specific than those whose hosts serve only as feeding sites. Furthermore, over evolutionary time primary (*rendezvous*) host range changes much more slowly than the choice of secondary hosts, which is more opportunistic (Hille Ris Lambers, 1980).

Whatever the consequences of the use of hosts as rendezvous sites, the evolution of host specificity in aphids is influenced by a wide range of factors, which may interact in ways that are still poorly understood. Indeed, even single factors are difficult to examine: we cannot easily distinguish between the direct effects of poor nutrition (which influences the *selection* on host choice) and the reduction in growth resulting from an aphid's reluctance to feed (which *is* host choice). Furthermore, in seeking explanations of host specificity we may not even be able to distinguish cause from effect: if we observe that an aphid's life cycle is closely synchronized with its host's phenology, is specificity a consequence of this precise adaptation? Or has the life cycle evolved to match the biology of the single host?

Bearing these limitations in mind, three of the factors discussed above seem capable of producing selection pressures strong enough to maintain host specificity: the close adaptation of aphids' life cycles to the host's phenology, the need for effective induction of galls and the use of the host as a rendezvous for the sexes.

8.5.2 Sympatric speciation, host races and biotypes

The choice of a host not only affects the chance of finding a mate, but may also determine mate choice. Thus, mating may be assortative with respect to variation in host range. This is one of the conditions for sympatric speciation by divergence of host races (Bush, 1975; Zwölfer and Bush, 1984) and Müller (1985, 1986) argues that this process has been important in the evolution of aphids.

The barriers to sympatric speciation, however, seem insurmountable, for two main reasons: first, the cost of specificity (failure to find a host) is such that extremely strong disruptive selection is required for sympatric divergence of host races; second, host races cannot become reproductively isolated if their host ranges overlap, as is the case in several putative examples of sympatric speciation. I have argued here that a more plausible interpretation is that many of the differences between sympatric races are conserved by asexual reproduction or self-fertilization (S. A. Ward, in preparation). With self-fertilization, what seem to be incipient species (Müller, 1985) may instead be temporarily distinct lines, which will ultimately either disappear or merge with the parent population, unless their karyotypes differ.

The relation between the selection for specialization in *rendezvous* hosts and sympatric divergence by selfing is a problematic one. Clearly self-fertilization is unimportant in aphids whose males migrate independently of the females that produce the oviparae; and, equally clearly, members of these species will fail to mate if they accept the wrong host. It is not clear, however, whether failure to mate can prevent generalization in species able to self-fertilize—do selfing aphids fail to mate if on the wrong host? This will depend on several features of their biology: the size of the colony produced by a single colonist; the sex ratio; the relative timing of the births of males and oviparae; dispersal over the host; and survival.

8.5.3 Applied significance

Whether we regard host races as incipient species or as temporarily distinct lines, they have considerable importance in agricultural entomology. This is because morphologically indistinguishable aphids may differ greatly in their pest status (Müller, 1986), so regional schemes for monitoring populations of aphid pests may yield unreliable forecasts of the levels of infestation of crops. Consideration of the effects of aphids' life cycles on biotype formation (in Section 8.4.3) suggests that regional monitoring may be more useful for predicting outbreaks of heteroecious aphids with alate males (which mate randomly with respect to their summer host range) than for species with assortative mating or frequent self-fertilization.

ACKNOWLEDGEMENTS

This chapter has benefited greatly from the constructive criticisms of W. J. Bailey, J. Ridsdill-Smith and P. W. Wellings. I am grateful also to Jenny Nagle for secretarial assistance. This work was supported by the Australian Research Grants Scheme.

References

Addicott, J. F. (1978) Competition for mutualists: aphids and ants. *Can. J. Zool.*, **56**, 2093–2096.

Åhman, I., Weibull, J. and Pettersson, J. (1985) The role of plant size and plant density for host finding in *Rhopalosiphum padi* (L.) (Hem.: Aphididae). *Swed. J. Agric. Res.*, **151**, 19–24.

Akimoto, S. (1988) Competition and niche relationships among *Eriosoma* aphids occurring on the Japanese elm. *Oecologia*, **75**, 44–53.

Argandoña, V. H., Luza, J. G., Niemeyer, H. M. and Corcuera, L. J. (1980) Role of hydroxamic acids in the resistance of cereals to aphids. *Phytochemistry*, **19**, 1665–1668.

Bernays, E. and Graham, M. (1988) On the evolution of host-specificity in phytophagous arthropods. *Ecology*, **69**, 886–892.

Börner, C. (1952) Europae Centralis Aphides: die Blattläuse Mitteleuropas. *Mitt. Thür. Bot. Ges.*, **3** (Weimar).

Bournoville, R. (1977) Variabilite des populations de deux especes d'aphides des legumineuses. *Ann. Zool. Ecol. Anim.*, **9**, 588–589.

Bush, G. L. (1975) Modes of animal speciation. *Ann. Rev. Ecol. Syst.*, **6**, 339–364.

Cartier, J. J., Isaak, A., Painter, R. H. and Sorensen, E. L. (1965) Biotypes of pea aphid *Acyrthosiphon pisum* (Harris) in relation to alfalfa clones. *Can. Entomol.*, **97**, 754–760.

Chambers, R. J., Sunderland, K. D., Wyatt, I. J. and Vickerman, G. P. (1983) The effects of predator exclusion and caging on cereal aphids in winter wheat. *J. Appl. Ecol.*, **20**, 209–224.

Colwell, R. K. and Fuentes, E. R. (1975) Experimental studies of the niche. *Ann. Rev. Ecol. Syst.*, **6**, 281–310.

Dahl, M. L. (1968) Biologische und morphologische Untersuchungen über den Formenkreis der schwarzen Kirschenlaus *Myzus cerasi* (F.) (Homoptera: Aphididae). *Dtsch. Entomol. Z.*, **15**, 281–312.

den Boer, P. J. (1968) Spreading of risk and stabilization of animal numbers. *Acta Biotheor.*, **18**, 165–194.

Dixon, A. F. G. (1970) Quality and availability of food for a sycamore aphid population. In *Animal Populations in Relation to their Food Resources* (ed. A. Watson), Blackwell, Oxford, 271–287.

Dixon, A. F. G. (1975) Seasonal changes in fat content, form, state of gonads and length of adult life in the sycamore aphid, *Drepanosiphum platanoidis* (Schr.) *Trans. R. Entomol. Soc. Lond.*, **127**, 87–99.

Dixon, A. F. G. (1976) Timing of egghatch and viability of the sycamore aphid, *Drepanosiphum platanoidis* (Schr.) at bud burst of sycamore, *Acer pseudoplatanus* L. *J. Anim. Ecol.*, **45**, 593–603.

Dixon, A. F. G. (1985) *Aphid Ecology*, Blackie, Glasgow.

Doyle, R. (1978) Settlement of planktonic larvae: a theory of habitat selection in varying environments. *Amer. Nat.*, **109**, 113–126.

Dritschilo, W., Krummel, J., Nafus, D. and Pimentel, D. (1979) Herbivorous insects colonising cyanogenic and acyanogenic *Trifolium repens*. *Heredity*, **42**, 49–56.

Eastop, V. F. (1972) Biotypes of aphids. In *Perspectives in Aphid Biology* (ed. A. D. Lowe), *Entomol. Soc. N.Z. Bull.*, **2**, 40–51.

Eastop, V. F. (1973) Deductions from the present day host plants of aphids and related insects. In *Insect/Plant Relationships* (ed. H. F. van Emden), Blackwell, Oxford, 157–178.

Feder, J. L., Chilcote, C. A. and Bush, G. L. (1988) Genetic differentiation between sympatric host races of the apple maggot fly *Rhagoletis pomonella*. *Nature*, **336**, 61–64.

Frazer, B. D. (1988) Predators. In *Aphids: their Biology, Natural Enemies and Control* (eds A. K. Minks and P. Harrewijn), World Crop Pests Vol. 2B, Elsevier, Amsterdam, 217–230.

Futuyma, D. J. and Mayer, G. C. (1980) Non-allopatric speciation in animals. *Syst. Zool.*, **29**, 254–271.

Gilbert, L. E. and Singer, M. C. (1975) Butterfly ecology. *Ann. Rev. Ecol. Syst.*, **6**, 365–397.

Hille Ris Lambers, D. (1950) Hostplants and aphid classification. *Proc. Int. Congr. Entomol.*, **8**, 141–144.

Hille Ris Lambers, D. (1980) Aphids as botanists? *Symp. Bot. Uppsala*, **22**, 114–119.

Jaenike, J. (1978) On optimal oviposition behaviour in phytophagous insects. *Theor. Popul. Biol.*, **14**, 350–356.

Kennedy, J. S., Booth, C. O. and Kershaw, W. J. S. (1959a) Host-finding by aphids in the field. I. Gynoparae of *Myzus persicae* (Sulzer). *Ann. Appl. Biol.*, **47**, 410–423.

Kennedy, J. S., Booth, C. O. and Kershaw, W. J. S. (1959b) Host-finding by aphids in the field. II. *Aphis fabae* Scop. (gynoparae) and *Brevicoryne brassicae* L.; with a reappraisal of the role of host-finding behaviour in virus spread. *Ann. Appl. Biol.*, **47**, 424–444.

Kieckhefer, R. W. and Lunden, A. O. (1983) Host preferences and reproduction of four cereal aphids (Hemiptera, Aphididae) on some common weed grasses of the northern plains. *Environ. Entomol.*, **12**, 986–989.

Klingauf, F. (1971) Die Wirkung des Glucosids Phlorizin auf das Wirtswahlverhalten von *Rhopalosipum insertum* (Walk.) und *Aphis pomi* De Geer (Homoptera: Aphididae). *Z. ang. Entomol.*, **68**, 41–55.

Klingauf, F. (1988) Host plant finding and acceptance. In *Aphids: their Biology, Natural Enemies and Control* (eds A. K. Minks and P. Harrewijn), World Crop Pests, Vol. 2A, Elsevier, Amsterdam, pp. 209–23.

Klingauf, F., Nöcker-Wenzel, K. and Klein, W. (1971) Einfluss einiger Wachskomponenten von *Vicia faba* L. auf das Wirtswahlverhalten von *Acyrthosiphum pisum* (Harris) (Homoptera: Aphididae). *Z. Pflkrankh. Pflschutz.*, **78**, 641–648.

Komazaki, S. (1986) The inheritance of egg hatching time of the spiraea aphid, *Aphis citricola* van der Goot (Homoptera, Aphididae) on the two winter hosts. *Kontyû, Tokyo*, **54**, 48–53.

Kondrashov, A. S. (1983a) Multilocus model of sympatric speciation. I. One character. *Theor. Popul. Biol.*, **24**, 121–135.

Kondrashov, A. S. (1983b) Multilocus model of sympatric speciation. II. Two characters. *Theor. Popul. Biol.*, **24**, 136–144.

Lammerinck, J. (1968) A new biotype of cabbage aphid on aphid resistant rape. *N. Z. J. Agric. Res.*, **11**, 341–344.

Leather, S. R. and Wellings, P. W. (1981) Ovariole number and fecundity in aphids. *Entomol. Exp. Appl.*, **30**, 128–133.

Levene, H. (1953) Genetic equilibrium when more than one niche is available. *Amer. Nat.*, **87**, 331–333.

Levins, R. and MacArthur, R. H. (1969) An hypothesis to explain the incidence of monophagy. *Ecology*, **50**, 910–911.

Lewontin, R. C. (1987) The shape of optimality. In *The Latest on the Best: Essays on Evolution and Optimality* (ed. J. Dupré), Bradford Books, MIT Press, Cambridge, MA, 151–159.

MacArthur, R. H. (1972) *Geographical Ecology*, Harper and Row, New York.

MacGibbon, D. B. and Benzenberg, E. J. (1978) Location of glucosinolate in *Brevicoryne brassicae* and *Lipaphis erysimi* (Aphididae). *N.Z. J. Sci.*, **21**, 389–392.

Markkula, M. and Roukka, K. (1970) Resistance of plants to the pea aphid *Acyrthosiphon pisum* Harris (Hom., Aphididae). I. Fecundity of the biotypes on different host plants. *Ann. Agric. Fenn.*, **9**, 127–132.

Maynard Smith, J. (1966) Sympatric speciation. *Amer. Nat.*, **100**, 637–650.

Maynard Smith, J. (1970) Genetic polymorphism in a varied environment. *Amer. Nat.*, **104**, 487–490.

McPheron, B. A., Smith, D. C. and Berlocher, S. H. (1988) Genetic differences between host races of *Rhagoletis pomonella. Nature*, **336**, 64–66.

Mittler, T. E. (1958) Studies on the feeding and nutrition of *Tuberolachnus salignus* (Gmelin) (Homoptera, Aphididae). *J. Exp. Biol.*, **35**, 74–84; 616–638.

Moericke, V. (1955) Über die Lebensgewohnheiten der geflügelten Blattläuse (Aphidina) unter besonderer Berücksichtigung des Verhaltens beim Landen. *Z. ang. Entomol.*, **37**, 29–91.

Möller, F. W. (1971) Bastardierungen innerhalb des Artenkomplexes um die grünstreifige artoffelblattlaus *Macrosiphum euphorbiae* (Thomas). *Beitr. Entomol.*, **21**, 531–537.

Moran, N. A. (1984) Reproductive performance of a specialist herbivore, *Uroleucon nigrotibium* (Homoptera), on its host and a non-host. *Oikos*, **42**, 171–175.

Moran, N. A. (1987) Evolutionary determinants of host specificity in *Uroleucon*. In *Population Structure, Genetics and Taxonomy of Aphids and Thysanoptera* (eds J. Holman, J. Pelikan, A. F. G. Dixon and L. Weisman), SPB Academic Publishing, The Hague, 29–38.

Moran, N. A. (1988) The evolution of host-plant alternation in aphids: evidence for specialization as a dead end. *Amer. Nat.*, **132**, 681–706.

Mosbacher, G. C. (1962) Über die Nahrungswahl bei *Dactynotus* Raf. (Aphididae). I. Die Wirtsspektren der Gruppe *D. jaceae* (L.) s. lat. und *D. cichorii* (Koch) s. lat. *Z. Ang. Entomol.*, **51**, 377–428.

Müller, F. P. (1962) Biotypen und Unterarten der 'Erbsenlaus' *Acyrthosiphum pisum* Harris. *Z. Pflkrankh. Pflschutz.*, **69**, 129–136.

Müller, F. P. (1966) Schädliche Blattläuse in den Tropen und Subtropen unter besonderer Berücksichtigung von Rassendifferenzierung. *Z. ang. Entomol.*, **58**, 76–82.

Müller, F. P. (1968) Eine rote holozyklische Rasse von *Metopolophium festucae*

(Theobald, 1917) (Homoptera: Aphididae). *Z. ang. Entomol.*, **61**, 131–141.

Müller, F. P. (1970) Zucht- und Übertragungsversuche mit populationen und Klonen der grünfleckigen Kartoffelblattlaus *Aulacorthum solani* (Kaltenbach, 1843). *Dtsch. Entomol. Z.*, **17**, 259–270.

Müller, F. P. (1971) Isolationsmechanismen zwischen sympatrischen bionomischen Rassen am Beispiel der Erbsenblattlaus *Acyrthosiphon pisum* (Harris) (Homoptera, Aphididae). *Zool. Jb. Syst.*, **98**, 131–152.

Müller, F. P. (1976) Hosts and non-hosts in subspecies of *Aulacorthum solani* (Kaltenbach) and intraspecific hybridizations (Homoptera: Aphididae). *Symp. Biol. Hung.*, **16**, 187–190.

Müller, F. P. (1980) Wirtspflanzen, Generationenfolge und reproduktive Isolation intraspezifischer Formen von *Acyrthosiphon pisum*. *Entomol. Exp. Appl.*, **28**, 145–157.

Müller, F. P. (1982) Das Problem *Aphis fabae*. *Z. Ang. Entomol.*, **94**, 432–446.

Müller, F. P. (1983) Untersuchungen über Blattläuse der Gruppe *Acyrthosiphon pelargonii* im Freiland-Insektarium. *Z. ang. Zool.*, **70**, 351–368.

Müller, F. P. (1985) Biotype formation and sympatric speciation in aphids (Homoptera: Aphidinea). *Entomol. Gener.*, **10**, 161–181.

Müller, F. P. (1986) Genetic and evolutionary aspects of host choice in phytophagous insects, especially aphids. *Biol. Zbl.*, **104**, 225–237.

Müller, F. P. and Hubert-Dahl, M. L. (1979) Wirtswechsel, Generationenfolge und reproduktive Isolation von *Ovatus crataegarius* (Walker, 1850) und *O. insitus* (Walker, 1849). *Dtsch. Entomol. Z.*, **26**, 241–253.

Müller, F. P. and Steiner, H. (1985) Das Problem *Acyrthosiphon pisum* (Homoptera: Aphididae). *Z. ang. Zool.*, **72**, 317–334.

Nielson, M. W., Schonhoerst, M. H., Don, H. *et al.* (1971) Resistance in alfalfa to four biotypes of the spotted alfalfa aphid. *J. Econ. Entomol.*, **64**, 506–510.

Oster, G. F. and Wilson, E. O. (1978) *Caste and Ecology in the Social Insects*, Princeton University Press, Princeton, NJ.

Paterson, H. E. H. (1981) The continuing search for the unknown and unknowable: a critique of contemporary ideas on speciation. *S. Afr. J. Sci.*, **77**, 113–119.

Petterson, J. (1970) Studies on *Rhopalosiphum padi*. I. Laboratory studies on olfactory responses to the winter host, *Prunus padus*. *Landbruks-Hogesk. Ann.*, **36**, 381–399.

Pimm, S. L. (1978) Sympatric speciation: a simulation model. *Biol. J. Linn. Soc.*, **11**, 131–139.

Rausher, M. D. (1984) The evolution of habitat preferences in subdivided populations. *Evolution*, **38**, 596–608.

Read, D. P., Feeny, P. P. and Root, R. B. (1970) Habitat selection by the aphid parasite *Diaeretiella rapae* (Hymenoptera: Braconidae) and hyperparasite *Charips brassicae* (Hymenoptera: Cynipidae). *Can. Entomol.*, **102**, 648–671.

Redfearn, A. and Pimm, S. L. (1988) Population variability and polyphagy in herbivorous insect communities. *Ecol. Monogr.*, **58**, 39–55.

Rice, W. R. (1984) Disruptive selection on habitat preference and the evolution of reproductive isolation: a simulation study. *Evolution*, **38**, 1251–1260.

Risebrow, A. and Dixon, A. F. G. (1987) Nutritional ecology of phloem-feeding insects. In *Nutritional Ecology of Insects, Mites, Spiders and Related Invertebrates* (eds F. Slansky Jr and J. G. Rodriguez), Wiley, New York, 421–448.

Rohde, K. (1979) A critical evaluation of intrinsic factors responsible for niche restriction in parasites. *Amer. Nat.*, **114**, 648–671.

Service, P. M. (1984) Genotypic interactions in an aphid-host plant relationship: *Uroleucon rudbeckiae* and *Rudbeckia laciniata*. *Oecologia*, **61**, 271–276.

Shearer, D. (1976) Unpublished Ph.D. Thesis, University of East Anglia, Norwich, UK.

Smith, B. D. (1966) Effect of the plant alkaloid sparteine on the distribution of the aphid *Acyrthosiphum spartii* (Koch). *Nature*, **212**, 213–214.

Stephens, D. W. and Krebs, J. R. (1986) *Foraging Theory*, Princeton University Press, Princeton, NJ.

Tamaki, G. and Allen, W. W. (1969) Competition and other factors influencing the population dynamics of *Aphis gossypii* and *Macrosiphoniella sanborni* on greenhouse chrysanthemums. *Hilgardia*, **39**, 447–505.

Tamaki, G., Fox, L. and Chauvin, R. L. (1980) Green peach aphid: orchard weeds are host to fundatrix. *Environ. Entomol.*, **9**, 62–66.

Taylor, L. R. (1977) Migration and the spatial dynamics of an aphid, *Myzus persicae*. *J. Anim. Ecol.*, **46**, 411–423.

Thomas, K. H. (1962) Die Blattläuse des Formenkreises *Brachycaudus prunicola* (Kalt.). *Wiss. Z. Univ. Rostock*, **11**, 325–342.

Thomas, K. H. (1968) Die Blattläuse aus der engeren Verwantschaft von *Aphis gossypii* Glover und *A. frangulae* Kaltenbach unter besonderer Berücksichtigung ihres Vorkommens an Kartoffel. *Entomol. Abh. Mus. Tierk. Dresden*, **35**, 337–389.

Tingey, W. M. and Sinden, S. L. (1982) Glandular pubescence, glycoalkaloid composition, and resistance to the green peach aphid, potato leaf hopper, and potato flea beetle in *Solanum berthaultii*. *Am. Potato J.*, **59**, 95–106.

Traynier, R. M. M. and Hines, E. R. (1988) Probes by aphids indicated by stain induced fluorescence in leaves. *Entomol. Exp. Appl.*, **45**, 198–201.

van Emden, H. F. (1972) Aphids as phytochemists. In *Phytochemical Ecology* (ed. J. B. Harborne), Academic Press, London, 25–43.

van Emden, H. F. (1978) Insects and secondary plant substances—an alternative viewpoint with special reference to aphids. In *Biochemical Aspects of Plant and Animal Coevolution* (ed. J. B. Harborne), Academic Press, New York, 309–323.

van Emden, H. F., Eastop, V. F., Hughes, R. D. and Way, M. J. (1969) The ecology of *Myzus persicae*. *Ann. Rev. Entomol.*, **14**, 197–270.

Ward, S. A. (1987a) Optimal habitat selection in time-limited dispersers. *Amer. Nat.*, **129**, 568–579.

Ward, S. A. (1987b) Cyclical parthenogenesis and the evolution of host range in aphids. In *Population Structure, Genetics and Taxonomy of Aphids and Thysanoptera* (eds J. Holman, J. Pelikan, A. F. G. Dixon and L. Weisman), SPB Academic Publishing, The Hague, 39–44.

Ward, S. A. (1991) Assessing functional explanations of host specificity. *Am Nat.* (in press).

Ward, S. A. and Dixon, A. F. G. (1982) Selective resorption of aphid embryos and habitat changes relative to life-span. *J. Anim. Ecol.*, **51**, 859–864.

Ward, S. A. and Dixon, A. F. G. (1984) Spreading the risk, and the evolution of mixed strategies: seasonal variation in aphid reproductive biology. *Adv. Invert. Reprod.*, **3**, 367–386.

Ward, S. A., Dixon, A. F. G. and Wellings, P. W. (1983a) The relation between fecundity and reproductive investment in aphids. *J. Anim. Ecol.*, **52**, 451–461.

Ward, S. A., Wellings, P. W. and Dixon, A. F. G. (1983b) The effect of reproductive investment on pre-reproductive mortality in aphids. *J. Anim. Ecol.*,
52, 305–313.

Ward, S. A., Leather, S. R. and Dixon, A. F. G. (1984) Temperature prediction and
the timing of sex in aphids. *Oecologia*, **62**, 230–233.

Weber, G. (1983) Untersuchungen zur ökologischen Genetik der grünen Pfirsich-
blattlaus *Myzus persicae*. Ph.D. Dissertation, Göttingen.

Weber, G. (1985a) On the ecological genetics of *Sitobion avenae* (F.) (Hemiptera,
Aphididae). *Z. Ang. Entomol.*, **100**, 100–110.

Weber, G. (1985b) Genetic variability in host plant adaptation of the green peach
aphid, *Myzus persicae*. *Entomol. Exp. Appl.*, **38**, 49–56.

Weber, G. (1985c) On the ecological genetics of *Metopolophium dirhodum* (Walker)
(Hemiptera, Aphididae). *Z. Ang. Entomol.*, **100**, 451–458.

Weber, G. (1986) Ecological genetics of host plant exploitation in the green peach
aphid, *Myzus persicae*. *Entomol. Exp. Appl.*, **40**, 161–168.

Wellings, P. W., Ward, S. A., Dixon, A. F. G. and Rabbinge, R. (1989) Yield loss
assessment. In *Aphids: their Biology, Natural Enemies and Control* (eds A. K.
Minks and P. Harrewijn), World Crop Pests, Vol. 2C, Elsevier, Amsterdam,
pp. 49–64.

Wensler, R. J. D. (1962) Mode of host selection by an aphid. *Nature*, **195**, 830–831.

Wink, M., Hartmann, T., Witte, L. and Rheinberger, J. (1982) Interrelationship
between quinolizidine alkaloid producing legumes and infesting insects:
exploitation of an alkaloid-containing phloem sap of *Cytisus scoparius* by the
broom aphid *Aphis cytisorum*. *Z. Naturforsch.*, **37**, 1081–1086.

Yamaguchi, Y. (1985) Sex ratios of an aphid subject to local mate competition with
variable maternal condition. *Nature*, **318**, 460–462.

Zucker, W. V. (1982) How aphids choose leaves: the roles of phenolics in host
selection by a galling aphid. *Ecology*, **63**, 972–981.

Zwölfer, H. and Bush, G. L. (1984) Sympatrische und parapatrische Artbildung.
Z. Zool. Syst. Evolut.-Forsch., **22**, 211–233.

9

Oviposition and the defence of brood in social wasps

J. Philip Spradbery

9.1 INTRODUCTION

Social insects are characterized by three criteria: there is an overlap between at least two generations of adults, individuals cooperate in caring for the young and there is some reproductive division of labour (Wilson, 1971). Although social insects comprise a slender 1.5% of insect species, their abundance and impact on ecosystems are such that they are the dominant arthropods of most terrestrial habitats (Wilson, 1971; Jeanne and Davidson, 1984).

Although the bees, ants and termites have attracted the most research effort as judged by numbers of specialists and published works, the wasps provide the most stunning examples of the various stages in the evolution of social behaviour from solitary beginnings. Virtually every step is represented by the contemporary wasp fauna. Wasp studies have led to some of the most important discoveries in insect sociobiology such as nutritional control of caste, the use of behavioural characters in taxonomy and phylogeny, and the significance of the reciprocal exchange of food and dominance behaviour (Wilson, 1975).

There are three sub-families of social wasps within the Vespidae: Stenogastrinae, Polistinae and Vespinae (Carpenter, 1982). The Stenogastrinae are small, delicate wasps, often called 'hover wasps', which are found in the tropics of Southeast Asia and New Guinea. They build small nests from mud, vegetable matter or a mixture of both and are considered primitively social (Carpenter, 1988). The Polistinae occur worldwide with two major biological groupings, the independent nest-founding species and the swarming species. The Vespinae include the familiar social wasps (*Vespula, Dolichovespula*) of Europe and the yellowjackets of North America. While the majority of vespine species occur predominantly in temperate regions, the hornets (*Vespa* spp.) are also found in the tropics of Southeast Asia and New Guinea. Morphological differences between queens (inseminated, egg-laying females) and workers (uninseminated and usually sterile females) is most marked in the Vespinae.

Many wasps (notable exceptions being species of swarming Polistinae) undergo a non-social phase during their life cycle when nests are initiated. This solitary stage is frequently characterized by very heavy mortality, although these losses are more than compensated by those successful colonies which produce huge numbers of reproductives (Jeanne and Davidson, 1984). Nests can be extremely large, up to 1.3 million adults in the polistine, *Agelaia vicina* in Brasil, with an estimated 33m² of comb area (Jeanne, 1991). The perennial colonies of the introduced European wasp, *Vespula germanica*, in Australia and New Zealand are probably the largest insect societies ever recorded. Nests and brood can weigh up to 450 kg with 4 million cells and up to 300 000 adults (Spradbery, 1973a).

Reproductive strategies and the mechanical and behavioural repertoire which wasps exhibit to ensure survival of their offspring provides the theme for this chapter. Care of the young is not unique to social insects: butterflies guard their eggs (Rothschild, 1979), while some dung beetles have highly developed levels of parental care (Edwards and Aschenborn, 1989). But, among the social wasps, the extraordinary range of physical, physiological and behavioural adaptations which have evolved to protect the reproductive investment is unrivalled among insects.

9.2 OVIPOSITION

To realize their potential reproductivity, wasps require appropriate resources such as proteinaceous food, cells in which to oviposit and the social conditions which favour ovarian development and oviposition. During the course of this chapter, I will touch upon these and other aspects of oviposition, emphasizing those of special interest or importance.

9.2.1 Fecundity and reproductivity

The fecundity of wasps (which have polytrophic ovarioles) varies with different species and can change with different episodes of the life cycle. The Stenogastrinae and Polistinae have only 3 + 3 ovarioles and produce little more than six eggs at any one time. Among the Vespinae, *Vespula* and *Dolichovespula* species have 6 + 6 ovarioles and the hornets (*Vespa* spp.) vary from 7 + 7 through 8 + 8, 9 + 9, 10 + 10 and even a 10 + 12 in the giant Japanese hornet, *Vespa mandarinia* (Kugler *et al.*, 1976). The fecundity of some vespine species is truly remarkable with up to 300 eggs laid daily by the single queen in *V. vulgaris* and *V. germanica* (Spradbery, 1971).

Fig. 9.1 Queen of *Vespula germanica* ovipositing in cell. In this species, more than 300 eggs may be laid daily. (Photograph by J. Green.)

Fecundity can be drastically reduced as a result of social interactions which occur among multiple foundress females in the Polistinae. Here, social hierarchies are established which result in one (or few) females becoming or remaining egg-layers, with ovarian regression and oocyte resorption in the remainder (see Section 9.4.1 below).

9.2.2 Egg placement

Wasp nests are characterized by cells which are used to hold and incubate eggs and for rearing the immatures. Solitary wasps oviposit on paralysed prey or glue their eggs directly to the inside wall of the cell chamber, sometimes via a delicate suspensory filament. This is done before or after provisioning the cell with food. In the social wasps, both Polistinae and Vespinae glue their eggs directly to the wall of the cell in similar fashion to many solitary wasps. They insert the tip of the abdomen into the cell (Fig. 9.1), extrude an egg from the oviduct and fix it to the wall of the cell with a glandular adhesive (Fig. 9.2(a)).

By contrast, the Stenogastrinae exhibit a most unusual method of egg-laying. These wasps first collect a small quantity of a viscous secretion from the tip of the abdomen with the mandibles. They use this droplet as a tool to manipulate the egg as it emerges from the oviduct and also to attach the egg to the base of a cell (Turillazzi, 1985a). Sometimes a further quantity of the secretion is added on top of the egg. Before the egg hatches, further supplies of the abdominal secretion are often added to

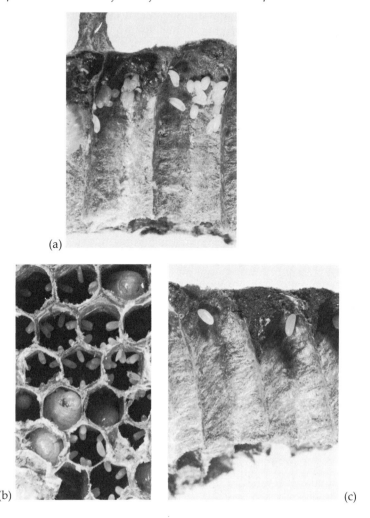

Fig. 9.2 Multiple eggs in cells of *Vespula germanica* comb (a, b) compared with the normal single egg per cell (c). Excess eggs in cells is a feature of perennial colonies with several egg-laying queens present. (Photographs by J. Green.)

the egg cells. The secretion is possibly analogous to the glandular adhesive used by other wasps to glue eggs in cells and the source is most likely Dufour's gland (Hansell, 1982a). A further remarkable feature of this stenogastrine secretion is its possible function as an ant repellent. It seems that eggs or larvae covered in the material are not preyed on by ants and even marauding hornets frequently leave such eggs and young larvae untouched (Turillazzi, 1985b).

9.2.3 Egg loss

Apart from the loss of deposited eggs through predation and other forms of colony destruction, there is also an apparent wastage of eggs resulting from a variety of intrinsic factors unique to the social wasps.

When several foundresses establish a nest, a process of dominance behaviour results in there being only one egg-laying female. If a subordinate wasp manages to lay eggs these can be recognized and eaten by the dominant female (see below, for further details). When egg eating occurs there is likely to be a trophic advantage accruing to the consumer although the regular production of trophic eggs, which is so characteristic of many ant societies, does not occur regularly in social wasps. However, in some *Polistes* species and *Mischocyttarus drewseni*, the first larvae to hatch are fed with eggs from other cells while eggs can provide a regular source of food if foraging is prevented or curtailed by inclement weather (Jeanne, 1972; Röseler, 1991).

If the queen of a *Vespula* colony should die prematurely, the ovaries of the normally sterile workers develop and the workers begin laying haploid (unfertilized, male-producing) eggs. This situation is invariably accompanied by multiple oviposition in cells such that a dozen or more eggs can be found in a single cell as hundreds of workers exercise their reproductive rights in the absence of the queen. A very similar situation occurs in overwintering colonies of *Vespula germanica* in Australia where dozens to hundreds of newly-recruited queens lay many eggs in individual cells (Fig. 9.2(b)) due to the limited number of available, empty cells (Spradbery, 1973b, and unpublished observations). While only one egg in each cell can survive to maturity, the excess eggs, while being reproductively wasted, would undoubtedly be trophically recycled.

9.3 PROTECTION FROM EXTERNAL FORCES

The social wasp colony with its eggs and larvae, adults and occasional stores of honey, represents a concentrated source of food, rich in protein and highly attractive to other organisms. Protection from outsiders, be they predatory ants, bats or badgers, parasitic flies, moths or wasps, even marauding hornets (see Fig. 9.15(b)) and usurping conspecifics, has been a major driving force in the evolution of sociality and adaptiveness of nest structure in wasps.

Apart from safety in numbers, from small groups of nest-founding females to the tens of thousands of adults in many wasp colonies, passive defence by the use of physical structures and barriers characterize the wasps. Chemical repellents and alarm communication are additional features of colony defence in wasps. The variety of structural and behavioural defence mechanisms is testimony to the adaptiveness of vespid wasps.

9.3.1 Physical barriers

Nests of vespid wasps provide protection from physical and biotic adversity and a focal point for social activities. The variety of nest architecture resulting from the building activities of wasps, both solitary and social, has long fascinated the layman and specialist alike. The morphological equipment and the manipulative and sensory skills necessary for the building and provisioning of nests and the ability to orientate to such structures, underlines the remarkable degree of nervous organization in the aculeate Hymenoptera.

The great majority of solitary Vespidae employ earth in the construction of cells for oviposition and brood care. In the most primitive of the social Vespidae, the Stenogastrinae which inhabit the rain forests of Southeast Asia and New Guinea, some species use clay, others a mixture of mud and vegetable fibres and a few make nests of a coarse, paper-like material. Most other social wasps (Polistinae, Vespinae) construct nests from woody fibres scraped from trees, poles, etc. It appears that the heavy mud cells and poor quality carton of the Stenogastrinae has restricted the evolution of large nests and complex societies in this sub-family (Hansell, 1984). The Polistinae and Vespinae, on the other hand, have evolved a more advanced wood-pulp technology which allows the construction of large, complicated structures.

As nests become bigger, the increase in number of inhabitants results in increasing specialization, better coordination among colony members and increasing control over the internal environment of the nest (Jeanne, 1977). In the temperate Vespinae for example, the large nests are maintained at a virtually constant temperature of 30°C, such homeostasis resulting in optimum conditions for raising immatures and insulating the colony from the vagaries of a temperate climate (Spradbery, 1973a).

Nests, especially in tropical areas, are frequently used to store food in the form of honey droplets in the cells. Another form of food storage is made possible by rearing brood in open cells. Most wasp larvae produce secretions from their labial gland (which later produce the silk for the cocoon). This secretion is rich in sugars and amino acids (Hunt, 1982). Adult vespine and some polistine wasps feed off this secretion and, in the case of the queen (which lays up to 300 eggs per day), the larval secretions probably provide the bulk of the protein required for ovogenesis (Spradbery, 1973a; Hunt *et al.*, 1987).

The origins and divergence of nesting behaviour in wasps must have been primarily to protect the brood from predation—especially ant predation (Fig. 9.3). Adult wasps are probably rarely taken by predators

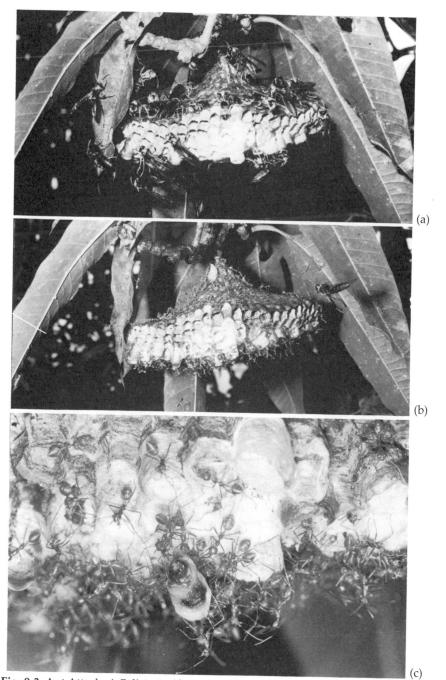

(a)

(b)

(c)

Fig. 9.3 Ant Attack. A *Polistes tepidus* nest with more than 30 adult wasps. (a) overrun by the tree ant, *Oecophylla* sp., (b) during an overnight raid (Papua New Guinea). Detail (c) ants removing wasp larvae from comb. (Photographs by author.)

compared with the brood, which represents a large and attractive food source. With eggs and larvae exposed in open cells they are accessible and detectable to predators and with no chance of escape. The relatively long tenure of nests at permanent sites must also increase the risk of discovery (Jeanne, 1975).

(a) Nests site

Although most wasps build aerial nests suspended from tree trunks, blades of grass, tiny rootlets and man-made artifacts, several species build their nests in protected, pre-formed cavities such as holes in trees, under cave-like overhangs or inside cavity walls, attics and even empty coconut shells (Fig. 9.4). A species of swarming *Ropalidia* in New Guinea frequently nests inside the galvanized pipes (Fig. 9.5) that now make up the majority of hand rails and fence posts in the cities—a more robust modern alternative to the usual nesting site of bamboo poles.

To reduce the incidence of detection by ants, many wasps build their nests on leaves rather than twigs, frequently along the midrib of the leaf with its increased cryptic advantage. A familiar sight in both tropical and temperate countries these days is the large number of nests suspended from buildings, especially the eaves and gutters, where they gain additional protection from rain and sun. It is recognized that ants forage less on buildings than surrounding ground structures and vegetation (Jeanne, 1975). As we shall see later, some wasps gain protection by building their nests in trees inhabited by pugnacious but non-predatory species of ants.

Where nest sites are particularly attractive for some reason or are in short supply, large numbers of nests may be built close together. The congregations of colonies are undoubtedly attractive to predators and possibly to parasites. The Australian paper wasp, *Ropalidia plebeiana*, builds exposed nests under overhanging cliffs and their modern counterpart, concrete bridges. Although there is usually plenty of space under the bridges, this wasp builds hundreds of nests in dense aggregations (Fig. 9.6) (Itô *et al.*, 1988). In this case there is likely to be some protection to be gained by staying close together and avoiding marginal predation —an example of the 'selfish herd' effect of Hamilton (1971).

(b) Nest architecture

An entire chapter could be devoted to wasp nest architecture in relation to specialization and evolution in the context of brood protection (see Jeanne, 1975, for a comprehensive review).

Nests consist of two elements; the individual cells for rearing brood (the 'reproductive function' of Jeanne, 1977) and the purely structural ('non-

Fig. 9.4 *Ropalidia* sp. nest in coconut shell. (Photograph by author.)

reproductive') components which are of engineering importance. A few of the more striking examples of architectural diversity in relation to defence of eggs and larvae will be described here.

There are basically two types of nest structure: the first is open, exposed and without envelopes. Most familiar are the nests of the paper wasps, *Polistes* and *Ropalidia*, with a single pillar or petiole attached to the substrate (see Figs 9.11 and 9.12). This narrow support is more easily protected from ants than cells which are built directly onto the substrate as in some Stenogastrinae of Southeast Asia (Fig. 9.7). Other stenogastrines build their cells without a suspensory pillar but on rootlets, which achieves the same effect as a defensive pillar (Fig. 9.8) (Spradbery, 1975, 1989).

The second basic nest structure is the closed nest with a surrounding envelope which restricts access to a single entry/exit hole which can be readily and economically manned by guards (Figs. 9.7 and 9.9). The entrance hole in some species is physically 'closed' at the threat of danger when the wasps plug the entrance with their heads, as in many South

Fig. 9.5 *Ropalidia* sp. nest inside steel pipe handrail. Note guard wasps around entrance. (Photograph by author.)

American polistine species (Chadab-Crepet and Rettenmeyer, 1982). It is among this group that the versatility and sheer size of wasp nests becomes evident; detailing the variation in nests has spawned its own lexicon of descriptive terms (for example *stellocyttarous calyptodomous*, which describes the nests of the common *Vespula* species with their combs supported by pillars and connected to an envelope).

While a single comb or several combs suspended one below the other, or even side by side, are reasonably familiar wasp nest structures, some tropical *Ropalidia* construct spiral ramps of comb, apparently solving many of the engineering problems inherent in building a covered yet expandable nest (see 9.4(a)) (Kojima and Jeanne, 1986; Spradbery and Kojima, 1989). The variations in nest structure can also be a useful character in the taxonomy of wasps (Richards and Richards, 1951; Sakagami and Yoshikawa, 1968).

Sometimes the entrance hole is extended into a tube-like turret which undoubtedly enhances the protection afforded by a single entrance. The turret is a spectacular addition to many embryo nests of *Vespa* species when the solitary queen in establishing the nest (Fig. 9.7(e)). Turrets also occur in the larger nests of some South American swarming Polistinae such as *Angiopolybia pallens* (Jeanne, 1975) and *Parachartergus fraternus*

(a)

(b)

Fig. 9.6 Nesting aggregation of *Ropalidia plebeiana*. (a) Road bridge habitat in New South Wales, Australia. (b) Detail of aggregation. (Photographs by author and Y. Itô, from Itô *et al.* 1988.)

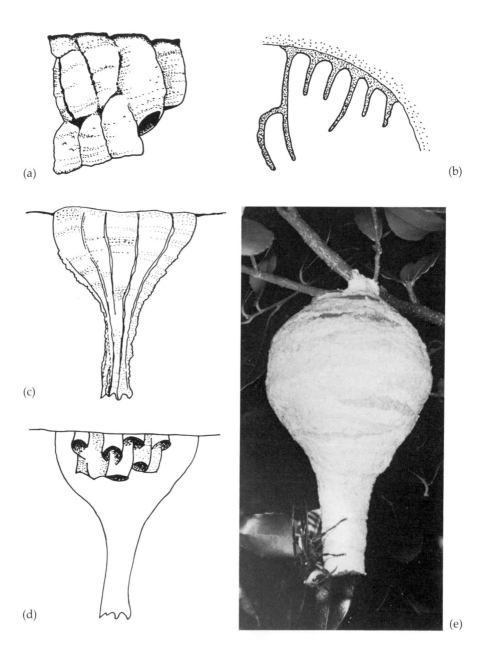

Fig. 9.7 Nest structures in social wasps. (a) *Parischnogaster striatula*—an exposed mud and carton nest with the cells built directly onto the substrate; (b) section of nest (redrawn from Ohgushi *et al.*, 1983); (c) *Eustenogaster calyptodoma*—brittle carton nest, with cells protected by an envelope; (d) to illustrate cells (redrawn from Ohgushi *et al.*, 1986); (e) Embryo nest of the hornet, *Vespa analis*. Note long turret leading to entrance hole. (Photograph by M. Matsuura.)

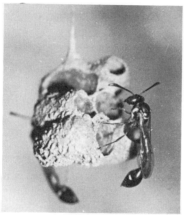

Fig. 9.8 Mud nests of New Guinea stenogastrine wasps attached to rootlets (photographs by author) (a) *Stenogaster concinna* (b) *Anischnogaster iridipennis*.

(Fig. 9.9(b)) (see also Richards, 1978, Plates 1–4). The turrets could also serve an environmental control purpose by retaining warmed air within the nest (R. L. Jeanne, unpublished data).

(c) Ant guards

With ants being the dominant predators in the lives of tropical wasps, the development of specific anti-ant structures would not be surprising. Two such modifications have been described, both within the Stenogastrinae, but made from very different raw materials.

In an unnamed *Stenogaster* species in the Philippines, cells are made from decayed wood and built onto a slender root above which a pair of discs occur (Fig. 9.10(a)) (Williams, 1919). The discs reminded Williams of the metal plates which are fastened to the mooring lines of ships as rat guards. Their function in the wasp is likely to provide similar protection from ants.

In Indonesia, *Parischnogaster nigricans serrei* builds linear nests with up to 40 cells in a row, attached to small roots or even pieces of wire (Turillazzi and Pardi, 1981). Most nests have one or several rings of a gelatinous, pearly white substance around the nest support, 1–6 cm from the nest and always between the nest and the substrate. Pagden (1962) had already suggested that these rings were sticky ant guards on nests of *P. jacobsoni* in Malaysia. Turillazzi and Pardi (1981) observed the 'construction' of such ant guards, which is begun one or two days after the first cell is built. The glutinous substance is exuded from the tip of the abdomen (possibly from Dufours gland) from whence it is collected by the wasp's hind legs, transferred to the middle legs and thence to its mouth. From here, the material is applied to the support or existing ant guard material until a circular pad is built up (Fig. 9.10(b)). Ants seemed incapable of traversing the secretion and, when it was spread onto a piece of wire and presented to ants, they made no attempt to pass over the treated area.

(d) Camouflage

Another strategy to avoid detection by predators, especially vertebrate ones, is the concealing of a nest by camouflage. Some wasps add secretory products to nests which darken them as well as provide some degree of water repellancy, resulting in a variegated appearance of the nest carton. The envelopes of *Metapolybia* spp. on tree trunks make nests difficult to see while some *Mischocyttarus* make nests which look remarkably like dead twigs (Jeanne, 1977).

In New Guinea, *Ropalidia cristata* builds nests on lichen-covered trees and rocks. The wasps collect the surrounding lichen and apply little pieces to the comb and more especially to the pupal cocoons, which are vivid white and would otherwise be easily detected (Spradbery and Kojima, 1989). Similar behaviour also occurs in some South American Polistinae (Evans and Eberhard, 1970).

So far in this chapter, protection from outside influences has been achieved by passive defence mechanisms. As nests get bigger, the capacity to employ more direct, active defence becomes a more viable option.

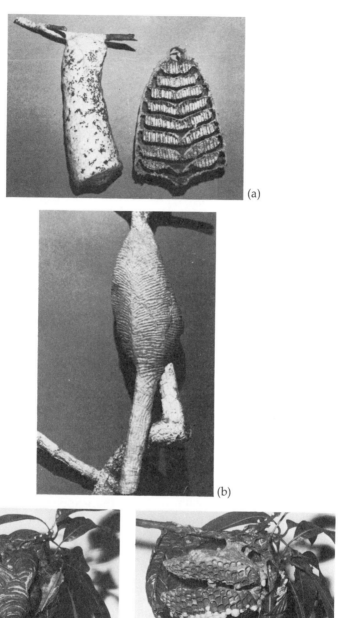

Fig. 9.9 Protective nest structures in social wasps (photographs by author) (a) *Chartergus* sp. with stout outer envelope and a single entrance hole. (b) *Parachartergus fraternus* with turret-like entrance. (c, d) Nest of the hornet, *Vespa affinis*, nest with single entrance (with and without envelope).

9.3.2 Behavioural responses

When nests of social wasps are approached by aliens, be they predators, parasites or conspecifics, the inhabitants invariably react in a defensive manner. If the threat increases, such behaviour may result in offensive reactions and stinging episodes, especially if many adults are present. Defensive behaviour can be viewed as an escalating preparation for attack as progressive threshold levels are surpassed.

How do wasps perceive danger, what is their response and how do they communicate danger? This section deals with the behavioural or active defence of the nest and brood.

(a) Adult response to sights and sounds

Exposed nests probably rely more heavily on visual cues to indicate threat than enclosed ones, although all wasp nests, even subterranean colonies, respond to vibrations, be it jarring of the nest itself or the surrounding substrate.

In the exposed nests of *Mischocyttarus drewseni* in Brazil, the approach of large objects causes the adults on the comb to turn and face the intruder, spread their wings and adopt the typical threatening posture of vespid wasps with raised bodies and forelegs (see Fig. 9.11). If the object gets closer, the adults bend their abdomens to one side and buzz their wings while producing a characteristic odour. Only wasps which are able to see the approaching danger perform in this way but, as soon as wing buzzing commences, the remaining adults are alerted—presumably through vibrations of the nest structure created by the wing buzzing behaviour (Jeanne, 1972; Chadab, 1980). This audio communication is probably accompanied by chemical cues. Also, there appears to be a very definite positive correlation in this species between the number of wasps on the nest and the degree of aggression. Jeanne (1982) also found similar behaviour in the South American paper wasp, *Polistes canadensis*.

Although wasp colonies are no match for the Army ants (*Eciton* spp.), wasps will actively defend against ant predation by biting the approaching ants (West-Eberhard, 1969; Post and Jeanne, 1981) or grabbing them, flying off and dropping them at some distance from the nest (Chadab-Crepet and Rettenmeyer, 1982).

In *Polybia occidentalis*, which constructs gourd-like enclosed nests hanging from trees and shrubs, the defence response (= pre-attack behaviour) is initiated by jarring the nest (Jeanne, 1981a). Jeanne distinguished two levels of response to potential predators: alarm recruitment and attack. The recruitment of wasps from within the nest occurs in response to physical movement of the nest, which results in a one-second burst of wing buzzing by the inhabitants. The wasps then pour out of the single

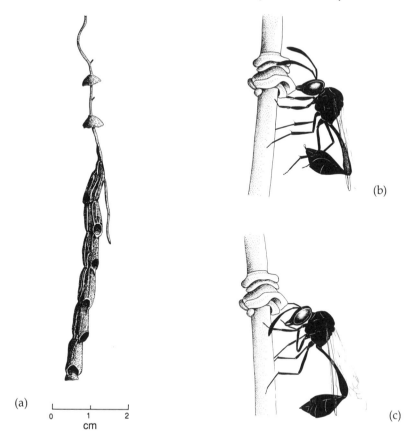

Fig. 9.10 (a) Structural umbrella-like 'rat guards' in a stenogastrine wasp (from Williams, 1919). (b) ant guard made from an abdominal secretion in *Parischnogaster nigricans serrei* (redrawn by Anne Hastings from Turillazzi and Pardi, 1981). In the first sketch, a blob of exudate has been expressed at the tip of the abdomen.

entrance hole and fan out over the envelope of the nest, displaying their typical threat posture. If no further threat is perceived, the wasps fold their wings after about 15–30 s and begin moving back inside the nest, the first wasps that responded being the last to return after approximately 60–75 s. When Jeanne disturbed a *P. occidentalis* nest very gently, there was no immediate response, but by increasing the level of disturbance by tapping or shaking the nest, wasps suddenly appeared at the entrance hole, suggesting a response to an alarm signal generated by the wasps themselves rather than to the mechanical disturbance. In experiments in which Jeanne (1981a) injected the contents of the *P. occidentalis* venom gland into a nest via a hypodermic syringe, the wasps responded with typical alarm recruitment behaviour, further supporting a chemically-mediated response (see Section 9.3.3 (b)).

When the threat continues or increases, the wasps will ultimately attack the would-be aggressor. There is little doubt that visual cues play a major part in attack behaviour, although the role of chemical emissions and olfactory response is now becoming more apparent (see below). When a nest of wasps is sufficiently disturbed, the inmates fly at the attacker in response to visual cues such as movement or dark objects against light backgrounds (Jeanne, 1981a). In the larger vespine species, movement is a far stronger releaser than dark colours (Maschwitz, 1964). Interestingly, returning foragers do not normally engage in defensive activities but are stimulated into aggressive behaviour after they re-enter the nest (Gaul, 1953). See Akre and Reed (1984) for detailed review of vespine defensive behaviour.

When populous nests are disturbed by large predators, group stinging or mobbing often occurs, an effective response when there are large numbers of defending wasps. The pain inflicted by wasps which inject venom into vertebrate enemies has probably been experienced at one time or another by most of us. There is little doubt that venom injection is a powerful deterrent to would be predators—evolution has ensured that stinging insects produce highly effective venoms that hurt.

(b) Alarm signals generated by larval wasps

While it is clear that adult wasps help defend their colonies by producing and responding to sounds or vibrations of the nest structure, recent studies by Jacob Ishay and his coworkers in Israel of the hornet, *Vespa orientalis*, adds larvae-produced sounds to the vocabulary of wasp language. Shaudinischky and Ishay (1968) first described the sounds produced by hornet larvae. They use their mandibles to scrape the walls of their cells, thereby producing a specific, rhythmic sound, audible to the human ear, and which probably acts as a hunger signal. Such solid-borne sounds, using the nest structure as a sounding board, are likely to be a highly effective method of communication in wasp nests. Certainly the nest structure is eminently suitable for both sound production and reception. Such sounds or vibrations decay very rapidly, enabling additional communication immediately after transmission of a particular signal.

Barenholz-Paniry and Ishay (1988) recorded the sounds made by mature larvae in response to major disturbances. The mechanical disturbance of the nest structure caused a remarkable response by the larvae—a completely synchronous scratching of the cell walls with their mandibles at a tempo of 50 beats per minute. The perfect synchrony enhances the sound the larvae produce and it seems most likely that this sound is a warning signal to the adult hornets in the event of mechanical disturbance of the nest.

(c) Ant–wasp associations

Although ants are major predators of wasps, there are several South American polistine species which build their nests in trees where ants also nest (Richards and Richards, 1951; Richards, 1978). The ants in these associations are species of *Dolichoderus* and *Azteca*, non-predatory ants but with highly pugnacious behaviour which apparently keeps the area clear of the predaceous ant species (Evans and Eberhard, 1970). Indeed, some species of Polistinae have never been found nesting except in association with *Azteca* ants (Hamilton, 1972; Richards, 1978).

9.3.3 Chemical defences

There have been several references in earlier sections of this chapter to the possibility of there being different forms of chemical defence systems in wasps. Two classes of chemical defence will be reviewed here: the passive use of barriers, and the alarm or recruitment pheromones.

(a) Chemical barriers

In exposed nests containing brood, especially nests in the early stages of colony establishment and more particularly with a single foundress in attendance, the risks of predation are extremely high. Every time the female forages for food or building materials the nest must be left unattended and vulnerable to predation. Many polistine wasp species have evolved a chemical defence mechanism which at least protects their nests from enemy number one—the marauding ants.

In the Brazilian wasp, *Mischocyttarus drewseni*, nests are generally established by single females and suspended by a long, thin pillar up to 3 cm long down which ants must descend to gain access to nest and brood. It was found that the foundress females frequently rubbed the tip of their abdomen along the surface of the suspensory pillar (Fig. 9.12a) (Jeanne, 1970). Jeanne rubbed the tip of the wasp's abdomen onto glass capillaries which, when offered to ants (*Monomorium pharaonis*), led to far fewer of them traversing the capillary to gain access to a piece of meat attached to the end of the tube than would have been expected for unmarked capillaries. The terminal sternite of *M. drewseni* and most other non-swarming Polistinae has a tuft of hairs on an unsclerotized part of the cuticle which is associated with well-developed abdominal glands. The tuft of hairs is called 'van der Vecht's organ' after its discoverer (van der Vecht, 1968). In living wasps, the organ is generally moist, indicating

Fig. 9.11 Nest of the Australian paper wasp *Polistes humilis*. Note the narrow, shiny black pedicel and the alarm posture of the adult females with bodies raised and wings outstretched. (Photograph by J. Green.)

vigorous glandular activity. The hairs of the organ are not unlike a brush and seem ideally suited to painting or smearing substances onto a substrate (Fig. 9.13).

Since Jeanne's (1970) work, most other species of wasp which build exposed nests suspended by pillars have been found to apply chemical ant repellents to the nest support. Turillazzi and Ugolini (1979) confirmed the use of ant repellent in three species of European *Polistes* and Post and Jeanne (1981) in *Polistes fuscatus* in the USA. Kojima (1983) described similar behaviour in *Ropalidia fasciata* on the Okinawa Islands. The incidence of pillar-rubbing behaviour is far higher (two to seven times in *R. fasciata*) in single-female nests than multi-foundress ones. The repellent is smeared onto the pillar most often just before leaving the nest on a foraging trip with the efficacy of the ant repellancy being lost after 2–4 h following its application. Rubbing behaviour becomes most intense when nests contain mature larvae and less so with only eggs or early-stage larvae present. With the appearance of the first adult progeny, the frequency of application of ant repellent decreases as the additional workforce play a more active defence role in protecting the nest against ants.

Some components of the ant repellent produced by van der Vecht's gland in *Polistes fuscatus* have been isolated and the active component identified as methyl palmitate (Post *et al.*, 1984b).

Although not confirmed as being a chemical repellent, the enclosed nests of the South American *Nectarinella championi* have sticky droplets around the narrowed nest entrance which may well deter ants or other arthropod predators from entering the nest (Schremmer, 1977).

Fig. 9.12 (a) *Mischocyttarus drewseni* female applying ant repellent to the suspensory pillar of the nest (photo by R. L. Jeanne, from *Science*, © 1970 by the American Association for the Advancement of Science). (b) *Polistes humilis* applying repellent to nest pillar (photograph by author).

(b) Chemical alarms

As mentioned above, the wing buzzing and abdomen wagging, shaking and bending behaviour of *M. drewensi* is accompanied by a strong odour (Jeanne, 1972). The same odour was also detected at the site of a stinging episode and lasted for a minute or two. Jeanne suspected that this substance might be an alarm pheromone. In Brazil, *Polybia occidentalis* constructs large, enveloped nests. Jeanne (1981a) injected the contents of a venom sac (dissected from a female wasp) and other parts of the wasp's body into the nest of *P. occidentalis* via a hypodermic syringe. Only the venom sac and the sting apparatus elicited a strong response (a rapid movement) with large numbers of the nest inhabitants responding. Similar experiments with *Polistes canadensis* again demonstrated that the venom sac and sting elicit defensive behaviour among wasps on the nest (Jeanne, 1982). Apparently, the alarm pheromone component of the wasp's venom reduces the threshold for an attack response although visual cues, such as a moving target, are also necessary to unleash a stinging attack by these wasps.

In their study of two North American species of *Polistes*, Post *et al.* (1984a) found that the alarm pheromones in the venom were not species specific.

Not surprisingly, similar alarm pheromones have been demonstrated in the underground nesting Vespinae of Europe. Both *Vespula germanica* and *V. vulgaris* have efficient chemical alarm systems which lead to prompt defence against potential enemies (Maschwitz, 1964). In the *Dolichovespula* species, which build aerial nests in trees and shrubs, If the nest is disturbed mechanically there is an immediate rush of wasps to investigate the disturbance, exactly as in *Polybia occidentalis* in Brazil (Jeanne, 1981a). When Maschwitz (1984) crushed whole wasps or parts of wasps including dissected-out glands and offered them to a *D. saxonica* nest, the venom gland and sac were the only components to stimulate the wasps to rush out of the nest and behave aggressively. Similar responses have been described for *Vespula squamosa* in the USA when offered extracts from venom sacs (Landolt and Heath, 1987). The response includes recruitment of wasps from a nest and their orientation flight to the source of the venom extract. Stinging behaviour was focused at or close by the source of the venom extract (filter paper discs attached to black spheres or boxes). The same reaction was recorded for *P. occidentalis* by Jeanne (1981a), who found that crushed venom sac, especially on a moving target, stimulated the wasps to attack with considerable ferocity.

The venom of the European hornet, *Vespa crabro* has been extracted and a variety of volatile components have been identified (Veith *et al.*, 1984). One of these components, 2-methyl-3-ene-2-ol, caused the typical alarm reaction elicited by whole venom sacs.

9.3.4 When all else fails

Despite the array of protective structures and behaviours employed by wasps to defend their brood from the effects of parasites, predators and

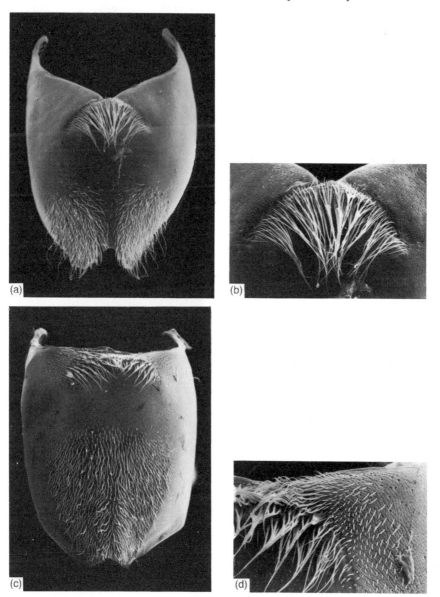

Fig. 9.13 Brush-like van der Vecht's organ which is used to apply ant repellent to wasp nests. (SEMs by Helen Geier and Colin Beaton.) (a–b) *Polistes humilis* (c–d) *Ropalidia plebeiana*.

Fig. 9.14 A parasitic ichneumonid wasp *Arthula* sp. (below) on a nest of *Polistes* in New Guinea (photograph by author).

climatic vicissitudes, when colonies become heavily parasitized (see Fig. 9.14) or nests are destroyed by predators (see Fig. 9.15), there are still a few strategies remaining to circumvent total disaster.

In the tropical *Polistes canadensis*, several (up to 38!) separate combs can be built in a group by one colony. Jeanne (1979) found that combs of this species were frequently infested with a predaceous tineid moth. When infestations were light or confined to a particular part of a comb, the parasitized portion of combs was cut around and excised by the wasps. However, if parasitism became too heavy to employ this removalist technique, the wasps simply abandoned the comb and withdrew to one or more of the remaining combs in the group.

When conditions in a nesting cavity become inadequate due to space restrictions or flooding, etc., hornets (*Vespa* spp.) regularly move to a new, more suitable piece of real estate up to 180 m away (Matsuura, 1984). At this stage, the foundress queen is still mobile (unlike mature vespine colonies where the queen becomes physogastric and loses her capacity to fly (Spradbery, 1973a)). Examples of nest relocation by hornets have been recorded in England (Nixon, 1985) and Japan (Matsuura, 1984). Even after construction of the new nest, the workers will continue to tend the brood of the original nest until they pupate and emerge as adults up to a month later.

Following severe disturbance or loss of a nest due to predation, invasion by parasites, fire or weather-related catastrophe, many wasp species retain the capacity and mobility to evacuate the nest and move in a coordinated way (= swarm) to a new site. This 'absconding swarm', which results from abandonment of the old nest, generally contains several queens (egg-laying females) and workers which relocate to the new locality by following scented trails laid down by groups of scout wasps (Jeanne, 1981b; West-Eberhard, 1982). The glands of the sixth abdominal segment (as distinct from the ant-repellent producing seventh abdominal gland) are used to mark trails in the swarming species (Jeanne, 1981b).

Thus, the ultimate response to a damaged nest is to flee: when all else fails the sacrifice of some brood is preferable to the total loss of the colony, adults and all.

9.4 INTERNAL CONFLICT AND COMPETITION

Although the threat from external factors remains a powerful force in the success or demise of wasp nests, within colonies, especially newly-formed ones, there can be considerable conflict as females struggle for control of the oviposition resource (= empty cells) in their attempts to exercise their reproductive privileges.

There are few recorded examples of groups of egg-laying females living side by side in reproductive and social harmony. Among them are the swarm-founding Polistinae of South and Central America. Here, several queens (egg-laying and inseminated females) with a retinue of workers establish new nests with great rapidity. To fill the new cells with eggs, many ovipositing females (polygyny) are needed. However, when such nests are firmly established, there is often a reduction in the number of queens to a single one (functional monogyny) (West-Eberhard, 1978).

Other examples of multi egg-layers coexisting on polygynous colonies include the unusual perennial colonies of the European wasp, *Vespula germanica*, which has been accidentally introduced to Australasia and other countries in recent time (Edwards, 1980). In the large, overwintering nests of this species as many as 600 or more laying queens are found (unpublished observations). The queens do not behave aggressively towards each other—in stark contrast to the normal situation in spring when nest-founding queens of *V. germanica* fight to the death during conflicts over site and embryonic nest ownership (Spradbery, 1973a,b). In the New Guinea hornet, *Vespa affinis*, nests are established by several queens which appear reproductively and socially equal with this apparent compatibility continuing throughout the life of the colony (Spradbery, 1986).

However, in the great majority of wasp societies there is, at least at some stage in the life cycle, a period of intense social conflict and reproductive competition. This final section reviews how wasps control oviposition in nests and how the typical matriarchal dictatorship is established and maintained.

9.4.1 Behavioural dominance

In the independent founding (as distinct from swarming) species of Polistinae, nests are frequently established by several females (= pleometrosis). Multi-foundress nests may have up to 20 females (e.g. *Ropalidia marginata* in India: Gadagkar *et al.*, 1978). Reutilization of old nests by several queens which overwinter on the parental combs is common in *Ropalidia fasciata* in Japan (Itô *et al.*, 1985) and *Polistes humilis* in Australia (Itô, 1986, and personal observations). Sometimes a nest is established by a group of females, but more often one female begins a nest and is soon joined by others. By whatever route a multi-foundress nest is formed, the scene is set for the inevitable struggle to decide who lays the eggs.

The behaviour which results in one female becoming the sole or principal egg-layer was first observed in the European paper wasp, *Polistes gallicus*, by Heldmann (1936) and later studied in classical detail by Pardi (1948). Essentially, interacting females on the new nest display dominant or submissive behaviour (Fig. 9.16). Sometimes the reactions are mild, but if two females are of equivalent status they generally exhibit extreme force. Such domestic violence can lead to death or dismember-

Fig. 9.15 *Polistes* nest attacked by a hornet (*Vespa tropica*) in New Guinea. (Photograph by author.)

ment of the combatants. In *Ropalidia plebeiana*, dominant females snip off the wings of their rivals with catastrophic results for the vanquished (Yamane *et al.*, 1991). The behavioural repertoire used in dominance reactions also includes the exchange of regurgitated food, antennal fencing, biting or akinesis and withdrawal by a subordinate.

In small groups of foundress females, the net result of these social interactions is the rapid development of an essentially linear dominance hierarchy or pecking order. Coinciding with this development is a drastic reduction in the overt expressions of dominance as the interactions become ritualized with a concomitant reduction in the time and energy expended on dominance behaviour. Furthermore, the dominant wasp(s) generally occupies a specific location on the nest, gains a distinct trophic advantage during frequent adult–adult food exchanges and forages herself little if at all, thus conserving her resources for ovarian development. If the dominant female should be lost, the next wasp in the pecking order takes over as principal egg-layer and, during the course of next establishment, several females may contribute to the reproductivity of the nest (Pardi, 1948). When the first female progeny emerge from nests there is often a revival of overt dominance behaviour.

The physiological factors which determine dominance and status are linked to ovarian development, which in turn is controlled by the activity of endocrine glands (corpora allata) and their production of ovary-stimulating juvenile hormone (Röseler *et al.*, 1980; Röseler, 1991, for review).

9.4.2 Egg eating and egg guarding

Lower ranked females quite often manage to lay eggs, especially in a newly-founded nest. The dominant female can usually recognize these eggs and invariably eats them ('differential oophagy' of Gervet, 1964). Recognition of 'alien' eggs is related to the age of the egg and presumably its odour and can last for as little as 40 min in *Polistes canadensis* (West-Eberhard, 1969) or up to a day in *P. dominulus* (Gervet, 1964). It seems that oophagy is practised by the dominant female, which immediately lays an egg in the newly vacant cell—another advantage of superior ovarian development. Oophagy is also practised by the vespine species *Dolichovespula* (Greene, 1979).

Because of the risks of egg eating by rivals, some wasps guard the cells in which they have recently oviposited. These egg-guarding vigils can last up to 3 h in *Metapolybia aztecoides* (West-Eberhard, 1981).

9.4.3 Control over cell construction

Because an empty cell is the ovipositon resource and stimulates egg-laying, control of oviposition can also be synonymous with the control of cells—either their construction or in keeping existing cells filled. To ensure that empty cells are rarely available, two strategies are employed by the dominant wasp or queen. In many species of *Polistes*, the queen is the major or sole initiator of cell construction and is thus in a commanding position to lay an egg in the newly-formed cell base (West-Eberhard, 1969; Dew, 1983). At other times, the queen ensures that there is no accumulation of empty cells by ovipositing in cells soon after they are vacated by emerging adults.

In *Dolichovespula maculata* in North America, the queen patrols the edge of combs where the new cells are added. The queen also plays a major role in trimming the ragged pupal cappings immediately after adult emergence, thus guaranteeing her option to oviposit in freshly-prepared, empty cells (Greene, 1979). Although the workers initiate new cells in the hornet, *Vespa orientalis*, they only build them to half their normal depth until the queen lays an egg in the cell (Ishay, 1973). An accumulation of empty cells in the hornet's nest eventually inhibits the building of new cells (Ishay *et al.*, 1986).

9.4.4 Pheromones—the chemical control of oviposition

Using physical acts of dominance to suppress ovarian development and egg-laying by rivals, be they co-foundresses or daughters, is an energy-consuming process with limited spatial influence. Control of social interaction via chemical means would add a new perspective with a potentially wider sphere of influence. (For review see Spradbury, 1991.)

The first signs of such chemical control have been observed in the swarming Polistinae, in which queen recognition by workers is based on chemical cues (Naumann, 1970; West-Eberhard, 1977). Here, the ritualized postures which declare queen status can be more effectively reinforced by the use of chemical odours. For example, in *Protopolybia exigua*, the workers surround queens and form courts, there is no overt aggressive behaviour, ovarian suppression in workers is highly effective and there is thus no need for oophagy (Simoes, 1977).

The culmination of chemical control of reproduction in social wasps is the use of pheromones in the Vespinae, where a single queen heads a colony of up to 5000 sterile worker daughters. If the queen is lost, the ovaries of the normally sterile workers develop almost immediately and they begin ovipositon within a week. Nevetheless, in the more primitive (and numerically smaller) species, there is an apparent combination of physical dominance with some chemical control as in *Dolichovespula* (Greene, 1979). Oophagy and cell guarding by queens has been observed with some ovarian development and occasional oviposition by workers. When workers attempt to lay eggs, they are frequently physically mauled by the queen—a very polistine trait. Similar behaviour has been reported

Fig. 9.16 Dominance behaviour in *Polistes humilis*. The dominating wasp (left) is chewing the antenna of the subordinate with her mandibles. The subordinate is responding with an extremely submissive posture: head down against the nest surface. (Photograph by author.)

in species of the *Vespula rufa* group which make relatively small nests with up to 400 workers present (Landolt *et al.*, 1977; Reed and Akre, 1983). Incomplete suppression of ovary development in workers of this species group frequently leads to competition for vacant cells between the queen and her workers.

Royal courts in which groups of workers surround a queen, licking her and stroking her with their antennae, is a feature of some hornet species (Fig. 9.17) (Ishay *et al.*, 1965; Matsuura, 1984). Although royal courts are a feature of honeybee colonies and associated with pheromone ('queen substance') dissemination (Free, 1987), their role in *Vespa* colonies has not been experimentally resolved. In *Vespa orientalis*, queens certainly produce ovary-suppressing pheromones, but their dissemination is not necessarily via the royal courts. When *Vespula germanica* workers were housed with a *V. orientalis* queen the workers' ovaries were suppressed without any direct contact with the hornet queen (Ishay *et al.*, 1986).

In the *Vespula* spp. (*V. germanica*, *V. vulgaris*), which produce the largest vespine colonies with thousands of workers, the effectiveness of queen-produced pheromones in suppressing worker ovarian development is supreme. There is no physical contact between queen and workers, except occasional reciprocal feeding, yet ovary suppression is wholly effective until the queen dies or becomes physiologically inept at the end of the annual colony cycle. Experimental evidence to date suggests a low volatile (non-gaseous) pheromone(s) which is most likely applied to the comb structure as the queen moves around the nest laying eggs in vacant and newly-built cells. By swapping combs 'contaminated' by a queen, it is possible to suppress worker ovary development in groups of workers which otherwise have no contact with the queen (Akre and Reed, 1983, and personal observations). The glandular source and chemical identity of such pheromones remain unknown but are the subject of current research by the author.

9.4.5 Alternatives to social bondage

What happens to those females which help establish nests as co-foundresses but fail to become egg-layers in the dominance hierarchy shake out? By remaining on the nest, their reproductive future is bleak because the chances of taking over the principal egg-laying role (= replacement queen) are, in reality, quite remote. The incumbent queen, by remaining on the nest and not participating in risky foraging activities and being at the receiving end of most trophallactic exchanges, continues to maximize her survival and reproductivity.

Nevertheless, subordinate co-foundresses surreptitiously sneak in the occasional egg which could remain undetected by the queen. But at this early stage of the colony cycle, those eggs will most likely develop into workers, not reproductives (which are produced towards the end of the

Fig. 9.17 Royal court in the Japanese hornet, *Vespa analis* (photograph by M. Matsuura). Queen surrounded by workers, some of which are licking and antennating the queen's body.

season). Another major problem facing the 'failed' co-foundress is that, with the emergence of the first generation of workers, these supernumary co-foundresses are generally driven off the nest. Such refugee females may attempt to usurp other nests and take over as the primary egg-layer. Usurpation is common among Stenogastrinae (Hansell, 1982b), in Polistinae (Kasuya, 1982; Makino, 1989) and even among spring-nesting queens of vespine species (Nixon, 1986). Such intraspecific usurpations were the likely setting for the evolution of interspecific social parasitism which is so widespread in the Vespidae (Matthews, 1982).

Several wasp species overwinter as queens on the parental nest or nearby and in the following spring they can reutilize the nest. In one of these species, an Australian paper wasp *Ropalidia plebeiana* (see Fig. 9.6), a most remarkable tactic is sometimes employed to overcome reproductive competition (Itô *et al.*, 1988). Here, the wasps on a single nest settle into two or more groups, each headed by a primary egg-layer. Once loose territories are established on the comb, the cells are chewed away until the different parts of the nest become physically separated. This unique method of colony fission can result in several new colonies being produced from the original parental nest with a consequent elimination of energetically wasteful dominance behaviour.

9.5 CONCLUDING REMARKS

By examining details of just one element in the social repertoire of wasp communities—oviposition and protection of the oviposition resource—a formidable array of behavioural patterns is evident. The evolution of reproductive strategies in the social Vespidae has been influenced by selective pressures from without and within the community. Although external forces such as climate, predation and parasitism, plus competition from conspecifics, have influenced nest design and defence strategies, the social conflict and tension within the community have had the most potent impact on wasp social biology; from the primitively social wasps such as *Polistes*, where reproductive conflict is expressed in overt competitive combat, to the highly social Vespinae which have evolved chemical recognition and dissemination systems which ensure complete control over the reproductive function of nestmates. In *Vespula*, this control is underlined by the gross morphological distinctions between queens and workers in which the latter caste cannot even be inseminated by males due to size and structural differences in their genitalia.

In a world experiencing dramatic ecological and climatic change, how might the social wasps fare under changing selective pressures? Two examples will suffice to illustrate the extremes of such change. In Papua New Guinea, local agricultural practice involves slashing and burning the rain forest to extend the vegetable gardens that provide much of the country's staple food. The stenogastrine wasps, which seem to live a precarious existence in untrammelled rain forest, cannot survive such dramatic changes to their habitat. Indeed, the type locality for *Stenogaster concinna* in southern Papua New Guinea has been destroyed since its discovery and description 15 years ago (Spradbery, 1975).

A change of habitat and climate can be achieved by the introduction of organisms to new geographical areas as well as change in local conditions. One of the most common social wasps of Europe, *Vespula germanica*, has been accidentally introduced to many countries in recent years, including north America, Chile and Argentina, South Africa, New Zealand and Australia. In Australia, the wasp has spread dramatically since its introduction to the mainland ten years ago and now occupies more than 500 000 km² of southeastern Australia. A feature of its successful colonization has been the phenomenon of overwintering and subsequent growth of nests many times larger than found in Europe—one with an estimated biomass of half a tonne (Spradbery, 1973a). In such nests, many laying queens are found—quite contrary to the normal annual colony in which a single queen prevails. Productivity, as measured by numbers of queens (= reproductive units) produced per colony in these perennial nests is staggering, with production continuing throughout the winter months. During a second or third summer season, queen production in such colonies can lead to unprecedented population increase. The exploitation of the Australian environment by *V. germanica* exceeds normal expectations and underlines how some insects can respond to new conditions and environmental challenges.

ACKNOWLEDGEMENTS

I am grateful to Robert L. Jeanne for his valuable comments on an earlier draft of the manuscript. Although photo credits are included in legends to figures, I acknowledge the constant help in matters photographic provided by John P. Green. The following gave permission for the use of original figures: Japanese Journal of Entomology (Fig. 9.4(a)); Y. Itô and Society of Population Ecology (Fig. 9.6); R. Ohgushi, Japanese Journal of Entomology and Kanazawa University (Fig. 9.7 (a–d)); Hawaiian Sugar Planter's Association (Fig. 9.10(a)); S. Turillazzi and Monitore Zoologico Italiano (Fig. 9.10(b,c)); R. L. Jeanne and The American Association for the Advancement of Science (Fig. 9.12(a)) and M. Matsuura (Fig. 9.7(e), 9.17).

References

Akre, R. D. and Reed, H. C. (1983) Evidence for a queen pheromone in *Vespula* (Hymenoptera: Vespidae). *Can. Entomol.*, **115**, 371–377.

Akre, R. D. and Reed, H. C. (1984) Vespine defence. In *Defensive Mechanisms in Social Insects* (ed. H. R. Hermann), Praeger, New York, 59–94.

Barenholz-Paniry, V. and Ishay, J. S. (1988) Synchronous sounds produced by *Vespa orientalis* larvae. *J. Acoust. Soc. Amer.*, **84**, 841–846.

Carpenter, J. M. (1982) The phylogenetic relationships and natural classification of the Vespidae (Hymenoptera). *Syst. Entomol.*, **7**, 11–38.

Carpenter, J. M. (1988) The phylogenetic system of the Stenogastrinae (Hymenoptera: Vespidae). *J. N.Y. Entomol. Soc.*, **96**, 140–175.

Chadab, R. (1980) Early warning cues for social wasps attacked by Army ants. *Psyche*, **86**, 115–123.

Chadab-Crepet, R. and Rettenmeyer, C. W. (1982) Comparative behaviour of social wasps when attacked by Army ants or other predators and parasites. In *The Biology of Social Insects* (eds M. D. Breed, C. D. Michener and H. E. Evans), Westview Press, Boulder, CO, 270–274.

Dew, H. E. (1983) Division of labour and queen influence in laboratory colonies of *Polistes metricus* (Hymenoptera: Vespidae). *Z. Tierpsychol.*, **61**, 127–140.

Edwards, P. B. and Aschenborn, H. H. (1989) Maternal care of a single offspring in the dung beetle, *Kheper nigroaeneus*: the consequences of extreme parental investment. *J. Nat. Hist.*, **23**, 17–27.

Edwards, R. (1980) *Social Wasps. Their Biology and Control*, Rentokil Library, East Grinstead.

Evans, H. E. and Eberhard, M. J. W. (1970) *The Wasps*, University of Michigan Press, Ann Arbor, MI.

Free, J. B. (1987) *Pheromones of Social Bees*, Chapman and Hall, London, 218 pp.

Gadagkar, R., Gadgil, M. and Maharal, A. S. (1978) Observations on population ecology and sociobiology of the paper wasp *Ropalidia marginata* (Lep.) (Family Vespidae). *Symp. Ecol. Anim. Pop. Zool. Survey of India, Calcutta*, 1–22.

Gaul, A. T. (1953) Additions to vespine biology. XI. Defence flight. *Bull. Brooklyn Entomol. Soc.*, **48**, 35–37.

Gervet, J. (1964) Le comportement d'oophagie différentielle chez *Polistes gallicus* L. (Hymen. Vesp.). *Insectes Soc.*, **11**, 343–382.

Greene, A. (1979) Behavioural characters as indicators of yellowjacket phylogeny (Hymenoptera: Vespidae). *Ann. Entomol. Soc. Amer.*, **72**, 614–619.

Hamilton, W. D. (1971) Geometry for the selfish herd. *J. Theor. Biol.*, **31**, 295–311.

Hamilton, W. D. (1972) Altruism and related phenomena mainly in social insects. *Ann. Rev. Ecol. Syst.*, **3**, 193–232.

Hansell, M. H. (1982a) Brood development in the subsocial wasp *Parischnogaster mellyi* (Saussure) (Stenogastrinae: Hymenoptera). *Insectes Soc.*, **29**, 3–14.

Hansell, M. H. (1982b) Colony membership in the wasp *Parischnogaster striatula* (Stenogastrinae). *Anim. Behav.*, **30**, 1258–1259.

Hansell, M. (1984) How to build a social life. *New Sci.*, May 1984, 16–18.

Heldmann, G. (1936) Über das Leben auf Waben mit mehreren überwinterten Weibchen von *Polistes gallica* L. *Biol. Zentralbl.*, **56**, 389–400.

Hunt, J. H. (1982) Trophallaxis and the evolution of eusocial Hymenoptera. In *The Biology of Social Insects* (eds M. D. Breed, C. D. Michener and H. E. Evans), Westview Press, Boulder, CO, 201–205.

Hunt, J. H., Jeanne, R. L., Baker, I. and Grogan, D. E. (1987) Nutrient dynamics of the swarm-founding social wasp species, *Polybia occidentalis* (Hymenoptera: Vespidae). *Ethol.*, **75**, 291–305.

Ishay, J. (1973) The influence of cooling and queen pheromone on cell building and nest architecture by *Vespa orientalis* (Vespinae, Hymenoptera). *Insectes Soc.*, **70**, 243–252.

Ishay, J., Ikan, R. and Bergmann, E. D. (1965) The presence of pheromones in the Oriental hornet, *Vespa orientalis*. *J. Insect. Physiol.*, **11**, 1307–1309.

Ishay, J., Rosenzweig, E. and Pechaker, H. (1986) Comb building by worker groups of *Vespa crabro* L., *V. orientalis* L., and *Paravespula germanica* Fabr. (Hymenoptera, Vespinae). *Monit. Zool. Ital.*, **20**, 31–51.

Itô, Y. (1986) Spring behaviour of an Australian paper wasp *Polistes humilis synoecus*: colony founding by haplometrosis and utilization of old nests. *Kontyû*, **54**, 191–202.

Itô, Y., Iwahashi, O., Yamane, S. and Yamane, S. (1985) Overwintering and nest reutilisation in *Ropalidia fasciata* (Hymenoptera, Vespidae). *Kontyû*, **53**, 486–490.

Itô, Y., Yamane, S. and Spradbery, J. P. (1988) Population consequences of huge nesting aggregations of *Ropalidia plebeiana* (Hymenoptera: Vespidae). *Res. Popul. Ecol.*, **30**, 279–295.

Jeanne, R. L. (1970) Chemical defense of brood by a social wasp. *Science*, **168**, 1465–1466.

Jeanne, R. L. (1972) Social biology of the neotropical wasp *Mischocyttarus drewseni*. *Bull. Mus. Comp. Zool. Harv. Univ.*, **144**, 63–150.

Jeanne, R. L. (1975) The adaptiveness of social wasp nest architecture. *Q. Rev. Biol.*, **50**, 267–287.

Jeanne, R. L. (1977) Ultimate factors in social wasp nesting behaviour, *Proc. IUSSI Congr. Wageningen*, 164–168.

Jeanne, R. L. (1979) Construction and utilisation of multiple combs in *Polistes canadensis* in relation to the biology of a predaceous moth. *Behav. Ecol. Sociobiol.*, **4**, 293–310.

Jeanne, R. L. (1981a) Alarm recruitment, attack behaviour, and the role of the alarm pheromone in *Polybia occidentalis* (Hymenoptera: Vespidae). *Behav. Ecol. Sociobiol.*, **91**, 143–148.

Jeanne, R. L. (1981b) Chemical communication during swarm emigration in the social wasp *Polybia sericea* (Olivier). *Anim. Behav.*, **29**, 102–113.

Jeanne, R. L. (1982) Evidence for an alarm substance in *Polistes canadensis*. *Experientia*, **38**, 329–330.

Jeanne, R. L. (1991) Social biology of the swarm-founding Polistinae. In *Social Biology of The Vespid Wasps* (eds K. G. Ross and R. W. Matthews), Cornell University Press, Ithaca, N.Y.

Jeanne, R. L. and Davidson, D. W. (1984) Population regulation in social insects.

In *Ecological Entomology* (eds C. B. Huffaker and R. L. Rabb), John Wiley and Sons, New York.

Kasuya, E. (1982) Take-over of nests in a Japanese paper wasp *Polistes chinensis antennalis* (Hymenoptera: Vespidae). *Appl. Entomol. Zool.*, **17**, 427–431.

Kojima, J. (1983) Defence of the pre-emergence colony against ants by means of a chemical barrier in *Ropalidia fasciata* (Hymenoptera: Vespidae). *Jpn J. Ecol.*, **33**, 213–223.

Kojima, J. and Jeanne, R. L. (1986) Nests of *Ropalidia* (*Icarielia*) *nigrescens* and *R* (*I*) *extrema* from the Philippines, with reference to the evolutionary radiation in nest architecture within the subgenus *Icarielia* (Hymenoptera: Vespidae). *Biotropica*, **18**, 324–336.

Kugler, J., Orion, T. and Ishay, J. (1976) The number of ovarioles in the Vespinae (Hymenoptera). *Insectes Soc.*, **23**, 525–533.

Landolt, P. J. and Heath, R. R. (1987) Alarm pheromone behaviour of *Vespula squamosa* (Hymenoptera: Vespidae). *Fla Entomol.*, **70**, 222–225.

Landolt, P. J., Akre, R. D. and Greene, A. (1977) Effects of colony division on *Vespula atropilosa* (Sladen) (Hymenoptera: Vespidae). *J. Kans. Entomol. Soc.*, **50**, 135–147.

Makino, S. (1989) Usurpation and nest rebuilding in *Polistes riparius*: two ways to reproduce after the loss of the original nest (Hymenoptera: Vespidae). *Insectes Soc.*, **36**, 116–128.

Maschwitz, U. (1964) Gefahrenalarmstoffe und Gefahrenalarmierung bei sozialen Hymenopteren. *Z. Vergl. Physiol.*, **47**, 596–655.

Maschwitz, U. (1984) Alarm pheromone in the long-cheeked wasp *Dolichovespula saxonica* (Hym., Vespidae). *Dt Entom. Z.*, **31**, 33–34.

Matsuura, M. (1984) Comparative biology of the five Japanese species of the genus *Vespa* (Hymenoptera: Vespidae). *Bull. Fac. Agric. Mie. Univ.*, **69**, 1–131.

Matthews, R. W. (1982) Social parasitism in yellowjackets. In *Social Insects in the Tropics* (ed. P. Jaisson), University of Paris-Nord, Paris, 193–202.

Naumann, M. G. (1970) The nesting behaviour of *Protopolybia pumila* in Panama (Hymenoptera: Vespidae), Ph.D. Thesis, University of Kansas, Lawrence.

Nixon, G. E. J. (1985) Secondary nests in the hornet *Vespa crabro* L. (Hym., Vespidae). *Entomol. Mon. Mag.*, **121**, 189–198.

Nixon, G. E. J. (1986) Piratical behaviour in queens of the hornet *Vespa crabro* L. (Hym.) in England. *Entomol. Mon. Mag.*, **122**, 233–238.

Ohgushi, R. and Yamane, S. (1983) Supplementary notes on the nest architecture and biology of some *Parischnogaster* species in Sumatera Barat (Hymenoptera: Vespidae). *Sci. Rep. Kanazawa Univ.*, **28**, 69–78.

Ohgushi, R., Yamane, S. and Abbas, N. D. (1986) Additional descriptions and records of Stenogastrine nests collected in Sumatera Barat Indonesia with some biological notes (Hymenoptera: Vespidae). *Kontyû*, **54**, 1–11.

Pagden, H. T. (1962) More about *Stenogaster*. *Malay Nat. J.*, **16**, 95–102.

Pardi, L. (1948) Dominance order in *Polistes* wasps. *Physiol. Zool.*, **21**, 1–13.

Post, D. C. and Jeanne, R. L. (1981) Colony defence against ants by *Polistes fuscatus* (Hymenoptera: Vespidae) in Wisconsin. *J. Kans. Entomol. Soc.*, **54**, 599–615.

Post, D. C., Downing, H. A. and Jeanne, R. L. (1984a) Alarm response to venom by social wasps *Polistes exclamans* and *P. fuscatus* (Hymenoptera: Vespidae). *J. Chem. Ecol.*, **101**, 1425–1433.

Post, D. C., Mohamed, M. A., Coppel, H. C. and Jeanne, R. L. (1984b) Identification of ant repellent allomone produced by social wasp *Polistes fuscatus* (Hymenoptera: Vespidae). *J. Chem. Ecol.*, **10**, 1799–1807.

Reed, H. C. and Akre, R. D. (1983) Comparative colony behaviour of the forest yellowjacket, *Vespula acadica* (Sladen) (Hymenoptera: Vespidae). *J. Kans. Entomol. Soc.*, **561**, 581–606.

Richards, O. W. (1978) *The Social Wasps of the Americas excluding the Vespinae*, Br. Mus. (Nat. Hist.), London.

Richards, O. W. and Richards, M. J. (1951) Observations on the social wasps of South America (Hymenoptera: Vespidae). *Trans. R. Entomol. Soc. Lond.*, **102**, 1–170.

Röseler, P. F. (1991) Reproductive competition during colony establishment. In *Social Biology of the Vespid Wasps* (eds K. G. Ross and R. W. Matthews), Cornell University Press, Ithaca, NY.

Röseler, P. F., Röseler, I. and Strambi, A. (1980) The activity of corpora allata in dominant and subordinated females of the wasp *Polistes gallicus*. *Insectes Soc.*, **27**, 97–107.

Rothschild, M. (1979) Female butterfly guarding eggs. *Antenna*, **3**, 94–95.

Sakagami, S. F. and Yoshikawa, K. (1968) A new ethospecies of *Stenogaster* wasps from Sarawak, with a comment on the value of ethological characters in animal taxonomy. *Ann. Zool. Soc. Jpn*, **41**, 77–84.

Schaudinischky, L. and Ishay, J. (1968) On the nature of the sounds produced within the nest of the Oriental hornet *Vespa orientalis* F. (Hymenoptera). *J. Acoust. Soc. Amer.* **44**, 1290–1301.

Schremmer, F. (1977) Das Baumrinden-nest der neotropischen Faltwespe *Nectarinella championi* umgeben von einem leimring als Ameisen-abwehr (Hymenoptera: Vespidae), *Ent. Germ.*, **3**, 344–355.

Simoes, D. (1977) Etologia e diferenciação de casta em algumas vespas socialis. Ph.D. Thesis, University of São Paolo, Ribeirão Preto, São Paulo, Brazil, 182 pp.

Spradbery, J. P. (1971) Seasonal changes in the population structure of wasp colonies (Hymenoptera: Vespidae). *J. Anim. Ecol.*, **40**, 501–523.

Spradbery, J. P. (1973a) *Wasps. An Account of the Biology and Natural History of Social and Solitary Wasps*, Sidgwick and Jackson, London; University of Washington Press, Seattle, 408 pp.

Spradbery, J. P. (1973b) The European wasp *Paravespula germanica* (F.) (Hymenoptera: Vespidae) in Tasmania, Australia. *Proc. VII Int. Congr. IUSSI, London*, 375–380.

Spradbery, J. P. (1975) The biology of *Stenogaster concinna* van der Vecht with comments on the phylogeny of Stenogastrinae (Hymenoptera: Vespidae). *J. Austr. Entomol. Soc.*, **14**, 309–318.

Spradbery, J. P. (1986) Polygyny in the Vespinae with special reference to the hornet *Vespa affinis picea* Buysson (Hymenoptera: Vespidae) in New Guinea. *Monit. Zool. Ital.*, **20**, 101–118.

Spradbery, J. P. (1989) The nesting of *Anischnogaster iridipennis* (Smith) (Hymenoptera: Vespidae) in New Guinea. *J. Austr. Entomol. Soc.*, **28**, 225–38.

Spradbery, J. P. (1991) Evolution of queen number and queen control in the Vespidae. In *Social Biology of the Vespid Wasps* (eds K. G. Ross and R. W. Matthews), Cornell University Press, Ithaca, NY, in press.

Spradbery, J. P. and Kojima, J. (1989) Nest descriptions and colony populations of eleven species of *Ropalidia* in Papua New Guinea (Hymenoptera: Vespidae). *Jpn J. Entomol.*, **57**, 632–53.

Turillazzi, S. (1985a) Egg deposition in the genus *Parischnogaster* (Hymenoptera: Stenogastrinae). *J. Kans. Entomol. Soc.*, **58**, 749–752.

Turillazzi, S. (1985b) Function and characteristics of the abdominal substance secreted by wasps of the genus *Parischnogaster* (Hymenoptera: Stenogastrinae). *Monit. Zool. Ital.*, **19**, 91–99.

Turillazzi, S. and Pardi, L. (1981) Ant guards on nests of *Parischnogaster nigricans serrei* (Buysson) (Stenogastrinae). *Monit. Zool. Ital.*, **15**, 1–7.

Turillazzi, S. and Ugolini, A. (1979) Rubbing behaviour in some European *Polistes* (Hymenoptera: Vespidae). *Monit. Zool. Ital.*, **13**, 129–142.

van der Vecht, J. (1968) The terminal gastral sternite of female and worker social wasps (Hymenoptera: Vespidae). *Proc. K. Ned. Akad. Wet.*, *Amsterdam*, **71**, 411–422.

Veith, H. J., Koeniger, N. and Maschwitz, U. (1984) 2-Methyl-3-butene-2-ol, a major component of the alarm pheromone of the hornet *Vespa crabro*. *Naturwissenschaften*, **71**, 328–329.

West-Eberhard, M. J. (1969) The social biology of polistine wasps. *Univ. Mich. Mus. Zool. Misc. Publ.*, **140**, 1–101.

West-Eberhard, M. J. (1977) The establishment of the dominance of the queen in social wasp colonies. *Proc. VIII Int. Congr. IUSSI, Wageningen*, 223–227.

West-Eberhard, M. J. (1978) Temporary queens in *Metapolybia* wasps: nonreproductive helpers without altruism? *Science*, **200**, 441–443.

West-Eberhard, M. J. (1981) Intragroup selection and the evolution of insect societies. In *Natural Selection and Social Behavior* (eds R. D. Alexander and D. W. Tinkle), Chiron, NY, 3–17.

West-Eberhard, M. J. (1982) The nature and evolution of swarming in tropical social wasps (Vespidae, Polistinae, Polybiini). In *Social Insects in the Tropics* (ed P. Jaisson), University of Paris-Nord, Paris, 97–128.

Williams, F. X. (1919) Philippine wasp studies. Part II. Descriptions of new species and life histories. *Bull. Exp. Stn Hawaii Sugar Planters Assoc.*, **14**, 19–186.

Wilson, E. O. (1971) *The Insect Societies*, Belknap Press of Harvard University Press, Cambridge, MA, 548 pp.

Wilson, E. O. (1975) *Sociobiology The New Synthesis*, Belknap Press of Harvard University Press, Cambridge, MA, 697 pp.

Yamane, S., Ito, Y. and Spradbery, J. P. (1990) Comb cutting in Ropalidia plebeiana, a new method of colony fission in social wasps (Hymenoptera: Vespidae). *Insectes Soc.* (in press).

10

Competition in dung-breeding insects

James Ridsdill-Smith

10.1 INTRODUCTION

This chapter examines the reproductive behaviour of dung-breeding flies and dung beetles. The utilization by these insects of dung for their offspring and the effects of intraspecific and interspecific competition, both on the ability of dung-breeding insects to oviposit and on the fitness of their offspring, are discussed.

Dung has particular features as a habitat for insects. First, it comes in discrete patches, which for cattle dung may be 1000–2000 ml in volume (240 g dry wt), while sheep dung ranges from pellets of 0.1 g dry wt each to lumps of 6 g dry wt (Lumaret and Kirk, 1987). Second, it is ephemeral and dries out so that it can be used by most insects for only 1–4 weeks. A well-developed succession of species occurs in dung, from those adapted to very fresh dung which arrive within a minute of it being dropped, to those better adapted to drier dung which arrive several days later (Mohr, 1943). All dung-breeding insects are attracted to fresh dung by its odour, which disappears when the dung crusts over. They lay eggs and develop as larvae in the dung, and most pupate or pupariate in soil near the dung.

The abundance of insects depends in part on the number of eggs laid, on the mortality of the eggs and subsequent offspring, and on the dispersal of adults. Insect predators and parasites are important regulating factors, and their effectiveness depends on their searching behaviour and capture rate (Hassell and May, 1985). Population growth may also be resource limited (Nicholson, 1954; Tilman, 1982). The mechanisms by which animals obtain resources are likely to be behavioural, including their foraging tactics (Tilman, 1982). The foraging of species between resources that are ephemeral and finely divided, such as fruit, carrion or dung, is likely to result in a patchy distribution (Atkinson and Shorrocks, 1984; Hassell and May, 1985; Hanski, 1986). Competition occurs when resources become limiting, and would be expected to be particularly important in populations living in dung. Where individuals of the same species utilize the same resource, intraspecific competition will occur; and where individuals of different species utilize the same resource, interspecific competition will occur. Competition is an interaction between individuals, and since there is frequently variation in the behaviour of these individuals it can influence competition.

This chapter begins with a discussion of competition, the nature of dung as a resource, and an overview of reproductive behaviour of a range of dung insects. The second half considers intraspecific and interspecific competition as well as the effects of individual variation in reproductive behaviour.

10.2 THE NATURE OF COMPETITION

One of the important regulating factors in determining the size of a population is competition. Although weather determines many aspects of the population dynamics of the bush fly, *Musca vetustissima*, including rates of development and levels of activity, the major negative feedback in its dynamics is the effect of its own density (Hughes and Sands, 1979).

Intraspecific competition can be viewed in the following ways: (a) the individuals influence each other; (b) competition results in a decrease in the contribution that the individuals could have made had there been no competitors, and is usually measured in terms of reduced survival, growth or fecundity; (c) the resource for which the individuals are competing is in limited supply, such as food or space to live; and (d) the effects of competition are density dependent (Begon and Mortimer, 1981). The nature of competition between individuals may be direct or indirect. Direct competition occurs where individuals respond not to each other's presence, but to the level of resource depletion that each produces; whereas indirect competition occurs where individuals respond directly to each other, rather than to the level of resource depletion each has caused (Begon *et al.*, 1986).

Interactions between two species can result in positive, negative or neutral effects on either species. Competition can involve one of two types of interaction. In the first of these, each species negatively affects the mortality, fecundity or growth of the other. This is symmetric interspecific competition, and although a variety of processes may cause the interactions, competition will only occur when the resource is in limiting supply. In the second interaction there is a negative effect on one species, but no detectable effect on the other. This is asymmetric competition. Lawton and Hassell (1981) reviewed the evidence for interspecific interactions in 35 pairs of species of insects, where the magnitude of the reciprocal interactions are available, and concluded that asymmetric competition occurred in 23. Moon (1980b) suggested that a number of dung-breeding pest flies could be controlled through asymmetric competition by the use of dung beetles to remove the dung resource. This principle has been the basis for an attempt to control dung-breeding flies by the introduction of scarabaeine dung beetles to Australia.

Interspecific interactions will be discussed in this chapter in relation to the reproductive behaviour of dung-breeding insects, with particular reference to several species of flies and to dung beetles of the family Scarabaeidae (Scarabaeinae and Aphodiinae).

10.3 THE NATURE OF THE RESOURCE

The dung resource, which is food and living space for the insects described within this chapter, is made up of undigested plant residues, products of the gut flora and fauna including live and dead microorganisms, and secretions and cell debris from the gut mucosa of the dung producer (Greenham, 1972). On dung with a high nitrogen content, dung beetles lay more eggs (Lee and Peng, 1982; Macqueen *et al.*, 1986; Ridsdill-Smith, 1986). Flies reared on dung with a high nitrogen content are larger (Greenham, 1972; Sands and Hughes, 1976; Palmer and Bay 1983; Matthiessen *et al.*, 1984; Macqueen *et al.*, 1986). This can be considered a response to dung quality. Changes in dung quality that influence egg production of the dung beetle, *Euoniticellus intermedius*, can be predicted best from the dry matter content of the dung fluid (Aschenborn *et al.*, 1989). While the liquid portion of the dung forms food for adult flies and dung beetles, larvae also feed on the solid portion (Hammer, 1941; Halffter and Edmonds, 1982). Dung is not readily assimilated, as illustrated by the larvae of *Aphodius rufipes*, which use only 2–3% of the energy they ingest and pass large quantities of dung through their guts (Holter, 1975).

Cattle dung retains sufficient moisture in summer to be used by dung beetles for breeding and feeding, but sheep pellets may lose 50% of their water in 3 h in summer and are relatively little used by dung beetles at this time (Lumaret and Kirk, 1987). In winter, dung loses water more slowly, and pads are available to insects for longer. The combined effects of dung-feeding larvae and adults tunnelling through the dung, and feeding on the dung fluid, will reduce the time for which dung is suitable as a place for them to live. High densities of adult dung beetles or fly larvae can completely remove the dung fluid by shredding the dung pad to dry flakes. Dung beetles also bury dung from the pad to construct their brood masses, reducing the amount of dung available for other insects.

In addition to physical changes in dung, the quality of dung will change with the seasons, reflecting the feeding of the dung producer. Seasonal changes in pasture growth, which are influenced by temperature and rainfall, are the major factors influencing the characteristics of dung as measured by its effects on bush fly, *Musca vetustissima* (Greenham, 1972; Sands and Hughes, 1976). For example, fewer eggs are laid in poor dung. Thus, in the bush fly the proportion of females ovipositing fall from 80–90% in good dung to 40–50% in poor dung (Ridsdill-Smith *et al.*, 1986; Ridsdill-Smith and Hayles, 1989). In winter rainfall regions, egg production of the dung beetle, *Onthophagus binodis*, on dung from cattle grazing dry summer annual pasture averaged 7% of that on dung from green spring annual pasture (Ridsdill-Smith, 1986). In summer rainfall regions, egg production of *Euoniticellus intermedius* on dung from cattle grazing dry winter pastures averaged 30% of that on dung from green summer pastures (Macqueen *et al.*, 1986).

Dung quality influences larval growth rate and size of emerging adult flies. Size of bush fly is greater on dung collected from cattle feeding on green spring pasture than on dung collected from cattle feeding on dry senescent pasture in summer (Fig. 10.1). Strong seasonal patterns in size of pupae and adults have been recorded for bush fly (Greenham, 1972; Sands and Hughes, 1976; Hughes and Sands, 1979; Matthiessen *et al.*, 1984; Macqueen *et al.*, 1986), dung fly, *Scathophaga stercoraria* (Sigurjonsdottir, 1984), and for horn fly, *Haematobia irritans* (Kunz, 1980; Palmer and Bay, 1983). Larval mortality of the small bush flies produced from poor dung is greater than of larger flies produced from good dung (Hughes and Sands, 1979; Macqueen *et al.*, 1986; Ridsdill-Smith *et al.*, 1986; Ridsdill-Smith and Hayles, 1989). Smaller progeny are produced by the dung beetle, *Onthophagus gazella*, from dung of low than high protein level (Lee and Peng, 1982).

When dung is collected regularly through the season, the relative responses of species to dung quality may be different. Thus, for dung collected from pasture in Queensland, the seasonal patterns of buffalo fly and bush fly sizes are similar, but mortality of buffalo fly does not vary seasonally, while that of bush fly increases to 100% on dung collected in

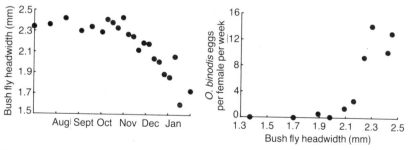

Fig. 10.1 Seasonal changes in dung quality measured by bush fly size (from Matthiessen *et al.*, 1984). Comparison of quality of the same dung assayed by *O. binodis* and by bush fly.

winter (Macqueen *et al.*, 1986). Egg production of the dung beetle, *O. binodis*, is much lower on dung collected in southwestern Australia in summer than in spring, while egg production of another dung beetle, *Onitis alexis*, is somewhat less affected by the seasonal changes in dung quality (Ridsdill-Smith, 1986). Size of bush flies and numbers of dung beetle eggs were measured on dung collected from a pasture throughout a season, and while the size of emerging flies ranged from 2.4 mm in spring dung to 1.3 mm in summer dung, *O. binodis* only oviposited in dung where the headwidth of emerging flies was 2.0 mm or more (i.e. the top half of the size range of flies) (Fig. 10.1).

10.4 REPRODUCTIVE BEHAVIOUR OF DUNG INSECTS

Dung insects aggregate at pads where they meet with their own species, select mates, lay their eggs, and their larvae feed. Thus, competition will occur between dung insects both for the resource as a place to select mates and oviposit, and for the resource as food for the adults and larvae. There is considerable variation in the way different dung insects utilize the dung, and in order to understand how competition will influence them, it is necessary to consider both their reproductive behaviour and the way in which they utilize the dung. The life histories of a range of different dung breeding flies and dung beetles are summarized in Table 10.1, and their behaviour described in this section.

10.4.1 Flies

Females require protein to develop their ovaries. The protein is obtained by predation on other species of flies (dung fly), by biting and feeding on the blood of the dung producer (horn fly), or by feeding on secretions around the face or from wounds of the dung producer (bush fly). It is this protein-seeking behaviour that determines the economic importance of these species as pests or beneficial insects. Thus bush fly is a nuisance pest of people, horn fly is a biting pest of cattle, and dung fly can be considered a beneficial insect.

Typically, mating will occur near to the time that the female fly oviposits on the dung pad. Bush fly does not mate on the pad, but nearby. In the laboratory, male bush fly pass sperm in the first 10 min during mating, and the accessory gland substances, which are transferred during the latter stages of mating, are both oviposition stimulants and cause the females to be non-receptive to further mating for a day (Tyndale-Biscoe, 1971). Although sperm from one mating can fertilize several batches of eggs, bush fly females mate more than once (Hughes *et al.*, 1972). The eggs are laid singly in the dung, but the female deposits all the

Table 10.1 Life history traits of selected dung insects.

	Flies			Dung beetles		
	Musca vetustissima	*Scathophaga stercoraria*	*Haematobia irritans*	*Aphodius rufipes*	*Onthophagus binodis*	*Copris lunaris*
Adult size (mg)	2–4 (fresh wt)	2–10 (fresh wt)	1–4 (fresh wt)	20–50 (dry wt)	6–38 (dry wt)	156–380 (dry wt)
Pre-oviposition feeding	Animal secretions, blood[1,2]	Predator on flies[4]	Biting host (blood)[9]	Dung[13]	Dung[16]	Dung[20]
Mating location	Away from pad	Pad[6]	Host[10]	Pad[14]	Near brood mass[17]	Near brood cake[20]
Egg location	Pad	Pad	Pad	Pad or soil near pad	Buried brood mass	Buried brood ball
Eggs per female per pad	25[2]	50[7]	18[11]	7[13,14,15]	10[16,19]	5[20]
Adult time in dung	2 h	5 h[5]	5 min[11]	5 days[15]	7–14 days[18]	5 days[20]
Egg to adult development time in dung	2 weeks[3]	3–4 weeks[8]	2 weeks[12]	6–8 weeks[15]	12–14 weeks[18]	13 weeks[20]

References: [1]Vogt and Walker (1987), [2]Hughes *et al.* (1972), [3]Hughes and Sands (1979), [4]Sasaki (1984), [5]Parker (1970a), [6]Parker (1970b), [7]Parker (1970c), [8]Amano (1983a), [9]Harris *et al.* (1974), [10]Zorka and Bay (1980), [11]Hammer (1941), [12]Thomas *et al.* (1974), [13]Klemperer (1980), [14]Landin (1961), [15]Holter (1979), [16]Ridsdill-Smith *et al.* (1982), [17]Cook (1990), [18]Ridsdill-Smith (unpublished data), [19]Cook (1988), [20]Klemperer (1982a).

mature eggs from her ovary in one pad usually in 30 min. Gravid female bush flies are stimulated to oviposit by the presence of other females, and they cluster at a favoured spot and eggs are laid in large clumps; an egg-laying pheromone may be involved (Hughes *et al.*, 1972). For example, significantly more females lay batches of eggs when densities of female bush flies at pads are greater. With less than 10 females per pad 63% of bush flies oviposited, compared with 77% with 48 females per pad (data from Ridsdill-Smith *et al.*, 1986). Dissection of adult flies shows that they complete 2–3 oviposition cycles during their lifetime, and so they will oviposit on only 2–3 different dung pads (Vogt, 1987). The eggs hatch in a few hours, and as the larvae grow they disperse throughout the pad, searching out areas where the dung is moist and undisturbed. In a two-week feeding period, when 1000–2000 larvae are present, they can shred the dung to dry flakes, destroying the food resource both for themselves and for other species. In the pad they are exposed to numerous predators as well as destruction of their food supply; mortality is high, averaging 94% in the field (Vogt *et al.*, 1987).

Horn flies mate on the dung producer. Mated females are not receptive to other males for a number of days following mating (Harris *et al.*, 1968), during which time they presumably lay their egg batch. Adult horn flies leave their dung-producing host for only 5 min to oviposit on dung before moving back to the same animal (Hammer, 1941; Kunz *et al.*, 1970). Females lay their eggs in groups of 2–3 on the sides of pads, or on vegetation under the dung, the whole batch of mature oocytes being laid on the one pad (Hammer, 1941). Horn fly larvae can also shred dung, the carrying capacity of a 1 kg pad being about 2000 larvae (Palmer and Bay, 1984). Immature mortality in the dung is high, averaging over 90% in the field (Thomas and Morgan, 1972).

The dung fly mates on and around the dung pad, and shows mate guarding, neither of which have been observed in other dung breeding flies. Male *Scathophaga stercoraria* arrive at the pad before the females and are highly mobile, moving around on both the dung and the pasture immediately surrounding the pad, grasping females as they arrive (Parker, 1970b). In dung fly, sperm from one mating can fertilize up to four batches of eggs, but with mated females the last male to mate before oviposition fertilizes about 80% of the succeeding batch of eggs irrespective of the previous number of matings (Parker, 1970c). Copulation takes 30–35 min, but after the male uncouples, he remains mounted and repels other males from interfering with the female for 15 min while she lays her eggs (Hammer, 1941; Parker, 1970a; Borgia, 1980). The eggs are laid separately on the surface of the dung, but all mature oocytes are laid on one pad. During her life a female oviposits on about five occasions in five pads (Parker, 1970c).

10.4.2 Dung beetles

The adults of all aphodiine and scarabaeine beetles obtain the protein they require to develop their ovaries feeding on dung (Halffter and Edmonds, 1982). The time from adult emergence till the first egg is laid ranges from one week with *Aphodius rufipes* (Klemperer, 1980), to about 8 months with *Copris lunaris* (Klemperer, 1982a). Species which have a highly seasonally restricted pattern of breeding probably have a reproductive diapause (Halffter and Matthews, 1966), as for example occurs in *Kheper nigro-aeneus*, which breeds once a year (Edwards, 1988).

Mating in *Aphodius* dung beetles typically takes only 30–60 s (Landin, 1961). *A. rufipes* deposits eggs in irregular batches of 6–8 stuck loosely together in a small chamber in the soil below the dung, or sometimes in the dung pad (Landin, 1961; Holter, 1979; Klemperer, 1980). The larvae of *A. rufipes* feed in the dung pad, and so are susceptible to environmental factors and to competitors for the dung, including their own parents (Holter, 1979). Larval mortality in the field is variable, ranging from 12 to 95% probably depending on factors influencing the pad as a place to live, such as the weather (Holter, 1979). If the pad dung resource becomes limiting, *A. rufipes* larvae may tunnel through the soil and feed on the large buried dung brood masses of the scarabaeine dung beetle, *Geotrupes spiniger*, incidentally killing the larvae of the larger beetle (Klemperer, 1980).

In *Onthophagus binodis*, the courtship behaviour involves the male and female making contact head to head, and the male suddenly jerking up his head and pronotum by pushing up his front legs. This is thought to disperse pheromones from exocrine glands situated on the under surface of the two profemurs (Cook, 1990). Similar jerking activity is observed in *Typhoeus typhoeus* males (Palmer, 1978). The pheromones are likely to be used for close range sexual attraction. In head to head encounters, the larger horned *O. binodis* males also butt the females, which may enable both the male and female to assess the size of the other partner. Copulation probably occurs on or near the buried brood mass during its construction and lasts approximately 9 min, during which a spermatophore is passed (Cook, 1990). Females construct tunnels about 10–20 cm deep in the soil under the pad, packing lumps of dung taken from the pad into the blind end to form a brood mass, weighing about 6 g (Ridsdill-Smith *et al.*, 1982). One egg is laid in each brood mass, which forms the complete food supply needed for the development of one larva. *O. binodis*, like all Scarabaeinae, has a single ovariole, and matures only one oocyte at a time (Halffter and Matthews, 1966; Tyndale-Biscoe, 1978). One pair of *O. binodis* buries brood masses, at a maximum rate of 1 or 2 per female per day (Ridsdill-Smith *et al.*, 1982; Cook, 1988), in branching tunnels in the soil, each female continuing to oviposit until the pad dung is no longer suitable. A female may produce up to 10 eggs before leaving the pad. Males may cooperate with the females in the construction of brood masses, and the tunnel entrance or the tunnel is nearly always occupied by one member of the pair, giving the appearance that the male is guarding the tunnel (Cook, 1990). More eggs are produced when females cooperate with males than in the absence of males (Cook, 1988; Cook, 1990). A mated *O. binodis* female can produce brood masses and eggs in the absence of a male (Cook, 1988). During its life *O. binodis* could oviposit in nine pads (Ridsdill-Smith *et al.*, 1982). Field mortality of eggs and larvae of *O. binodis* is about 50% (T. J. Ridsdill-Smith, unpublished data). The pattern of dung utilization shown by *O. binodis* is the most common pattern in the Scarabaeinae (Halffter and Edmonds, 1982), in which adults feed in the pad, and only bury dung when breeding. In other species, like *Onthophagus ferox*, from southwestern Australia, the adults bury dung in uncompacted balls close to the surface on which they feed, and construct more compact brood masses deeper in the soil for breeding (T. J. Ridsdill-Smith, unpublished data). *Onitis* species vary in the rate at which they bury dung, in the time at which dung removal starts after arrival at the pad, and in the depth at which brood masses are buried in the soil (Edwards and Aschenborn, 1987; Doube *et al.*, 1988). These patterns of dung utilization will affect any competition that occurs.

In courtship behaviour, the male *Copris lunaris* strokes the female with his legs and drums on her elytra with his anterior tibiae. Coupling occurs fairly quickly, and lasts for 5 min (Klemperer, 1982a). Mated females can lay fertile eggs nearly a year after mating (Klemperer, 1982a). Male and female *C. lunaris* carry dung down a tunnel and construct a cake of dung weighing about 100 g; the male is then excluded from the dung cake chamber (Klemperer, 1983). The female makes one brood ball complete with egg every 1–2 days until the dung cake is finished, with a normal fecundity of 5 eggs (Klemperer, 1982a), although *C. lunaris* females will continue to oviposit if dung of suitable texture is added to the chamber and fecundity can be increased to 12 eggs per female (Kirk and Feehan, 1984). The brood ball is so called because it is a free-standing ball, in contrast to the brood mass of *O. binodis* which is packed into a tunnel. In *C. lunaris*, oviposition is followed by a period of brood care by the mother which lasts for about 90 days (Klemperer, 1983). In this remarkable behaviour the female tends the brood balls during the summer, without further food, until the next generation emerges. There is one generation a year, and the female breeds in one pad per year. Mortality of larvae of brood-caring beetles is decreased by the female keeping the brood ball free of fungus, and attacking intruders near the brood balls including other dung-feeding arthropods (Klemperer, 1982b, 1983). In *Copris diversus*, mortality of immatures fell from 68% to 24% in the presence of a female (Tyndale-Biscoe, 1984). *Copris* species bury the dung required for a dung cake within 1–2 days of arriving at a pad (Doube *et al.*, 1988). At pads where beetles are not breeding, either sex of *C. lunaris* can bury 20–50 g of uncompacted dung in a shallow tunnel for adult feeding (Klemperer, 1982a).

The presence in the same pad of several individuals of the dung beetle *Phanaeus daphnis*, a species which produces brood balls like *C. lunaris*, may stimulate mating and oviposition, since more eggs per female are produced in the laboratory when more than one pair of newly-emerged adults is present (Halffter and Lopez, 1977). Female *P. daphnis* produce twice as many eggs when males are present than when ovipositing alone, but in the related *Phanaeus mexicanus* the presence of a male did not enhance egg production of a female (Halffter and Lopez, 1977). The role of the male in assisting the female in producing brood balls is thus not always the same. In another group of beetles, including species of the genus *Oniticellus*, the dung is removed from the pad, and brood balls are constructed in a chamber formed within the dung pad (Davis, 1989). They produce brood balls in or near the pad, rather than buried in the soil as in *Copris* spp.

A further group of the Scarabaeinae have adopted another method of utilizing dung for breeding. They remove pieces of dung from the pad and form it into spherical balls which are then rolled by the beetles away from the pad, and usually buried at some distance from the pad (Halffter and Edmonds, 1982). As with other species, a single egg is laid in each ball. These species are called the ball-rollers. In the ball-rolling species *Kheper nigroaeneus*, the male stands on his head and produces a pheromone from his abdomen (Tribe, 1975). This is not done at the dung pad, but at the spot where he has buried a dung ball in which the female can oviposit after mating (Edwards and Aschenborn 1988). *K. nigroaeneus* produces one egg each breeding occasion (Edwards and Aschenborn, 1989), with one or sometimes two breeding periods a year (Edwards, 1988). *K. nigroaeneus* is a brood-caring species, where the female remains with the brood ball for a period of 84 days (Edwards, 1988), and egg to adult mortality in the field in the presence of a female is 31% (Edwards and Aschenborn, 1989). Dung is removed from the pad for adult feeding by this species also, and smaller balls rolled by single beetles are mainly for adult feeding, whereas larger balls rolled by pairs are used for breeding (Edwards and Aschenborn, 1988).

10.5 INTRASPECIFIC COMPETITION

The effects of intraspecific competition can be seen in oviposition, growth and mortality rates. The dung-breeding flies lay all the mature oocytes in their ovaries in batches, and the number of eggs laid per female does not decrease at high densities. For example, similar number of eggs per female are produced at high and low densities of horn fly (Thomas and Kunz, 1985).

The adult flies emerging from pads at high densities are smaller than those emerging at low densities, for bush fly (Sands and Hughes, 1976; Ridsdill-Smith *et al.*, 1986; Ridsdill-Smith and Hayles, 1989), dung fly (Cotton, 1978; Amano, 1983b; Sigurjonsdottir, 1984), horn fly (Palmer and Bay, 1983), and buffalo fly (Doube and Moola, 1987). For all dung-breeding flies in which it has been tested, the approximate density at which size is reduced by competition is lower for size than it is for mortality (Table 10.2). Mortality is greater when the surviving flies are below a threshold size, which for the bush fly is a headwidth of 2 mm (Hughes and Sands, 1979; Ridsdill-Smith *et al.*, 1986; Ridsdill-Smith and Hayles, 1989), and for buffalo fly is a pupal weight of 1.2 mg (Doube and Moola, 1987).

The effect of mortality factors on populations can be quantified through the use of '*k*' values. The *k* value is the difference between the logarithms of the population before and after the mortality acts (Varley *et al.*, 1973). This can be applied to competition by taking the initial density (*B*) to be the numbers before the action of competition and the final density (*A*) to be the numbers after the effect of competition, where $k = \log B - \log A$, or $k = \log(B/A)$. The '*k*' value increases as the mortality rate—and therefore the ratio of initial to final density—increases. When the *k* values are plotted against the log of density the shape of the curve shows the exact nature of the density-dependent response (Varley *et al.*, 1973; Begon and Mortimer, 1981). Where the slope is less than 1 there is undercompensation in the competition and mortality does not increase as fast as the increase in density. With a slope of 1 there is exact compensation. This means that the number of survivors remains constant as density increases, so that above a threshold a fixed number of individuals get adequate resources while the rest get none; with increasing density an increasing number die (Begon and Mortimer, 1981). Where the slope is greater than 1 and tending to the vertical there is overcompensation in the

Table 10.2 Effect of larval crowding on size and mortality of flies in cattle dung.

Species	*Density (larvae/1000 ml pad) at which competition causes:*	
	a reduction in size	*an increase in mortality*
Face fly (Moon, 1980a)*	1000	2000
Dung fly (Amano, 1983b)	700	1400
Horn fly (Palmer and Bay, 1983)	500	4000

* Musca autumnalis

competition. This means that any increase in density causes a rapid increase in mortality. All individuals get an equal share of the resource but it is less than they each need; in an extreme situation all die (Begon and Mortimer, 1981). The relationship between k values and density can also be used to show the effects of increasing density on growth and fecundity, where B is the growth or fecundity in the absence of competition and A is the growth or fecundity in the presence of competition (Begon *et al.*, 1986).

The combined effects of density and dung quality can also be examined using k values. The k values for bush fly size increase at lower densities in poor dung than in good dung (Fig. 10.2), and competition thus occurs at lower densities in poor than in good dung. There is a maximum and a minimum size for viable flies, and thus the k values will have an upper limit, as seen in Fig. 10.2 for flies in poor dung.

Density of dung beetles influences mean oviposition per female. The k values for *Aphodius rufipes* egg production per beetle per pad in the field increase with beetle density in a linear relationship with a slope of 0.71 (Holter, 1979). This indicates a slightly undercompensating density-dependent egg loss for *A. rufipes*. The k values for larval mortality in the pads indicates a highly variable but overcompensating density-dependent response (slope = 1.82) (Holter, 1979) typical of scramble competition (Varley *et al.*, 1973). When *Aphodius constans* lays eggs over a number of days the older larvae, which are larger, may eliminate the younger larvae produced from eggs laid later (Lumaret and Kirk, 1987).

The k values for *Onthophagus binodis* egg production per week in the laboratory at 27°C show low levels of competition and undercompensating density-dependent responses up to a density of 20–40 beetles per pad, and at higher beetle densities high levels of competition and over-compensating density-dependent responses (Fig. 10.3). *O. ferox* is active at lower temperatures than *O. binodis* and k values for egg production per

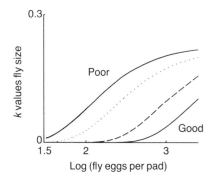

Fig. 10.2 The k values for bush fly size (headwidth) respond at lower egg densities in poor dung than good dung (from Ridsdill-Smith and Hayles, 1989).

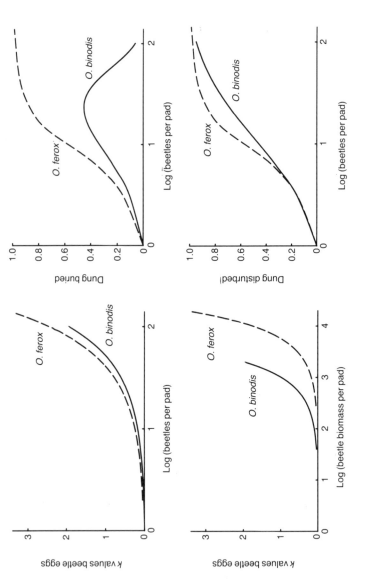

Fig. 10.3 The k values for dung beetle egg production respond at similar densities for two beetle species, but at different levels of their biomass (dry weight). The proportions of the dung buried vary between the two beetle species with density, but the proportions disturbed are similar (*O. binodis* data from Ridsdill-Smith *et al.*, 1982).

two weeks in the laboratory at 18°C show undercompensating density-dependent responses up to a density of 20–40 beetles per pad, and overcompensating density-dependent responses at higher densities (Fig. 10.3). *O. ferox* (140 mg dry wt) is a much larger beetle than *O. binodis* (20 mg dry wt) and, although the *k* values are similar with the same numbers of beetles per pad, when the curves are plotted using biomass of beetles per pad, *O. ferox* responds to density at much higher biomass levels than *O. binodis* (Fig. 10.3).

The type of competition that occurs is indicated by whether all beetles are affected to the same extent at high densities, or whether only some are affected. All female *O. ferox* dissected at high densities after the experiment shown in Fig. 10.3 are parous (definition in Tyndale-Biscoe, 1978), and thus all have oviposited. In another experiment, *O. binodis*, which had been ovipositing at low and high densities of 6 or 60 beetles per pad for 5 weeks, were also all parous (Table 10.3). All insects at the high density therefore obtained an equal share of the resource. The response of beetles is to their current density in the pad, and is not dependent on their previous history, as shown by the similar egg production of 10 *O. binodis* held at this density from their emergence compared with those previously held at a density of 120 beetles per pad (Table 10.3).

Total dung burial per pad increases with increasing density of *O. binodis* up to 20–40 beetles per pad, when dung disturbance is 60–80% of the pad, and then falls. In contrast the rate of dung burial by *O. ferox* increases in a sigmoid curve with increasing beetle density, flattening out when about 90% of the pad is buried owing to shortage of the food resource (Fig. 10.3). Differences in the amount of dung buried occur because, while breeding by both species is influenced by beetle density, *O. binodis* adults feed in the pad while *O. ferox* bury dung for adult feeding. However, the

Table 10.3 Rate of egg production by *O. binodis* can be predicted from density in the current pad and not density in the previous pad.

Experiment (each approx. 4 weeks)	Number of beetles per pad		Eggs per female per day	
	Low density	High density	Low density	High density
1 Observed	6	60	1.64†	0.22†
Predicted*	6	60	1.64	0.14
2 Observed	10	120	1.73†	0.09†
Observed	10 (after 4 weeks at 120)		1.42†	—
Predicted*	10	120	1.37	0.01

*Ridsdill-Smith *et al.* (1982).
†Surviving beetles dissected and all were parous.

total amount of dung disturbed increased with beetle density and a similar sigmoid curve is evident for both species (Fig. 10.3).

Individual *O. ferox* respond to the density of their own species, rather than to the level of resource depletion, indicating indirect interference competition. Average dung burial per beetle by *O. ferox* with two beetles per pad is 6.3, 4.5 and 6.4 ml after 3, 7 and 14 days in the same pad; with 32 beetles this falls to 2.1, 3.3 and 2.1 ml; while with 128 beetles the average dung burial per beetle is only 1.9, 1.1 and 0.5 ml after the same period of time. The rate per beetle did not change appreciably with time or with dung utilization, but changed dramatically with increase in beetle density. Fewer progeny are produced per female by *O. gazella*, and they are significantly smaller, at high rather than at low densities, as would be expected from interference competition (Lee and Peng, 1982).

Beetles can also reduce availability of the dung resource as living space. The dung shredded by high densities of *O. binodis* feeding in the pad is not available for breeding. At high densities of *O. ferox* relatively less of the dung buried is recovered as brood masses, presumably because, as dung in the pad becomes limiting, beetles are feeding on buried dung. Similarly, less of the dung buried at high densities by two *Copris* species is recovered as brood balls than at low densities (Giller and Doube, 1989).

10.6 INTERSPECIFIC COMPETITION

During the short wet season in Africa half a litre of elephant dung can attract 4000 scarabaeine beetles in 15 min (Heinrich and Bartholomew, 1979), and Doube (1987) reports large numbers of species attracted to a single pad. Clearly, there is not sufficient dung for all these beetles to breed, and interspecific competition would be expected to be intense. However, there are few data to show competition. Beetles introduced to Australia are more abundant than the same species in their native habitats. At Gouda, in the Cape region of South Africa, 100–200 beetles were trapped in a day. Sixteen species were found of which *Onthophagus binodis* made up 2% of the catch (Davis, 1987). However at Busselton, in southwestern Australia, a site with a similar climate, there were similar numbers of beetles, but there were only four species, of which *O. binodis* made up 75% of the catch (Ridsdill-Smith and Matthiessen, 1988). Interspecific competition may be the cause of the lower populations in Africa, but there are no experimental data to support this hypothesis.

A number of species of flies coexist in cattle dung, but there is only limited evidence for interspecific competition (Hammer, 1941; Valiela, 1969; Moon, 1980a). This is in contrast to the situation in carrion where competition between species causing mortality has been widely reported (Hammer, 1941; Lawton and Hassell, 1981; Blackith and Blackith, 1984;

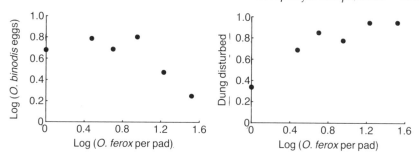

Fig. 10.4 Competition between dung beetles. The egg production per week of *O. binodis* (log(n + 1)) falls when pads contain more than 10 *O. ferox* (1 on log scale), when over 80% of the pad is disturbed.

Hanski, 1987). Carrion is far more unpredictable in time and space as a resource for larval feeding than dung, causing intraspecific competition between adults and between larvae for food (Nicholson, 1954). Inter-specific competition in carrion would also be expected to be great.

Competition between dung beetle species is frequently asymmetric. For example, there is a 55% reduction in egg production of 30 *O. binodis* per pad when 16 *O. ferox* are added to pads, and a 12% increase in egg production of 16 *O. ferox* when 30 *O. binodis* are added to pads. In another experiment, adding from 0 to 10 *O. ferox* had no impact on egg production of 20 *O. binodis* per pad, but there was a rapid fall when 16 and 32 *O. ferox* were added (Fig. 10.4). *O. ferox* reduces *O. binodis* egg production when just over 80% of the pad is disturbed indicating overcompensating density dependent competition (Fig. 10.4). Overcompensating density-dependent responses also occur with each of these species in a pad (Fig. 10.2). Giller and Doube (1989) note that dung burial by two large coprine species is mainly unaffected by the presence of a smaller onitine species, but dung burial by the onitine species is substantially reduced by the presence of 2–3 pairs of the coprine species. This is because the coprines are larger and would require more dung for each egg laid. They also have different dung burial behaviour, burying dung more quickly after arriv-ing at the pad (Giller and Doube, 1989). Another form of asymmetric competition, called cleptoparasitism, occurs with larvae of *Aphodius* spp. (Halffter and Matthews, 1966; Rougon and Rougon, 1980). The larvae of the small *Aphodius* can kill the larvae of larger dung beetles. However, brood-caring *Copris lunaris* females attack and drive off *Aphodius* larvae which intrude near their brood balls (Klemperer, 1982b).

The presence of different species of dung beetles in dung pads can lead to substantial mortality of larvae of a number of fly species through interspecific competition (Bornemissza, 1970; Blume *et al.*, 1973; Hughes *et al.*, 1978; Ridsdill-Smith, 1981), the level of mortality being greater when beetles are placed on dung straight after fly oviposition rather than 1 to 3 days later, even though dung burial after two weeks is no different (Hughes *et al.*, 1978; Moon *et al.*, 1980; Ridsdill-Smith *et al.*, 1987). The effect of interspecific interactions between the dung beetle, *O. binodis*, and bush fly on the oviposition and survival of the two species in the laboratory is different in good and poor dung. In good dung, the *k* values for fly size and fly mortality show an undercompensating density-dependent response to beetle density, and flies are influenced more by high fly densities than by high beetle densities (Ridsdill-Smith *et al.*, 1986) (Fig. 10.5). In good dung, *O. binodis* are breeding, and the *k* values for their egg production increase with increasing beetle density, as expected, but the *k* values for beetle egg production are also greater with higher densities, which is unexpected and shows interspecific competition (Fig. 10.5). Fly density influences beetle egg production more at low beetle densities, where beetles lay most eggs, than at high beetle densities,

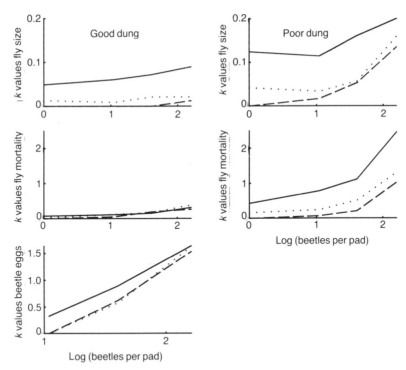

Fig. 10.5 Interspecific competition occurs between *O. binodis* and bush fly in good dung, but in poor dung the competition is asymmetric. Beetles oviposit in good dung and are influenced by bush fly, but in poor dung they do not oviposit and are not influenced by the presence of flies (from Ridsdill-Smith *et al.*, 1986). Solid line = 1267 fly eggs laid per pad; dotted line = 178 eggs; dashed line = 41 eggs per pad.

where beetles lay fewer eggs because of intraspecific competition. The shape of the k values indicates interference competition which is undercompensating and density-dependent. Flies interfere with beetle behaviour, since beetles leave pads earlier at higher fly densities than at lower fly densities, while beetle density does not affect the time at which beetles leave the pads (Ridsdill-Smith *et al.*, 1986). In poor dung, the k values for fly size and fly mortality show an undercompensating density-dependent response at low beetle densities, changing to an overcompensating density-dependent response at high beetle densities (Fig. 10.5). The beetles almost completely cease oviposition in poor dung, and so there would be no competition influencing their egg production. In the poor dung, the beetles cause greater fly mortality and competition is more asymmetric than in good dung. Dung quality affects both the type and the level of competition occurring. Fly mortality occurs when beetles are feeding or breeding and is influenced by dung disturbance as well as by dung burial (Moon *et al.*, 1980; Ridsdill-Smith *et al.*, 1987).

10.7 INDIVIDUAL VARIATION IN BEHAVIOUR

Some variability in behaviour will be adaptive to allow response to an unpredictable environment, and is quite common in insects. It enables an individual to react to a particular set of circumstances. For example, *Copris lunaris* most commonly produces brood balls in a chamber in the soil, but may depart from its normal behaviour and produce brood balls in a chamber next to the pad, similar to those seen in *Oniticellus* spp., while on other occasions it may roll and bury dung balls like *Scarabaeus* spp. (Klemperer, 1982a). Males of the ball-rolling *Kheper nigroaeneus* normally emit a pheromone when near a buried brood ball to attract females from a distance so that mating and oviposition can follow, but they may also emit a pheromone at times when no brood ball is present when only mating occurs (Edwards and Aschenborn, 1988).

One important source of variation in reproductive success is in individual size. The size range of most adult dung-breeding insects is considerable (Table 10.1), resulting from differences in availability of the resource. Various factors have been identified in this chapter that cause differences in size of the offspring, including intraspecific and interspecific competition, and seasonal changes in dung quality. Larger dung beetle adults are produced from larger brood masses by *Onthophagus binodis* (D. Cook, unpublished data), *Onthophagus gazella* (Lee and Peng, 1981) and *Copris diversus* (Tyndale-Biscoe, 1984). The horned males of *O. binodis* are not only larger than the hornless males, but can be clearly distinguished on the basis of body shape, including the presence of a pronotal horn (Cook, 1987). Virtually all the large horned male *O. binodis*

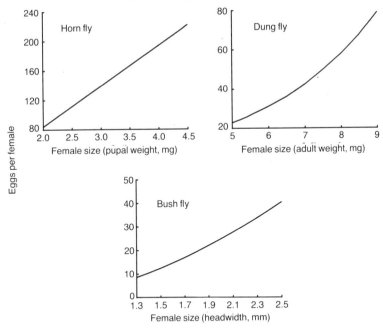

Fig. 10.6 Egg production of dung-breeding flies varies with female size as measured by pupal weight (mg) for horn fly (from Palmer and Bay, 1983), by adult weight (mg) for dung fly (from Borgia, 1981), and by headwidth (mm) for bush fly (from Vogt and Walker, 1987).

emerge from larger brood masses (D. Cook, unpublished data), and the sizes of these insects are therefore probably determined in a phenotypic manner, switched by nutrition. This structural dimorphism does not occur in *O. ferox*, where although males are larger than females, the shape of males and females is not different (Cook, 1987).

The effect of differences in female size and male size has to be considered separately, since their reproductive functions differ. Female fitness is selected for as a function of size. In the dung-breeding flies, larger females produce more eggs (Fig. 10.6). In the dung beetle, *O. binodis*, larger females produce larger brood masses than small females, but not more eggs (Cook, 1988) (Fig. 10.7), regardless of male size (D. Cook, unpublished data). Where both *O. gazella* parents are large, the brood masses produced are larger than those produced by smaller parents (Lee and Peng, 1981). Large *C. diversus* parents bury a larger dung cake than small parents and from it they produce more brood balls, but the weight of individual brood balls produced by either large or small parents is not significantly different (Tyndale-Biscoe, 1984). Presumably in all these cases the larger female is able to remove and bury more dung in each trip from the pad to the tunnel. Although female size can influence the size of the brood mass, other factors frequently override the female size effect, as in the ball-rolling beetle, *Kheper platynotus*, where there is no correlation between size of the ball and size of either the female or male beetle associated with it in the field (Sato and Imamori, 1987).

The impact of male size on reproductive behaviour is somewhat different, since it is the effect of male size on their mating behaviour that influences the population. Size of dung fly is highly variable in both sexes, and there is clear evidence of male competition for females. The gravid female dung fly chooses large males, both by moving towards larger males on the pad, and by moving to the prime oviposition site near the centre of the pad where they are mated by the larger males, which then guard the female while ovipositing (Parker, 1970b; Borgia, 1980, 1981, 1982). Large males are more active on the dung, and take over females from smaller males when two males are trying to mate with the same female, and also from a small male which has started mating (Parker, 1970b; Borgia, 1980, 1982). In a crowded situation females paired with large males are protected from damage and from being taken over by other males, and are thus able to oviposit more quickly (Borgia, 1981). The males do not discriminate between females on the basis of size (Borgia, 1981).

In dung beetles there are frequent records of male–male interactions where the larger male is successful (Halffter and Matthews, 1966). Male–male contests are usually for dung balls or at the entrance of breeding tunnels, and so represent competition for potential breeding resources and for females already associated with these resources. In the ball-rolling species, *Scarabaeus laevistriatus*, large beetles usually win over small beetles in contests for dung balls, although beetles with greater thoracic temperature are also more successful (Heinrich and Bartholomew, 1979). Male–male encounters have been observed with *C. lunaris* near the tunnel entrance where the males audibly stridulate (Klemperer, 1983). In

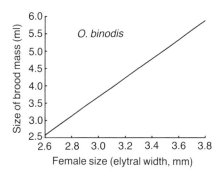

Fig. 10.7 Size of *O. binodis* brood mass (volume in ml) varies with size (elytral width in mm) of the female beetle (from Cook, 1988).

Typhoeus typhoeus large males invariably win contests for tunnels and for copulation with females (Palmer, 1978).

In *O. binodis* the mating behaviour of the large horned males differs from that of the small hornless males. It would be very interesting to know if there was any male–male competition, but no observations have so far been made on this. The courtship behaviour of the horned male includes head butting with the female and pushing up his front legs to expose the tibial glands, both of which would give the female a chance to assess the size of the male (Cook, 1990). After mating the horned male cooperates with the female to produce brood masses, assisting in the collection of dung and guarding the tunnel entrance presumably against other males (Cook, 1990). The smaller hornless males, in contrast, show little courtship behaviour, nor do they cooperate with females in brood mass production after mating. In the laboratory, in the absence of male–male competition, 82% of large horned males initiate courtship behaviour, but only 13% subsequently inseminate the females, while 97% of smaller hornless males initiate courtship behaviour, and 61% subsequently inseminate females (Cook, 1990). It is not clear from these observations if there is male or female choice or both, but it does appear that there is greater choosiness occurring between horned males and

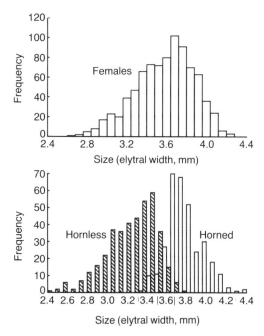

Fig. 10.8 Average *O. binodis* sizes in the field. Frequency distribution of females and both horned and hornless males (from Cook, 1988).

females than between hornless males and females. Female *O. binodis* produce significantly more brood masses per week with horned males than with hornless males, although the individual brood masses are not significantly larger (Cook, 1988), because it is female size that influences size of brood mass.

Horned and hornless *O. binodis* male encounters must occur frequently since the two morphs are present in the field in almost equal numbers throughout the year (Fig. 10.8) (Cook, 1988), and regularly coexist in the same pad (T. J. Ridsdill-Smith, unpublished data).

11.8 DISCUSSION AND CONCLUSIONS

Nicholson (1954), in his work with carrion flies, pointed out that competition influences the individual. Thus, in a competitive situation individuals with certain properties, whether these be tolerance to physical stresses or differences in behaviour, may be eliminated from the population in conditions where in the absence of competition they might well be able to survive. Intraspecific competition is common in dung-breeding insects. The flies lay their eggs in a few hours, and larval development is completed in 2–4 weeks. The main density response is in the size of the offspring, which influences populations through female size and fecundity. Dung beetles lay their eggs over a period of days and larval development is completed in 6–14 weeks. In the scarabaeine beetles, dung for the larvae is buried in the soil in brood masses, reducing moisture losses and temperature fluctuations. The main density response is in the rate of oviposition. *Aphodius rufipes*, which lays eggs in the soil, and whose larvae feed in the pad, has a density response both in rate of oviposition and, if larvae are overcrowded, in larval survival.

Interactions between species vary with physiological age of the insect and the food quality of the dung resource. In good dung there is interspecific competition between bush fly and *Onthophagus binodis*, which is not very encouraging news for those trying to control dung-breeding flies by introducing dung beetles. The mechanism appears to be interference competition. However, some of the larger dung beetle species which bury dung quickly are more effective in causing bush fly mortality (Ridsdill-Smith *et al.*, 1987). In poor dung, the feeding by adult dung beetles in the pad is not reduced by density effects, and beetles cause substantial fly mortality resulting from exploitation competition. Poor dung supports fewer bush fly larvae than good dung, and larvae are more susceptible to competition. Asymmetric competition is evident, with beetles causing high bush fly mortality, but flies having no effect on beetles. This is consistent with field observations that dung beetles suppress bush fly survival in poor dung produced in summer (Hughes *et al.*, 1978; Ridsdill-Smith and Matthiessen, 1988). Since different dung beetle species have different dung utilization behaviour, and respond in different ways to dung quality, it is necessary to study the reproductive behaviour of beetles and flies in good and poor dung to be able to predict the interactions occurring over a season.

An organism may respond to a variable environment in a number of ways: (a) the phenotype may be constant; (b) it may vary continuously, which is probably the most common situation; and (c) variation in the phenotype may be discrete with no intermediates, so that two or more phenotypes are seen (Levins, 1968; Stearns, 1982). Alternative phenotypes are exceedingly common in nature, and the existence of alternative behavioural patterns within a species is considered to be an indication of the presence of strong intraspecific competition (West-Eberhard, 1986). Well-known examples of alternative phenotypes in insects include the morphs of aphids (see Ward, Chapter 8, this volume), castes of social insects (see Spradbery, Chapter 9, this volume), alternative life history strategies in a wide range of insects (Dingle, 1986), including the dung beetle *Kheper nigroaeneus* (Edwards, 1988), and winged and wingless forms of grasshoppers (West-Eberhard, 1986). Southwood (1987) points out that in unpredictable environments there may be selection for high and low risk strategies that are evolutionarily stable in terms of individual behaviour. However, the variation in size of many dung-breeding insects would appear to be continuous, but in other species size (morph) appears to switch due to nutrition or the insect's environment. Dung as a resource is ephemeral, and individual pads may vary nutritionally, and may contain different densities of insects. As a result a range of different sized adults may be produced at one time. When these fly to fresh dung there will be insects of different sizes coexisting in the pad, and selection will favour the larger insects in the females through increased fecundity, and in male–male contests through possession of dung, oviposition sites or females. In dung insects conditions would seem to be favourable for the development of alternative reproductive behaviour patterns, and alternative reproductive behaviours are probably more common than has been previously expected. This should prove a productive area for future studies.

ACKNOWLEDGEMENTS

I am grateful for helpful comments on various drafts of this chapter by Drs Paul Wellings, Win Bailey, Dave Cook, Bernard Doube, Penny Edwards, Ilkka Hanski and Marina Tyndale-Biscoe.

References

Amano, K. (1983a) Studies on the biology of the yellow dung fly *Scatophaga stercoraria* L. (Diptera: Scatophagidae). I. Effects of temperature on the growth and development. *Ann. Rep. Plant Prot. N. Jap.*, **34**, 64–65.

Amano, K. (1983b) Studies on the intraspecific competition in dung-breeding flies. I. Effects of larval density on yellow dung fly *Scatophaga stercoraria* L. (Diptera: Scatophagidae). *Jap. J. Sanit. Zool.*, **34**, 165–175.

Aschenborn, H. H., Loughnan, M. L. and Edwards, P. B. (1989) A simple assay to determine the nutritional suitability of cattle dung for coprophagous beetles. *Entomol. Exp. Appl.*, **53**, 73–9.

Atkinson, W. D. and Shorrocks, B. (1984) Aggregation of larval Diptera over discrete and ephemeral breeding sites: the implications for coexistence. *Amer. Nat.*, **124**, 336–351.

Begon, M. and Mortimer, M. (1981) *Population Ecology—A Unified Study of Animals and Plants*, Blackwell Scientific Publications, Oxford.

Begon, M., Harper, J. L. and Townsend, C. R. (1986) *Ecology—Individuals Populations and Communities*, Blackwell Scientific Publications, Oxford.

Blackith, R. and Blackith, R. (1984) Larval aggression in Irish flesh-flies (Diptera: Sarcophagidae). *Ir. Nat. J.*, **21**, 255–7.

Blume, R. R., Matter, J. J. and Eschle, J. L. (1973) *Onthophagus gazella*: effect on survival of horn flies in the laboratory. *Environ. Entomol.*, **2**, 811–813.

Borgia, G. (1980) Sexual competition in *Scatophaga stercoraria*: size- and density-related changes in male ability to capture females. *Behaviour*, **75**, 185–206.

Borgia, G. (1981) Mate selection in the fly *Scatophaga stercoraria*: female choice in a male-controlled system. *Anim. Behav.*, **29**, 71–80.

Borgia, G. (1982) Experimental changes in resource structure and male density: size-related differences in mating success among male *Scatophaga stercoraria*. *Evolution*, **36**, 307–315.

Bornemissza, G. F. (1970) Insectary studies on the control of dung breeding flies by the activity of the dung beetle *Onthophagus gazella* F. (Coleoptera: Scarabaeinae). *J. Austr. Entomol. Soc.*, **9**, 31–41.

Cook, D. (1987) Sexual selection in dung beetles. I. A multivariate study of the morphological variation in two species of *Onthophagus* (Scarabaeidae: Onthophagini). *Austr. J. Zool.*, **35**, 123–32.

Cook, D. (1988) Sexual selection in dung beetles. II. Female fecundity as an estimate of male reproductive success in relation to horn size and alternative behavioural strategies in *Onthophagus binodis* Thunberg (Scarabaeidae: Onthophagini). *Austr. J. Zool.*, **36**, 521–532.

Cook, D. (1990) Differences in courtship, mating and postcopulatory behaviour between male morphs of the dung beetle *Onthophagus binodis* Thunberg (Coleoptera: Scarabaeidae). *Anim. Behav.*, **40**, 428–36.

Cotton, D. C. F. (1978) Laboratory evidence for intra-specific competition in the larvae of some Diptera living in cow-pats. *Sci. Proc. R. Dublin Soc. Ser. A*, **6**, 155–164.

Davis, A. L. V. (1987) Geographical distribution of dung beetles (Coleoptera: Scarabaeidae) and their seasonal activity in south-western Cape Province. *J. Entomol. Soc. South Afr.*, **50**, 275–285.

Davis, A. L. V. (1989) Nesting of Afrotropical *Oniticellus* (Coleoptera Scarabaeidae) and its evolutionary trend from soil to dung. *Ecol. Entomol.*, **14**, 11–21.

Dingle, H. (1986) The evolution of insect life cycle syndromes. In *The Evolution of Insect Life Cycles* (eds F. Taylor and R. Karban), Springer-Verlag, New York, Berlin and Heidelberg, 187–203.

Doube, B. M. (1987) Spatial and temporal organization in communities associated with dung pads and carcasses. In *Organization of Communities* (eds J. H. R. Gee and P. S. Giller), 27th Symp. Br. Ecol. Soc. 1986. Blackwell Scientific Publications, Oxford.

Doube, B. M. and Moola, F. (1987) Effects of intraspecific larval competition on the development of the African buffalo fly *Haematobia thirouxi potans*. *Entomol. Exp. Appl.*, **43**, 145–151.

Doube, B. M., Giller, P. S. and Moola, F. (1988) Dung burial strategies in some South African coprine and onitine dung beetles (Scarabaeidae: Scarabaeinae). *Ecol. Entomol.*, **13**, 251–261.

Edwards, P. B. (1988) Field ecology of a brood-caring dung beetle *Kheper nigroaeneus*—habitat predictability and life history strategy. *Oecologia*, **75**, 527–534.

Edwards, P. B. and Aschenborn, H. H. (1987) Patterns of nesting and dung burial in *Onitis* dung beetles: implications for pasture productivity and fly control. *J. Appl. Ecol.*, **24**, 837–851.

Edwards, P. B. and Aschenborn, H. H. (1988) Male reproductive behaviour of the African ball-rolling dung keetle *Kheper nigroaeneus* (Coleoptera: Scarabaeidae). *Coleopt. Bull.*, **42**, 17–27.

Edwards, P. B. and Aschenborn, H. H. (1989) Maternal care of a single offspring in the dung beetle *Kheper nigroaeneus*: the consequences of extreme parental investment. *J. Nat. Hist.*, **23**, 17–27.

Giller, P. S. and Doube, B. M. (1989) Experimental analysis of inter- and intra-specific competition in dung beetle communities. *J. Anim. Ecol.*, **58**, 129–142.

Greenham, P. M. (1972) The effects of the variability of cattle dung on the multiplication of the bushfly (*Musca vetustissima* Walk.). *J. Anim. Ecol.*, **41**, 153–165.

Halffter, G. and Edmonds, W. D. (1982) *The nesting behaviour of dung beetles* (Scarabaeinae)—*an ecological and evolutive approach*. Inst. Ecol. Mexico DF Publ. No. 10, 176 pp.

Halffter, G. and Lopez, Y. (1977) Development of the ovary and mating behaviour in *Phanaeus*. *Ann. Entomol. Soc. Amer.*, **70**, 203–213.

Halffter, G. and Matthews, E. G. (1966) The natural history of dung beetles of the subfamily Scarabaeinae (Coleoptera Scarabaeidae). *Folia Entomologica Mex.*, **12–14**, 1–312.

Hammer, O. (1941) Biological and ecological investigations on flies associated with pasturing cattle and their excrement. *Videnskabelige Medd. Dan. Naturhist. Foren.*, **105**, 1–257.

Hanski, I. (1986) Individual behaviour, population dynamics and community structure of *Aphodius* (Scarabaeidae) in Europe. *Acta Oecol. Oecol Gen.*, **7**, 171–187.

Hanski, I. (1987) Carrion fly community dynamics: patchiness seasonality and coexistence. *Ecol. Entomol.*, **12**, 257–266.

Harris, R. L., Frazar, E. D. and Schmidt, C. D. (1968) Notes on the mating habits of the horn fly. *J. Econ. Entomol.*, **61**, 1639–1640.

Harris, R. L., Miller, J. A. and Frazar, E. D. (1974) Horn flies and stable flies: feeding activity. *Ann. Entomol. Soc. Amer.*, **67**, 891–894.

Hassell, M. P. and May, R. M. (1985) From individual behaviour to population dynamics In *Behavioural Ecology* (eds R. M. Sibley and R. H. Smith), 25th Symp. Br. Ecol. Soc. 1984. Blackwell Scientific Publications, Oxford.

Heinrich, B. and Bartholomew, G. A. (1979) Roles of endothermy and size in inter- and intraspecific competition for elephant dung in an African dung beetle *Scarabaeus laevistriatus*. *Physiol. Zool.*, **52**, 484–496.

Holter, P. (1975) Energy budget of a natural population of *Aphodius rufipes* larvae (Scarabaeidae). *Oikos*, **26**, 177–186.

Holter, P. (1979) Abundance and reproductive strategy of the dung beetle *Aphodius rufipes* (L.) (Scarabaeidae). *Ecol. Entomol.*, **4**, 317–326.

Hughes, R. D. and Sands, P. (1979) Modelling bushfly populations. *J. Appl. Ecol.*, **16**, 117–139.

Hughes, R. D., Greenham, P. M., Tyndale-Biscoe, M. and Walker, J. M. (1972) A synopsis of observations on the biology of the Australian bushfly (*Musca vetustissima* Walker). *J. Austr. Entomol. Soc.*, **11**, 311–331.

Hughes, R. D., Tyndale-Biscoe, M. and Walker, J. (1978) Effects of introduced dung beetles (Coleoptera: Scarabaeidae) on the breeding and abundance of the Australian bushfly *Musca vetustissima* Walker (Diptera: Muscidae). *Bull. Entomol. Res.*, **68**, 361–372.

Kirk, A. A. and Feehan, J. E. (1984) A method for increased production of eggs of *Copris hispanus* L. and *Copris lunaris* L. (Coleoptera: Scarabaeidae). *J. Austr. Entomol. Soc.*, **23**, 293–294.

Klemperer, H. G. (1980) Kleptoparasitic behaviour of *Aphodius rufipes* (L.) larvae in nests of *Geotrupes spiniger* Marsh (Coleoptera Scarabaeidae). *Ecol. Entomol.*, **5**, 143–151.

Klemperer, H. G. (1982a) Normal and atypical nesting behaviour of *Copris lunaris* (L.): comparison with related species (Coleoptera Scarabaeidae). *Ecol. Entomol.*, **7**, 69–83.

Klemperer, H. G. (1982b) Parental behaviour in *Copris lunaris* (Coleoptera Scarabaeidae): care and defence of brood balls and nest. *Ecol. Entomol.*, **7**, 155–167.

Klemperer, H. G. (1983) The evolution of parental behaviour in Scarabaeinae (Coleoptera Scarabaeidae): an experimental approach. *Ecol. Entomol.*, **8**, 49–59.

Kunz, S. E. (1980) Horn fly production as affected by seasonal changes in rangeland forage conditions. *Southwest Entomol.*, **5**, 80–83.

Kunz, S. E., Blume, R. R., Hogan, B. F. and Matter, J. J. (1970) Biological and ecological investigations of horn flies in central Texas: influences of time of manure deposition on oviposition. *J. Econ. Entomol.*, **63**, 930–933.

Landin, B.-O. (1961) Ecological studies on dung-beetles (Col. Scarabaeidae). *Opusc. Entomol. Suppl.*, **32**, 1–227.

Lawton, J. H. and Hassell, M. P. (1981) Asymmetrical competition in insects. *Nature*, **289**, 793–795.

Lee, J. M. and Peng, Y.-S. (1981) Influence of adult size of *Onthophagus gazella* on manure pad degradation, nest construction and progeny size. *Environ. Entomol.*, **10**, 626–630.

Lee, J. M. and Peng, Y.-S. (1982) Influence of manure availability and nesting density on the progeny size of *Onthophagus gazella*. *Environ. Entomol.*, **11**, 38–41.

Levins, R. (1968) *Evolution in Changing Environments*. Princeton University Press, Princeton, NJ.

Lumaret, J. P. and Kirk, A. (1987) Ecology of dung beetles in the French Mediterranean region (Coleoptera: Scarabaeinae). *Acta Zool. Mex.* (NS), **24**, 1–55.

Macqueen, A., Wallace, M. M. H. and Doube, B. M. (1986) Seasonal changes in favourability of cattle dung in central Queensland for three species of dung-breeding insects. *J. Austr. Entomol. Soc.*, **25**, 23–29.

Matthiessen, J. N., Hayles, L. and Palmer, M. J. (1984) An assessment of some methods for the bioassay of changes in cattle dung as insect food using the bush fly *Musca vetustissima* Walker (Diptera: Muscidae). *Bull. Entomol. Res.*, **74**, 463–467.

Mohr, C. O. (1943) Cattle droppings as ecological units. *Ecol. Monogr.*, **13**, 275–298.

Moon, R. D. (1980a) Effects of larval competition on face fly. *Environ. Entomol.*, **9**, 325–330.

Moon, R. D. (1980b) Biological control through interspecific competition. *Environ. Entomol.*, **9**, 723–728.

Moon, R. D., Loomis, E. C. and Anderson, J. R. (1980) Influence of two species of dung beetles on larvae of face fly. *Environ. Entomol.*, **9**, 607–612.

Nicholson, A. J. (1954) An outline of the dynamics of animal populations. *Austr. J. Zool.*, **2**, 9–65.

Palmer, T. J. (1978) A horned beetle which fights. *Nature*, **274**, 583–584.

Palmer, W. A. and Bay, D. E. (1983) Effects of intraspecific competition and nitrogen content of manure on pupal weight survival and reproductive potential of the horn fly *Haematobia irritans irritans* (L.). *Prot. Ecol.*, **5**, 153–160.

Palmer, W. A. and Bay, D. E. (1984) A computer simulation model for describing the relative abundance of the horn fly *Haematobia irritans irritans* (L.) under various ecological and pest management regimes. *Prot. Ecol.*, **7**, 27–35.

Parker, G. A. (1970a) The reproductive behaviour and the nature of sexual selection in *Scatophaga stercoraria* L. (Diptera: Scatophagidae). I. Diurnal and seasonal changes in population density around the site of mating and oviposition. *J. Anim. Ecol.*, **39**, 185–204.

Parker, G. A. (1970b) The reproductive behaviour and the nature of sexual selection in *Scatophaga stercoraria* L. (Diptera: Scatophagidae). II. The fertilization rate and the spatial and temporal relationships of each sex around the site of mating and oviposition. *J. Anim. Ecol.*, **39**, 205–228.

Parker, G. A. (1970c) Sperm competition and its evolutionary effect on copula duration in the fly *Scatophaga stercoraria*. *J. Insect. Physiol.*, **16**, 1301–1328.

Ridsdill-Smith, T. J. (1981) Some effects of three species of dung beetles (Coleoptera: Scarabaeidae) in south-western Australia on the survival of the bush fly

Musca vetustissima Walker (Diptera: Muscidae) in dung pads. *Bull. Entomol. Res.*, **71**, 425–433.

Ridsdill-Smith. T. J. (1986) The effect of seasonal changes in cattle dung on egg production by two species of dung beetles (Coleoptera: Scarabaeidae) in south-western Australia. *Bull. Entomol. Res.*, **76**, 63–68.

Ridsdill-Smith, T. J. and Hayles, L. (1989) A re-examination of competition between *Musca vetustissima* Walker (Diptera: Muscidae) larvae and seasonal changes in favourability of cattle dung. *J. Austr. Entomol. Soc.*, **28**, 105–111.

Ridsdill-Smith, T. J. and Matthiessen, J. N. (1988) Bush fly *Musca vetustissima* Walker (Diptera: Muscidae) control in relation to seasonal abundance of scarabaeine dung beetles (Coleoptera: Scarabaeidae) in south-western Australia. *Bull. Entomol. Res.*, **78**, 633–639.

Ridsdill-Smith, T. J., Hall, G. P. and Craig, G. F. (1982) Effect of population density on reproduction and dung dispersal by the dung beetle *Onthophagus binodis* in the laboratory. *Entomol. Exp. Appl.*, **32**, 80–85.

Ridsdill-Smith, T. J., Hayles, L. and Palmer, M. J. (1986) Competition between the bush fly and a dung beetle in dung of differing characteristics. *Entomol. Exp. Appl.*, **41**, 83–90.

Ridsdill-Smith, T. J., Hayles, L. and Palmer, M. J. (1987) Mortality of eggs and larvae of the bush fly *Musca vetustissima* Walker (Diptera: Muscidae) caused by scarabaeine dung beetles (Coleoptera: Scarabaeidae) in favourable cattle dung. *Bull. Entomol. Res.*, **77**, 731–736.

Rougon, D. and Rougon, C. (1980) Le cleptoparasitisme en zone sahelienne: phenomene adaptatif d'insectes Coleopteres Coprophages Scarabaeidae aux climats arides et semi-arides. *C.R. Acad. Sci. Paris*, **291**, 417–419.

Sands, P. and Hughes, R. D. (1976) A simulation model of seasonal changes in the value of cattle dung as a food resource for an insect. *Agric. Meteor.*, **17**, 161–183.

Sasaki, H. (1984) A comparative study of the ecology of yellow dung-flies (Diptera: Scatophagidae). I. Predatory ability and predatory behaviour *Jap. J. Sanit. Zool.*, **35**, 325–331.

Sato, H. and Imamori, M. (1987) Nesting behaviour of a subsocial African ball-roller *Kheper platynotus* (Coleoptera Scarabaeidae). *Ecol. Entomol.*, **1**, 415–425.

Sigurjonsdottir, H. (1984) Food competition among *Scatophaga stercoraria* larvae with emphasis on its effects on reproductive success. *Ecol. Entomol.*, **9**, 81–90.

Southwood, T. R. E. (1987) Tactics, strategies and templets. *Oikos*, **52**, 3–18.

Stearns, S. C. (1982) The role of development in the evolution of life histories. In *Evolution and Development* (ed. J. T. Bonner), Springer-Verlag, Berlin, Heidelberg and New York, 237–258.

Thomas, G. D. and Kunz, S. E. (1985) Effects of season and density on the fecundity and survival of caged populations of adult horn flies (Diptera: Muscidae). *J. Econ. Entomol.*, **78**, 106–109.

Thomas, G. D. and Morgan, C. E. (1972) Field-mortality studies of the immature stages of the horn fly in Missouri. *J. Environ. Entomol.*, **1**, 453–459.

Thomas, G. D., Berry, I. L. and Morgan, C. E. (1974) Field development time of non-diapausing horn flies in Missouri. *Environ. Entomol.*, **30**, 151–155.

Tilman, D. (1982) *Resource competition and community structure*. Princeton University Press, Princeton, NJ.

Tribe, G. D. (1975) Pheromone release by dung beetles (Coleoptera: Scarabaeidae). *South Afr. J. Sci.*, **71**, 277–278.

Tyndale-Biscoe, M. (1971) Protein-feeding by the males of the Australian bushfly *Musca vetustissima* Wlk. in relation to mating performance *Bull. Entomol. Res.*, **60**, 607–614.

Tyndale-Biscoe, M. (1978) Physiological age-grading in females of the dung beetle *Euoniticellus intermedius* (Reiche) (Coleoptera: Scarabaeidae). *Bull. Entomol. Res.*, **68**, 207–217.

Tyndale-Biscoe, M. (1984) Adaptive significance of brood care of *Copris diversus* Waterhouse (Coleoptera: Scarabaeidae). *Bull. Entomol. Res.*, **74**, 453–461.

Valiela, I. (1969) An experimental study of the mortality factors of larval *Musca autumnalis* DeGeer. *Ecol. Monogr.*, **39**, 199–225.

Varley, G. C., Gradwell, G. R. and Hassell, M. P. (1973) *Insect Population Ecology*, Blackwell Scientific Publications, Oxford.

Vogt, W. G. (1987) Survival of female bush flies *Musca vetustissima* Walker (Diptera: Muscidae) in relation to reproductive age. *Bull. Entomol. Res.*, **77**, 503–513.

Vogt, W. G. and Walker, J. M. (1987) Potential and realised fecundity in the bush fly *Musca vetustissima* under favourable and unfavourable protein-feeding regimes. *Entomol. Exp. Appl.*, **44**, 115–122.

Vogt, W. G., Runko, S. and Walker, J. M. (1987) Estimation of egg to pupal survival of *Musca vetustissima* Walker (Diptera: Muscidae) in the field using genetic and radioactive markers. *Bull. Entomol. Res.*, **77**, 295–301.

West-Eberhard, M. J. (1986) Alternative adaptations, speciation and phylogeny (A Review). *Proc. Natl Acad. Sci. USA*, **83**, 1388–1392.

Zorka, T. J. and Bay, D. E. (1980) The courtship behaviour of the hornfly. *Southwest Entomol.*, **5**, 196–201.

11

Larval contribution to fitness in leaf-eating insects

Duncan Reavey and John H. Lawton

'I'm afraid I can't put it more clearly,' Alice replied very politely, 'for I can't understand it myself to begin with; and being so many different sizes in a day is very confusing.'
  'It isn't,' said the Caterpillar.
  Alice's Adventures in Wonderland, Lewis Carroll

11.1 THE LARVAL CONTRIBUTION TO FITNESS

11.1.1 How the Caterpillar might have replied to Alice

Considerable attention has been given so far to the effects on fitness of all kinds of adult behaviour. Now it is time to look at the larval stage. We have already seen how larvae can affect adult reproductive behaviour; for example, previous larval development can influence adult success in mate selection by determining adult body size and timing of reproductive maturity (Bailey, Chapter 3, this volume) and the future needs of larvae can constrain oviposition behaviour (Jones, Chapter 5, this volume). There are all sorts of other ways too in which the larval stage affects fitness. In this chapter we provide a framework of ideas which we believe are important in giving larvae the attention they deserve.

  Our discussion is broad. We look first at the timing of the larval stage and how it is constrained by the season and by the needs of the other life cycle stages. This is followed by consideration of possible life histories —different ways of reaching the same end point. Finally we examine larval behaviour and how this contributes to fitness. Most of the examples we use to lace the discussion are taken from the Lepidoptera; with roughly 150 000 described species, virtually all of which have phytophagous larvae, Lepidoptera are among the most diverse, if not the most diverse taxon of leaf-eating insects in ecosystems throughout the world (Strong *et al.*, 1984). In broader terms, however, the approach we take and the questions we ask are equally relevant to chrysomelid beetles, sawflies and other leaf-eating groups. But we must emphasize that the main thrust of our argument relates to insects with complete metamorphosis, feeding externally on their hosts, and exploiting one or only a small number of individual plants during larval development. We have drawn some of our examples and questions from organisms with rather different biologies —leaf miners living inside the host plant, grasshoppers grazing on thousands of individual plants, and so on. But other taxa, aphids for example, are so different that they deserve a chapter to themselves. We have neither the expertise nor the space to do justice to them here. Nor have we the knowledge to deal adequately with tropical insects of whatever taxon. Many of the ideas we develop apply equally well to tropical and temperate species; but some undoubtedly do not. Our distinctly temperate perspective needs to be borne in mind throughout the chapter.

Alice was right. The tangle of factors affecting when and where leaf-eating larvae occur and how they behave can be confusing. If she had spent more time with the Caterpillar it might have provided her with a framework something like the one we suggest here. One of our principal aims is to show once and for all that caterpillars cannot be dismissed as mere feeding machines.

11.1.2 Larval fitness as one inseparable component of an individual's fitness

Fitness is the proportionate contribution of an individual to future generations (Begon *et al.*, 1986) and hence is wholly relative. However, an individual can go a long way to maximizing fitness by maximizing its intrinsic rate of natural increase, r, where $r = \ln(R_0/T)$. R_0 is the net reproductive rate, that is, the mean number of eggs ultimately produced by an average egg in its average lifetime, and T is the generation time. R_0 is easily thought of as:

$$R_0 = \text{probability [newly laid egg surviving to oviposit (if female) or mate (if male)]} \times \text{expected fecundity} \times \tfrac{1}{2}$$

(The multiplier $\tfrac{1}{2}$ is there because the fecundity is shared with the male that has fertilized the eggs.) Selection acts to optimize the trade-off between the three variables—survival, fecundity and generation time —to maximize individual fitness.

Larval contribution to fitness cannot be analysed as a discrete component of the fitness of the individual (which is what matters), so strictly cannot be considered in isolation. There are carry overs to the larval stage from other life history stages and vice versa. For example, the larval stage might not necessarily occur at the best time of year for larval growth or survival; it must fit in with the constraints imposed by the needs of other stages. In the same way, an oviposition site that maximizes survival of the egg might not necessarily be the best site for the larva to survive its first instar. Indeed, it is easy to imagine an absurd number of possible trade-offs among and within all stages of the life cycle! The only conclusion then is that none of the individual stages is maximizing anything measurable; rather there is simply a trade-off among survival, fecundity and generation time over the *whole* life cycle that maximizes *r*. We believe that such a view is too pessimistic. While the situation is complex, it does appear possible to understand components of larval contribution to fitness in leaf-eating insects, at least in broad outline.

We will use the idea of fitness to help explain how the larvae of phytophagous insects behave. For example, many of the birch-feeding insects we are working with at York prefer young, early season foliage, though some can make use of older leaves if necessary; but others actually prefer older to younger leaves; and others appear to prefer young leaves but never normally encounter them (Fowler and Lawton, 1984). We will assume that this selection behaviour contributes to maximizing fitness. Whether a larva chooses the best from among the food available is very much a microscopic question. Before answering it we must first consider the macroscopic question of why it is feeding at that time of year in the first place. Why should a larva feed in June when a whole range of leaf qualities are present if it could feed in April when every leaf is young and of high quality? Therefore we will look first at patterns in the timings of life history stages. How do larval stages constrain the whole life cycle, and how do other stages in the life cycle constrain the timing and duration of larval stages?

11.2 THE TIMING OF THE LIFE CYCLE

11.2.1 Life history patterns

In climates with a marked seasonality the ultimate constraint on the way a life cycle can be organized is the pattern of the year itself, particularly the length and predictability of the dead season. Winter in north temperate regions must be passed in a dormant, resistant stage apparently generating a clear trend in overwintering stage with climate for North American butterflies (Hayes, 1982). Species undergoing diapause as pupa or adult tend to occur in warmer conditions; diapause as an egg predominates where it is cooler; diapause as a larva covers a wide range of temperatures but with predominance in the cool and the most extreme cold. Hayes' results are important in the present context because, if overwintering stages are constrained in some way as the results suggest, there may be knock-on effects on larval feeding periods. For example, diapausing overwintering larvae can resume feeding as soon as fresh foliage appears; the progeny of overwintering adults may be forced to feed later in the season because of the time required for eggs to be laid and hatch. Then, larvae might not be feeding at the 'best' time and overall fitness is a compromise between conflicting constraints at different stages in the life history. Intriguingly, despite Hayes' pioneering results, more general considerations suggest that the ability of any particular life cycle stage to undertake dormancy, or to tolerate harsh conditions, is not a major constraint on species' life histories (Tauber *et al.*, 1986) and hence on the timing of larval feeding. For example, more than half the British butter-flies overwinter as supposedly vulnerable larvae. Indeed, what is striking from a study of all 927 British macrolepidoptera is that just about *any* combination of the life history characteristics can and does occur (Gaston and Reavey, 1989) (Table 11.1). Eggs, larvae of various instars, pupae and adults can each serve as the overwintering stage; periods of active feeding by larvae vary markedly in length and are not related to adult size or to the type of food plant. It is not possible to find any very clear combinations of life history characteristics that define distinct groups of life history strategies and suites of characters are not strongly linked in regular ways. Our best guess at the moment, therefore, is that the timing of crucial events at other stages in the life histories of British Lepidoptera, such as the need to overwinter in a resistant stage, is not a strong constraint on when larvae feed. Hence, selection may well be able to optimize larval feeding periods, relatively untrammelled by the timing of other events in the life history.

11.2.2 Food as a constraint

For phytophagous insects, the feeding stages of the life cycle must be synchronized to the presence of suitable food (or foods, if larva and adult have different needs) and this constrains the non-feeding stages to other times of year. Variation from week to week in the quality of the food plant can be considerable (e.g. Dement and Mooney, 1974; Schultz *et al.*, 1982) and its effect on larval performance just as great; it may be enough to make synchrony of the larval stage with food plant phenology an overrid-ing requirement (Feeny, 1970; Mitter *et al.*, 1979; Schneider, 1980; West, 1985; Faeth, 1986; Potter and Kimmerer, 1986; Williams and Bowers, 1987; Du Merle, 1988; García-Barros, 1988; Rossiter *et al.*, 1988) and could constrain many species to a single generation each year (Schweitzer, 1979). Indeed, the timing of larval appearance in relation to leaf emerg-ence and foliage quality can be more important in host selection than phylogenetic relationships among foodplant species (Lechowicz and Mauffette, 1986). If the food resource is less restricted there is more scope to adjust the timing of the life cycle to avoid enemies or competitors or whatever.

The number of generations each year clearly has the potential to

Table 11.1 Relationships between body size, aspects of the life history and feeding biology for the 927 species of British macrolepidoptera (after Gaston and Reavey, 1989).

	Overwintering stage	Number of broods	Feeding specificity	Foodplant architecture	First month of larval feeding	Duration of larval stage
Body size	*Larger spp. tend to overwinter as adults; smaller spp. as pupae; eggs and larvae intermediate	Larger spp. have fewer broods	Larger spp. are more polyphagous	†n.s.	—	—
Overwintering stage		Not simple: no spp. overwinter as egg and have two broods, but no other simple trends	‡Not simple	Not simple	Depends on no. of intervening stages since diapause—but wide variation	—
Number of broods			n.s.	n.s.	n.s.	—
Feeding specificity				Not simple	Not simple	n.s.
Foodplant architecture					Tree feeders start earlier than herb feeders	n.s.

*When relationships are specified, as in that between overwintering stage and body size, the association is statistically significant.

†n.s. means no statistically significant association between the pair of variables.

‡A statistically significant association occurs but it is complex, with no simple pattern. Often the significance of the relationship is not understood.

markedly affect fitness; naively , for example, a species with two generations a year may have twice the fitness of a univoltine species. Food availability certainly sets the maximum number of generations possible in a season, at least for some species. Thus, polyphagous butterflies in North America tend to have more generations a year than oligophagous or monophagous species (Hayes, 1982), though this trend is not found for the British macrolepidoptera (Gaston and Reavey, 1989). On a more local scale, a limited range of available foodplants can certainly restrict a species to univoltinism, while a greater range allows multivoltinism. *Papilio zelicaon* (Papilionidae) is univoltine in areas of the USA where only native foodplants are available for a few months of the year, but it is multivoltine where newly introduced foodplants provide a food source for much longer (Sims, 1980). Foodplant choice by the tiger swallowtail *Papilio glaucus* (Papilionidae) determines the northerly limit at which a second generation is possible; this limit varies latitudinally over 600 miles depending on the foodplant species used (Scriber and Hainze, 1987). The yellow-ringed carpet *Entephria flavicinctata* (Geometridae) is single-brooded in northern England but double-brooded in Scotland. This is the only example among British macrolepidoptera of more generations occurring in the North than the South; the reverse occurs in 33 species. A reasonable explanation is based on the length of the growing season for the foodplants. In England *E. flavicinctata* feeds on *Saxifraga hypnoides*, a species of hills and mountains. In Scotland it uses *Sedum anglicum*, a coastal species reaching down to sea level, and it is in the rocky gullies and ravines by the sea that the moth occurs. Both foodplants are green throughout the year, but the growing season is considerably longer on the west coast of Scotland than in the hills of northern England. When the foodplant's growth season is longer, the potential growth season for the larva is longer, hence there is sufficient time for an extra generation.

11.2.3 Constraints other than food

Notice that we say food availability could set the *maximum* number of generations that are possible. There are many examples of similarly sized species in the same genus with the same food plants showing differences in voltinism (Slansky, 1974); examples among the British macrolepidoptera include the carpets *Xanthorhoe* spp. (Geometridae) and the thorns *Selenia* spp. (Geometridae) (Skinner, 1984). Although the reasons for these differences in life history among close relatives are not understood in detail, an obvious hypothesis is that other factors are superimposed on food availability so that a suitably timed single generation gives greater fitness than the maximum number of generations that foodplant availability allows (Slanksy, 1974; Tauber *et al.*, 1986). Table 11.2 shows that a single generation gives a greater *r* than two generations when:

Table 11.2 How differing survival through the year can determine the number of generations that is fitted in.

One generation

$$n_0 \xrightarrow{P(s)_1} n_{\text{mid season}} = \xrightarrow{P(s)_{\text{diapause}}} n_{\text{end of season}} =$$
$$P(s)_1 \times R_{01} \times n_0 \qquad\qquad P(s)_1 \times R_{01} \times n_0 \times P(s)_{\text{diapause}}$$

Two generations

$$n_0 \xrightarrow{P(s)_1} n_{\text{mid season}} = \xrightarrow{P(s)_2} n_{\text{end of season}} =$$
$$P(s)_1 \times R_{01} \times n_0 \qquad\qquad P(s)_1 \times R_{01} \times n_0 \times P(s)_2 \times R_{02}$$

---

n_0	number of individuals at the start of the season
$P(s)_1$	P(survival) to end of first generation
$P(s)_2$	P(survival) to end of second generation
$P(s)_{\text{diapause}}$	P(survival) to end of diapause replacing the second generation
R_{01}	basic reproductive rate of first generation
R_{02}	basic reproductive rate of second generation

Hence a single generation gives a greater rate of natural increase, r, than two generations when $P(s)_{\text{diapause}} > P(s)_2 \times R_{02}$.

P(survival in inactive stage during the second half of the season)

$> P$(survival through a second generation)
$\times R_0$ of the second generation.

Kukal and Kevan's (1987) study of the Arctic lepidopteran *Gynaephora groenlandica* (Lymantriidae) shows the influence that parasitism can have. This remarkable species has a generation time of about 14 years, during which 56% of mortality came from parasitism by an ichneumonid and a tachinid, compared to 13% from 'winter' mortality. Larval development takes so long because feeding is limited to three or four weeks a year; the season's feeding is completed and a hibernaculum constructed by the end of June even though suitable food is apparently available throughout July. Peak activity of the parasitoids coincides with inactivity of the larvae in July and the parasitoids attack feeding larvae but not hibernating larvae. In other words, what *G. groenlandica* appears to do is to trade feeding opportunities for parasitoid avoidance.

Parasitoids and predators appear to have major effects on the timing of other species' life histories too (Lawton, 1986a). Escape from an ichneumonid parasite that attacks pupae of the speckled wood *Parage aegeria* (Satyridae) in autumn could be the reason for overwintering as a third instar larva (Shreeve, 1986). Precise timing is important in Batesian mimicry where mimics emerging after their models are better protected than those emerging earlier, whether in the adult or larval stage (Rothschild, 1963; Waldbauer and Sheldon, 1971). Similarly, moths relying on a startle effect to distract a predator must avoid its habituation, so synchrony among species with contrasting patterns must be maintained (Sargent, 1978).

In brief, although synchrony with suitable larval food undoubtedly plays a crucial role in determining the overall fitness of leaf-eating insects, a number of other constraints, particularly interactions with natural enemies, obviously play a part in moulding life histories and in setting the larval contribution.

11.2.4 Fitting into the year

Diapause is an important way for insects to survive unfavourable seasons. Although there is wide flexibility across species in the overwintering life history stage, there is usually little or no intraspecific flexibility. Nearly all Lepidoptera diapause at an exact point in a single developmental stage (Tauber *et al.*, 1986) and this means that there can only be a whole number of generations per year; there cannot be 1¼ or 2⅖ or any other non-integer number. Among the British macrolepidoptera, exceptions are species with a 2+ year development (but these are apparently just as specific in their stages of diapause); there are also two exceptional species—the speckled wood *Parage aegeria* (Satyridae) (Shreeve, 1986) and brimstone moth *Opisthograptis luteolata* (Geometridae) (South, 1961) —which can fit 1 or 1½ generations into a year by overwintering as larva or pupa. Having a fixed, whole number of generations each year has interesting implications.

The principle that a decrease in generation time, *T*, always increases fitness if other parameters are unchanged does not necessarily apply to species fitting into a season of limited length and entering diapause at an exact point in their life cycle (Taylor, 1980). To see why, assume that there is a period of time towards the end of the season when individuals enter diapause. Only individuals within a certain range of age classes at the start of this period will diapause; the others die. Then, where the season is of limited length, reducing *T* can decrease fitness if the number of individuals falling within the range of critical age classes for diapause is reduced (Taylor, 1980; Blau, 1981). A longer generation time may put more individuals into the correct age range for diapause, and give greater fitness. In other words, there are circumstances where maximum larval growth rates (leading to short generation times) might not maximize fitness. This is the first of several reasons why selection may not act simply to maximize larval feeding rates.

It also matters that every year is different; in just about all environments few years are ever the same as the last. Some years will be difficult, with poor weather conditions or foodplant quality, and it might be difficult enough for an insect to reach a suitable state for diapause before conditions worsen and it is too late. But in better than usual years there is, literally, spare time. This could be filled in several ways, with the constraint that the individual must reach the precise diapause stage by the end of the season. Possibilities depend on the cues used by the insects to initiate and regulate their development. The spare time could simply be wasted. This is difficult to show but is reasonable for insects that maximize growth rate, reach a threshold weight, then undergo some sort of obligatory diapause. Alternatively, the spare time could be used to advantage. This requires a mechanism that informs the insect that spare time is available. Temperature and photoperiod in prediapause stages are common cues (e.g. Tauber *et al.*, 1986). One possibility is to squeeze in another generation. For example, at least 33 British macrolepidoptera are regularly bivoltine in the South but univoltine in the North, while 47 others have an additional generation in 'good' years (Skinner, 1984). These extra generations are often only partial, suggesting that not all individuals or populations are stimulated to produce them, or at least not in all localities. Yet the vast majority of species do not produce an additional generation, even in favourable years when there must be spare time available (Tauber *et al.*, 1986), though many show the capacity for another generation if conditions are manipulated in the laboratory. An alternative solution is to avoid diapause altogether. Of five tropical *Eurema* spp. (Pieridae), diapause as a non-reproductive adult was cued by lack of rainfall for one species and photoperiod for another, but three showed no diapause and behaved as opportunistic migrants when conditions worsened (Jones and Rienks, 1987).

Another possibility in better than average years is for the insect to increase its growth beyond the minimum that is sufficient in a poor season, perhaps by feeding for longer or on the better quality of food that might be available in a good year. If fecundity is correlated with weight at pupation (Hinton, 1981; Scriber and Slansky, 1981; but see Wiklund and Persson, 1983; Leather, 1988), extra larval feeding should increase adult fecundity and hence overall fitness. In other words, some species may simply continue feeding and growing until conditions become unfavourable (Atkinson and Begon, 1988), but it is not clear how generally this might apply. Alternatively, feeding rate could be adjusted so that development merely keeps pace with the year. Reducing the feeding rate in good years may give scope for behavioural alterations that could improve survival chances; for example, larvae may be able to spend more time hiding from enemies, perhaps by feeding only at night (Lance *et al.*, 1987). Indeed, reduced growth rates *per se* can actually reduce attack by enemies (Clancy and Price, 1987; Damman, 1987), contrary to the 'slower growth, higher mortality' dogma.

Plasticity in larval growth rate allows onset of diapause to be synchronized with deteriorating weather conditions. Larvae of the pale tussock moth *Calliteara pudibunda* (Lymantriidae) and the banded woolly bear *Pyrrharctia isabella* (Arctiidae) show a reduced growth rate, supernumerary instars and a longer development period when reared under long photophase conditions; under short photophase (and low temperature for the lymantriid) there is an intensive acceleration of development (Geyspits and Zarankina, 1963; Goettel and Philogène, 1978). This maintains the larval state, whatever the body weight, until pupation and diapause are cued by a shortening autumn day length. There are many other similar examples (Danks, 1987). Once again, it is clear that larvae of these species are not simply feeding at the maximum rate all the time. Although temperature and photoperiod are undoubtedly important as environmental cues used by many species to adjust their life histories to environmental conditions, they are not the only possibilities. Another possible cue for leaf eaters could come from plant growth hormones, or other hostplant chemicals, changing in occurrence and concentration in the foodplant. Changes in several plant hormones have direct effects on developmental rate and fecundity of the rangeland grasshopper *Aulocara elliotti* (Acrididae), and could have a role in synchronizing growth and reproduction with changes in the foodplant (Visscher, 1987).

There have been no systematic attempts to investigate how larvae of phytophagous insects adjust their feeding and overall rates of development to fit in with other requirements of the life history. What the above possibilities do indicate, however, is a second reason why many larvae are *not* mere feeding machines designed to maximize feeding and growth rates.

An altogether different and apparently relatively rare approach to making the most of 'good' years is to have flexibility in the overwintering stage; for example, *Ctenucha virginica* (Arctiidae) can overwinter in any of three larval instars (Fields and McNeil, 1988). Another possible benefit is a head start in growth for the following season. The unusually complex life cycle of the speckled wood butterfly has already been mentioned. When autumn conditions are warm, eggs laid in September can develop to pupae before overwintering and this can give a head start the following spring; in cooler years they cannot, though this has the advantage of parasite avoidance (Shreeve, 1986; but see Winokur, 1988). The five-spot burnet *Zygaena trifolii* (Zygaenidae) shows extreme flexibility, with overwintering possible in any of instars 3–8; this allows all sorts of development pathways to be followed in the wild, with intriguing polymorphism in the life cycles of siblings in controlled conditions (Wipking and Neumann, 1986). But in the main such extreme flexibility is rare. Why this is so is unclear, given the extreme interspecific variability in overwintering stages already noted for the British macrolepidoptera (Section 11.2.1). For whatever reason, most phytophagous insects in seasonal climates appear constrained to survive the unfavourable season in one particular stage of their life, with important effects on the life cycle and overall fitness.

11.2.5 Buffers in the life cycle

We have looked at overall constraints on the life cycle—factors influencing the number of generations and the set of life history characteristics. Each life cycle stage—ovum, successive larval instars, pupa and adult—contributes to r, but differ as to the weightings of each stage's contribution to P (survival) and to expected fecundity. While egg and pupa can directly contribute nothing material to fecundity and expose the individual to a mortality risk, it can be a good thing for them to extend beyond their minimum duration if these are the least disadvantageous ways of filling time until environmental conditions become suitable for the next active stage. And they can contribute to fitness if some eggs and pupae are more efficient at passing the time than others. Some life cycle stages are riskier than others, as examination of any insect life table shows (e.g. Varley *et al.*, 1973; Dempster, 1983; Strong *et al.*, 1984; Stiling, 1988). Unfortunately, data from such studies are rarely presented as, say, daily mortality rates, making it difficult to test the reasonable hypothesis that more vulnerable stages, experiencing higher instantaneous mortality rates, are passed more quickly (Williams, 1966; Edley and Law, 1988); unless, that is, the previous and subsequent stages are restricted to certain times of year and constrain the vulnerable stage to filling the time in between. When a buffer is needed to allow synchronization of the life cycle with the year, the least risky stages are the ones likely to provide this buffer. It is easy to take for granted that the larval stages are the most vulnerable, but life table data for Lepidoptera (e.g. Dempster, 1983; Strong *et al.*, 1984; Stiling, 1988) show that heavy mortality can fall upon any stage in the life history. We have not attempted detailed analyses of these data to calculate instantaneous mortality rates, but our impression is that, for many species, buffering by the larval stage is as plausible as any other. Take Goldschmidt's (1940) work on the races of the gypsy moth *Lymantria dispar* (Lymantriidae). Larvae of geographical races from regions where summers are short have a short development period and a more rapid growth rate than those of races experiencing longer summers. Shortening the larval period is apparently feasible but in 'long-summer' habitats the moth does not do it. Could this be an example of buffering by the larva?

Synchrony among individuals in a population could be achieved by the same sort of buffer. Synchrony is most important for reproducing adults, so varying the duration of the pupal stage—the stage most closely preceding the reproductives—may give optimal synchrony of adults. But there is no reason in principle why the duration of late larval instars cannot also be varied to ensure synchronous adult emergence. Selection may also favour synchrony of any life history stage as a means of swamping predators (e.g. Lloyd, 1984). Although poorly studied, if larval stages provide the buffer that ensures synchronization of later stages, this is a third reason why maximum larval growth rates are not necessarily expected.

11.3 TRADE-OFFS AMONG LIFE HISTORY STAGES

While each stage of the life cycle has a P(survival) and a potential contribution to fecundity that is measurable, it is their product over their whole life cycle that determines R_0. This in itself is a possible reason for apparent suboptimal behaviour in oviposition, larval feeding and so on, because what we see is a trade-off with other life history processes. Some of the obvious trade-offs are between optimum oviposition sites for eggs and the risks females take during laying; between egg survival and early larval survival (a good place to hide eggs may provide little or no food for larvae); between larval survival and larval growth (the best places to feed may carry high risks of predation); and between larval performance and pupal performance (the best feeding sites may be remote from safe pupation sites). We expand on these and related problems in the sections that follow.

11.3.1 The adult–ovum–larva trade-off

The trade-offs of oviposition are covered in detail by Jones (Chapter 5, this volume). The main feature is a trade-off between P(oviposition occurring), P(ovum surviving) and P(larva succeeding). Time available, predation risks and life span of the adult determine when it is best to oviposit and how selective the female can be. Oviposition behaviour of *Cactoblastis cactorum* (Phycitidae) (Robertson, 1987) is a striking example of this tangle of constraints and compromises; where eggs are laid (and where the larvae start out) not only depends on precision in selecting suitable foodplants for the larvae; just as important to the female is shelter from the wind during oviposition and the need to lay her first egg batch near the emergence site to reduce the energy cost of flight.

One component in the trade-off is the suitability of the oviposition site for the egg, the larva, and even the pupa. The site selected has a fundamental effect on larval growth and survival that declines only as larval mobility increases (see Thompson, 1988). At one extreme are the leaf miners, stem borers and gall formers that develop from egg to pupa within a single part of the plant. They are restricted throughout development to whatever growing conditions that single leaf, stem or gall has to offer. Of the external feeders, many are mobile only over small distances in early instars, but increasing body size brings greater mobility, and a greater choice of food within a single hostplant or between different individual hosts. At the other extreme are the highly mobile 'ballooning' first instar lepidopteran larvae. Examples include the winter moth *Operophtera brumata* (Geometridae) and gypsy moth *Lymantria dispar* (Lymantriidae), which show windborne dispersal from tree to tree until they find an acceptable foodplant species (Wint, 1983; Lance, 1983); winddispersed first instar larvae of *Ectropis excursaria* (Geometridae) settle down on foliage, which gives high first instar survival (Ramachandran, 1987).

The less mobile the larva, the more critical the adult's choice of oviposition site (Thompson, 1988). Whether adults of these species show a more precise, lengthier process of foodplant selection, and whether the larvae are less discriminatory, do not appear to have been considered. For species with immobile larvae the trade-off among egg survival, larval growth and larval survival will be heavily weighted on the needs of the larval stages. When larvae are mobile early on, oviposition need only be within striking range of the foodplants and this should allow a greater weighting on egg survival. Assessments of the advantages of laying on or away from foodplants are lacking but increasing egg survival is a plausible explanation for a whole range of observations. For example, larvae of *Hemileuca oliviae* (Saturniidae) are polyphagous and more or less randomly distributed on grasses, but ovipositing adults clearly select stiff, upright structures to provide physical support for the eggs (Bellows *et al.*, 1983); tropical *Euptychia* spp. (Satyridae) from disturbed habitats tend to oviposit away from their foodplants, but this is less frequent among forest species (Singer, 1984); and eggs of the silver-washed fritillary *Argynnis paphia* (Nymphalidae) are laid singly in crevices in the bark of trees though the foodplant is the low-growing perennial *Viola riviniana*—first instar larvae overwinter on the trunk (where they may be protected from enemies or bad weather) and only descend to the ground in spring (Carter and Hargreaves, 1986).

Offspring can begin to make the best of their situation even before they hatch from the egg. One particularly striking link between oviposition site and larval performance has recently been reported by Kennedy *et al.* (1987). Eggs of *Heliothis zea* (Noctuidae) maintained on the foliage of a wild tomato, *Lycopersicon hirsutum*, hatch into first instar caterpillars with elevated levels of resistance to potentially toxic 2-tridecanone present in the glandular trichomes of the plant. Small quantities of 2-tridecanone released from the host apparently induce the synthesis of detoxifying enzymes in the developing caterpillars; the enzyme(s) are lacking in larvae hatched from eggs laid on other species of hosts.

There are many more familiar examples where choice of oviposition site appears to maximize larval success, with success measured by growth rate and final size (Williams, 1983; Leather, 1985; Myers, 1985; Via, 1986; Forsberg, 1987) or survival (Rausher and Papaj, 1983; Weiss *et al.*, 1987). But in other studies, oviposition site choice leads to larval growth or survival which is well below what would be possible if a different plant species (Wiklund, 1975a; Courtney, 1981), a different plant location (Rausher, 1979), a different chemical cue (Fatzinger and Merkel, 1985) or a different location on the plant (Simberloff and Stiling, 1987) were chosen. Furthermore, there are many examples of captive larvae thriving on foodplants which are present in the field but which are usually ignored by adult females (Smiley, 1985). Are the ovipositing females getting it wrong?

All sorts of explanations for this apparently suboptimal behaviour have been put forward (Singer, 1984; Thompson, 1988). In some cases, female behaviour may indeed be suboptimal under present environmental conditions because habitats and plant species composition have recently changed. Getting it absolutely right at the present time is not essential for reproductive success. Alternatively, measurements of larval success are usually made in the laboratory and may be incomplete or inappropriate under field conditions. In particular, they may ignore mortality caused by enemies (predators, parasites and diseases) and competitors which may differ markedly on different species or parts of a hostplant. Thirdly, larval performance is only one of many considerations for optimal oviposition; what a searching female finds and accepts depends on factors like its own habitat preference (Root, 1973; Wiklund, 1975a; Courtney, 1982), the distribution of nectar and other food sources (Murphy *et al.*, 1984; Grossmueller and Lederhouse, 1987; Coleman and Jones, 1988; Jones and Coleman, 1988), the relative abundance of foodplants (Rausher, 1979; Holdren and Ehrlich, 1982), their apparency (Soberon *et al.*, 1988) and an urgency to lay (Jaenike, 1978; Courtney, 1982; Dennis, 1983). In other words, if there were an adult–ovum–larva trade-off, optimality would not require any of the usual, easily measured parameters of larval development to be maximized. We can find no studies that look more closely at these complex and interdependent trade-offs.

11.3.2 The larva–pupa–adult trade-off

Less well studied are trade-offs at the pupal stage. There must come a point when it is no longer worthwhile for a final instar larva to continue feeding, most likely when the costs of continued feeding outweigh the benefits. Although it has been generally accepted that increased body size increases fecundity of the female, favouring prolonged larval feeding,

Leather (1988) has recently questioned whether larger females necessarily lay more eggs under field conditions. Large body size could also increase the fitness of the male if a large body size is important for mobility, enhanced fertility (Haukioja and Neuvonen, 1985) or in sexual selection by females (e.g. Smith, 1984), although again, pressure for protandry or protogyny could push body size in the other direction (Singer, 1982; Mikkola, 1987). At the same time continued feeding could prolong the risks from predators or deteriorating weather. Hence there is a key trade-off between potential fecundity and survival (Roff, 1981, 1986), and one which could have different solutions for males and females (Lederhouse *et al.*, 1982). Yet the problem has received virtually no attention in the literature. The utilization efficiencies of four lepidopteran species feeding on sycamore decrease as the larvae grow; those with greater efficiencies descend to pupate later, but they suffer greater predation (Warrington, 1985). This example hints that trade-offs exist, but does no more than that.

The trade-off could extend into the adult stage if the length of the pupal stage is fixed. Time as a larva can be traded with time as a young adult according to the relative qualities (food availability, P(survival) and so on) of the larval and adult habitats (Singer, 1982).

Some idea of the subtleties involved in larval–pupal trade-offs and their effects on fitness can be gained by thinking about specific examples. A mobile larva can select its pupation site and some species are more choosy than others. The Papilioninae (Papilionidae) have two different approaches to effective concealment of the pupa (Wiklund, 1975b): Some species have a single pupal morph and these rely on reaching a particular, often quite specific, pupation site. Others are polymorphic and can make use of a variety of different sites; the appearance of the pupa, usually its colour, is environmentally controlled (Wiklund, 1975b; Sims and Shapiro, 1983). The contribution of the larva is potentially important in the first case because there could be a price in searching for suitable places to pupate. This could include a cost of increased mortality risk and a cost of lost body weight, and hence lower future fecundity. However, it could be more than balanced by the benefits of a well-chosen, safe, pupation site. There can certainly be a metabolic and time cost in moving long distances (Weiss *et al.*, 1987) but it is not clear whether this cost is anywhere near as great as the possible benefits, or whether it can restrict the choice of pupation sites. We know from the few data that exist on pupal mortality for Lepidoptera that loss in the pupal stage can approach 100% and it averages about 60% (White, 1986); there is a good chance that effective selection of the pupal site by the caterpillar can influence this to some degree.

Taking a broader view, 265 of 403 British macrolepidoptera covered by Carter and Hargreaves (1986) descend from their foodplant to overwinter on the ground surface (19% of species) or below it (48%). Perhaps not surprisingly, species overwintering as pupae tend to pupate below ground while those with other overwintering stages tend to stay on their foodplant or on the ground surface. Overwintering stages and over-wintering sites are not randomly associated ($\chi^2 = 20.05$, $n = 336$, $P < 0.0001$); species with a longer pupal period, typically those over-wintering as pupae, might be under more pressure to get it right. For example, a small fraction of mimosa webworm *Homadaula anisocentra* (Plutellidae) pupae survive in highly protected overwintering sites in urban environments in central Iowa at the northern edge of the species' range. Less well protected pupae perish at normal winter temperatures (Miller and Hart, 1987).

11.4 HOW TO MAKE IT AS A LEAF-EATING LARVA

Now that we have put the larval stage into the context of the year and of the other life cycle stages, and seen how the overall structure of the life cycle constrains and influences the larval contribution to fitness, we want to look more closely at what phytophagous larvae actually do. How does larval behaviour contribute to fitness? How efficient are larvae at making the best of a highly heterogeneous environment?

11.4.1 Life histories, and the growth rate–survival trade-off

From the wide variety of larval morphologies and behaviours it seems that there are many different ways of getting it right (Schultz, 1983a; Janzen, 1984). Within the time period set by constraints of the year and of the other life cycle stages, a larva must grow sufficiently and survive. Growth rate could be increased by feeding for longer or by feeding at the time of day when the plant's chemical composition is more favourable (Grison, 1952; Haukioja *et al.*, 1978; Raupp and Denno, 1983). Survival chances could be altered by more or less permanent concealment within rolled or tied leaves, mines, galls or stems (Price *et al.*, 1986; Hawkins and Lawton, 1987; Hawkins, 1988), by cryptic coloration and behaviour to suit, or by aposematism. Feeding as a specialist rather than a generalist could be another way of reducing pressure from predators (Brower, 1958; Bernays, 1988), though there is no clear explanation of why this should be so. But, throughout, fastest possible growth and highest possible survival are not necessarily compatible. The result is another trade-off.

That many larvae do not show the fastest possible growth is clear. Some prefer nutritionally poorer tissue (Raupp and Denno, 1983; Griswold and Trumble, 1985) but adaptive explanations have relied on guesswork.

Reducing the danger from predators and parasitoids is one possibility among many (Price *et al.*, 1980; Lawton, 1986a).

Feeding for only a small part of each day, as many species do, could hold the growth rate below the maximum, though there are pitfalls in interpreting 'resting' as wasted feeding time; it could well be essential if digestive processes do not keep pace with feeding (Porter, 1982, 1984). Nevertheless, there is every chance that a cryptic night feeder could increase its growth rate if it fed and moved by day as well as by night, but there would most likely be a cost in survival (Herrebout *et al.*, 1963; Heinrich, 1979; Schultz, 1983a; Fitzgerald *et al.*, 1988). Showing that such a trade-off actually exists will never be easy because the defences of larvae—characterized by daily activity patterns, coloration and so on —are often genetically fixed and hence extremely difficult to manipulate. For example, the optimal temperature for cryptic larvae of *Colias* spp. (Pieridae) closely matches the temperature range encountered when feeding and could well be a metabolic adaptation to predator pressure (Sherman and Watt, 1973). However, several examples suggest that trade-offs between growth rate and survival do exist. Whether a trade-off is optimal, with fitness truly maximized, is another question and, while there is a good chance it is, the possibility has never been rigorously tested.

Larvae of *Omphalocera munroei* (Pyralidae) prefer to feed on older, nutritionally poorer pawpaw leaves and also use them to form a protective shelter; younger, nutritionally superior but softer leaves cannot be used to form a shelter. Development took 20% longer on the older leaves but survival of natural-sized groups of larvae was only 2% on younger leaves without protective shelters compared to 42% on older leaves with shelters (Damman, 1987). Larvae of *Hemileuca lucina* (Saturniidae) feed gregariously in full sunlight but here they are very vulnerable to predatory wasps. Larvae disturbed by wasps moved quickly to the cooler interior of the plant where only mature leaves are available. After moving, larvae experienced lower temperatures and poorer food quality and therefore grew more slowly, but these disadvantages appeared to be more than compensated for by reduced risks of predation (Stamp and Bowers, 1988).

Feeding on poorer tissue could also enhance survival by reducing physical dangers; the risk of a herbivore being dislodged by blowing branches is less in some feeding positions than others and this could be the most important criterion in food choice (Dixon and McKay, 1970).

Some components of the life history are more flexible. For example, the ground colour of cryptic larvae can match the food eaten (e.g. Heinrich, 1979). One striking example of this is the camouflage adopted by *Nemoria arizonaria* (Geometridae): spring brood caterpillars mimic oak catkins; by the time of the summer brood, the catkins have fallen and the larvae mimic the twigs. The level of tannins in the diet, rather than temperature or photoperiod, appears to provide the cues for these developmental responses (Greene, 1989).

Not surprisingly, the pattern of daily behaviour can show marked plasticity if conditions demand. When populations of the gypsy moth are at low density, late instar larvae feed at night and spend the day resting away from the canopy in relatively well-protected sites in the litter or in fissures in the bark; at high density, larvae remain in the canopy, feed by day as well as by night and spend more time searching and less time feeding (Lance *et al.*, 1987); as risk of starvation increases, the larvae risk greater predation.

The larvae of many species of foliage-feeding insects change very little in appearance or behaviour during development, except by growing larger; in others, larval appearance, behaviour, feeding site and so on—the whole life history—change dramatically in different instars, presumably because selection pressures also change. Yet the selective advantages of major changes in larval life style have rarely been evaluated or subjected to experimental investigation. Changes can be as fundamental as that shown by a *Panotima* sp. (Pyralidae), a leaf surface grazer until the third instar when it becomes a stem borer (Lawton *et al.*, 1988). Of 1137 species of British microlepidoptera for which data are available (Emmet, 1988), 49 species of leaf miner become leaf tiers, 17 become open feeders and 14 species of stem borer become leaf tiers as larval development proceeds (K. J. Gaston and D. Reavey, unpublished data). What trade-offs are involved in these switches in feeding behaviour are quite unknown. Appearance, too, can be very different in successive instars, with cryptic coloration in some but aposematic coloration in others. For example, the larva of the alder moth *Acronicta alni* (Noctuidae) resembles a bird dropping when it is small but when fully grown it has conspicuous yellow and black banding (Owen, 1980; Carter *et al.*, 1988). The eggs of Ithomiinae (Nymphalidae) show high levels of toxic alkaloids derived from the adult and are aposematically coloured; the alkaloids are absent in the larvae and pupae and these stages are cryptic (Brown, 1985). Gregarious behaviour can be a defence for small larvae (e.g. Tostowaryk, 1972) but becomes unnecessary when larvae are large and less vulnerable (Stamp and Bowers, 1988). Unravelling the selective advantages of changes in larval appearance and behaviour during development would appear to offer considerable potential for future research.

The life history of a species is the final, intimate framework within which individual foraging behaviour—what the larva does minute by minute and hour by hour—contributes to overall fitness. It is to individual foraging behaviour that we must now turn.

11.4.2 Foraging behaviour and fitness: the case of birch-feeding Lepidoptera

By now it should be clear that the manner in which the microscopic details of an insect's feeding behaviour contribute to overall fitness may vary enormously from species to species, depending on such macroscopic constraints as the timing of the life cycle, the needs of other stages in the life history, whether the insect is cryptic or aposematically coloured, and how and where it feeds. Not surprisingly, at the moment there are no unifying models that explain caterpillar foraging behaviour on an hour-by-hour or minute-by-minute basis. But important elements are understood.

One obvious point is that foodplants are highly heterogeneous from the insect's point of view (e.g. Denno and McClure, 1983) in such things as leaf age, exposure to sunshine, accessibility to predators, herbivore damage and so on. So the right choice of where and when to feed could make a difference to larval contribution to fitness, at least in principle; for example, there is a vast literature showing that growth and survival are better if the leaf nitrogen level or leaf water content is higher (e.g. Bernays and Barbehenn, 1987; Mattson and Scriber, 1987; Tabashnik and Slansky, 1987). It is also clear that larvae have the ability to make choices (e.g. Hanson, 1983; Schoonhoven, 1987) and can show distinct preferences for a particular food, microclimate or whatever. Yet all too often preferences have been shown only in artificial laboratory situations, and with different outcomes under different experimental conditions (Risch, 1985; Marquis and Braker, 1987). More relevant is whether there is similar selectivity on the foodplant in the field. Most interesting and crucially important to fitness is whether the larvae make the *right* choices in the field. Because the literature on these problems is both large and diffuse, we have chosen to illustrate key ideas with examples drawn from the system with which we are personally most familiar, namely lepidopteran caterpillars feeding on birch *Betula*.

One important source of heterogeneity in the foodplant is in the patterns of herbivore damage and in the distribution and concentration of the so-called 'defence' compounds induced by this damage (Schultz, 1988). We will look at the topical subject of induced defences in more detail and see what effect these have on larval foraging behaviour and, ultimately, on fitness.

In response to leaf damage inflicted by herbivores, or to experimental artificial damage, there is a build-up in the levels of phenolics in the leaves of birch, both *Betula pendula* and *B. pubescens* (Haukioja, 1979; Haukioja and Niemelä, 1979; Lawton, 1986b; Hartley and Lawton, 1987 and references therein). Phenolics have the potential to impair the feeding, growth and other aspects of the performance of some, though not all, leaf-feeding

Table 11.3 Developmental and behavioural responses of birch-feeding herbivores to leaf damage. Marginal effects are indicated in paentheses.

Insect species	Birch species	Effects of leaf damage on development			Behavioural responses to damage from:			Reference
		Larval duration	Pupal weight	Larval survival in laboratory	Chewers	Artificial damage	Miners	
LEPIDOPTERA								
Lasiocampidae								
Eriogaster lanestris	B.pub.tort.	larval weight lower on damaged leaves during early growth, but not later						1
Thyatiridae								
Achlya flavicornis	B.pendula	No effect	Lower	No effect		Prefers	No pref.	2
Geometridae								
Archiearis parthenias	B.pub.tort.	Longer	No effect	No effect				1
Epirrita dilutata	B.pendula				No pref.	(Prefers)	Avoids	3
Epirrita dilutata	B.pub.pub.						No pref.	4
E. autumnata	B.pub.tort.	Longer	Lower	No effect				1,5,6
	B.pub.pub.	Longer	Lower					7
Operophtera brumata	B.pub.pub.						Some-times avoids	4
Apocheima pilosaria	B.pendula				No pref.	No pref.	Avoids	3
	B.pendula			No effect in field		No marked effect		8
	B.pub.pub.		Larval weight lower	Lower in field				9
Erannis defolaria	B.pendula				(Prefers)	Prefers	Avoids	3
Lymantriidae								
Orgyia antiqua	B.pendula						Avoids	10

Species	Birch						Ref.
Euproctis similis	*B.pendula*				No pref.	Avoids	3
Noctuidae							
Orthosia stabilis	*B.pub.pub.*					Avoids	4
Spodoptera littoralis	*B.pendula*				No pref.		10,11
	B.pendula				Prefers leaves adjacent to damaged ones over undamaged control		11
					Avoids damaged side of leaf		12
	B.pub.pub.					Avoids	10,11
Coleophoridae							
Coleophora serratella	*B.pendula*	Longer	No effect		Small movements in close vicinity		8,13
			in field				
	B.pub.pub.				More damage on previously damaged foliage		9
						Avoids	4
HYMENOPTERA							
Cimbicidae							
Trichiosoma lucorum	*B.pub.tort.*	No effect	No effect	No effect			1
Tenthredinidae							
Dineura virididorsata	*B.pub.tort.*	(Shorter)	No effect	No effect			1
Pristiphora sp.	*B.pub.tort.*	Longer	No effect	No effect			1
Pteronidea	*B.pub.tort.*	Longer	No effect	Lower			1

References: 1, Haukioja and Niemelä (1979); 2, D. Reavey (unpublished data); 3, Hartley and Lawton (1987); 4, G. R. Valladares (unpublished data); 5, Haukioja and Niemelä (1977); 6, Haukioja *et al.* (1985); 7, Haukioja and Hanhimäki (1985); 8, Bergelson and Lawton (1988); 9, Fowler and MacGarvin (1986); 10, Wratten *et al.* (1984); 11, Edwards *et al.* (1986); 12, Gibberd *et al.* (1988); 13, Bergelson *et al.* (1986).

insects (Bernays, 1978; Lindroth and Peterson, 1988). At the same time, damage also leads to a reduction in leaf water content (Hartley and Lawton, 1987) and reduced water content of foliage may also impair larval performance (Scriber and Slansky, 1981). Doubtless there are also other physical and chemical changes in damaged leaves. Which of these factors affects larval development is not important here; what matters is that damage induces deterioration in birch foliage and hence has the potential to impair larval performance and reduce fitness.

There are several ways in which this deterioration might affect herbivorous insects. Phenolics, acting in the main as feeding deterrents (Niemelä *et al.*, 1979; Hartley, 1987), low nitrogen levels or low water content can reduce a herbivore's growth rate. Slower growth could reduce the eventual fecundity or increase the period of exposure to enemies (Bergelson and Lawton, 1988). Caterpillars of seven lepidopteran and sawfly species grow more slowly or are smaller at pupation when fed on damaged rather than undamaged birch leaves (Table 11.3). Two show greater mortality when fed on damaged foliage. With these apparent disadvantages of feeding on damaged leaves, it would make sense for larvae of these species to be selective.

Most larvae certainly have an ability to discriminate. Laboratory tests have shown that caterpillars of eight birch-feeding lepidopterans prefer undamaged over damaged leaves (Table 11.3) but results are equivocal, depend on how the damage was caused and show no simple relationship to phenolic levels (Hartley, 1988). There are cases (and no doubt many more unreported) where no detectable preferences are shown and occasions when naturally damaged leaves are actually preferred. This is only to be expected if some other aspect of leaf quality predominates and the effects of damage are relegated to secondary importance. Note that the effects of damaged foliage on insect performance and its effects on behaviour do not necessarily correspond (Lawton, 1986b); the pale brindled beauty *Apocheima pilosaria* (Geometridae) shows lower pupal weight and higher larval mortality when reared on damaged birch leaves in the field (Fowler and MacGarvin, 1986) but has no tendency to avoid damaged leaves in choice tests (Hartley and Lawton, 1987) nor to show greater movement on trees with damaged leaves (Bergelson and Lawton, 1988). More generally, damaged birch leaves received less subsequent damage from natural but unidentified mixtures of herbivores in the field than did undamaged leaves in one experiment (Silkstone, 1987, but see G. R. Valladares and J. H. Lawton, unpublished MS). What are missing from all these studies are observations on the foraging behaviour of individual natural birch-feeding caterpillars of known age and species, coupled with rearing experiments in which animals are forced to develop on only one type of leaf. Only then will we really be able to determine the significance of what are, at present, a rather confusing series of observations.

It also makes sense to avoid damaged leaves if the damage gives a cue to searching predators and parasitoids (Heinrich and Collins, 1983). The result would be increased movement of the herbivore between leaves. The costs of this extra movement could be trivial, but, in theory at least, there could be increased metabolic cost or an increased risk if moving makes a larva more conspicuous or more likely to encounter an enemy (Schultz, 1983b; Fowler and Lawton, 1985). So far there is no evidence for this effect on birch. The presence of foliage damage had no marked effect on predation rates on two lepidopteran species; nor did it increase herbivore movement to an extent likely to alter mortality from predators (Bergelson and Lawton, 1988). One of the species studied was a case-bearing lepidopteran *Coleophora serratella* (Coleophoridae), a species which develops more slowly when reared on damaged leaves. Larvae responded to the damage they caused by moving away from its immediate area, though they were just as likely to move to another part of the same leaf as to other leaves (Bergelson *et al.*, 1986). These very local changes in leaf quality could force insects to change feeding site frequently, but whether this reduces fitness to any real extent (Gibberd *et al.*, 1988) is more questionable.

There is no substitute for time spent watching what larvae actually do in natural conditions. The behaviour can then be related to what is known of leaf chemistry. Phenolics are induced not only in insect-grazed leaves of *B. pendula*, but in adjacent, undamaged leaves too (Lawton, 1986b; Hartley and Firn, 1989). We now know that there is a clear increase in phenolics in undamaged leaves adjacent to damaged leaves up to two nodes along the branch towards its tip. However, there is no induction along the branch in the other direction (S. E. Hartley, unpublished results). If induced phenolics are important in determining where larvae should feed, we predict that feeding larvae should select, successively, leaves that are further from the tip of the branch, working their way from the tip of the branch towards the trunk. Late instar larvae of the yellow horned *Achlya flavicornis* (Thyatiridae) do not do this; rather, they do exactly the opposite, selecting leaves progressively nearer the tip. However, they consume the whole of the leaf in one or two feeding bouts so there may be little opportunity for defence cues to pass from the damaged leaf up the branch (Feichtinger and Reavey, unpubl.). Alternatively, the pattern of feeding may be constrained by unknown factors that are more important in determining fitness than are induced chemical changes in the leaves lying ahead of the feeding caterpillar.

11.4.3 Subtleties in foraging

The short summary in the previous section of work on birch herbivores highlights difficulties in relating short-term, microscopic foraging decisions of individual larvae to fitness. Here we briefly expand the discussion to embrace studies of other species to illustrate the rich possibilities that exist for future work on the foraging decisions made by leaf-eating larvae.

One striking aspect is the ability of at least some larvae to select the intake of protein and digestible carbohydrate in a ratio that is optimal for conversion into biomass; this has been shown for the corn earworm *Heliothis zea* (Noctuidae) among several other insects (Waldbauer *et al.*, 1984). But not only this. Towards the end of the larval stage there is an increasing need for carbohydrate over protein, and the intake of each is adjusted by *H. zea* as it nears pupation so that the ratio remains optimal (Cohen *et al.*, 1987; see also Zucoloto, 1987). This self-selection seems as clear on living foodplants as it does on artificial diets (Cohen *et al.*, 1988). In the same way, nymphs of the desert locust *Schistocerca gregaria* (Acrididae) vary their preference for foods of different water content; they regulate their water and dry matter intake by their behaviour (Lewis and Bernays, 1985). It is clear, too, that behaviour changes and larvae become less choosey as they become hungry (e.g. Jones, 1977), that is, as noted in Section 11.4.1, larvae are willing to take greater risks as the threat of starvation looms. Such a variation in preferences—perhaps from hour to hour—is a complication that has not really been taken on board in many studies of dietary choice in leaf-eating insects and deserves more attention.

One area that would be particularly profitable to study is how and why feeding preferences change as larvae grow (e.g. Hansen and Ueckert, 1970). Physical obstacles on the plant surface can become more or less of a hindrance and affect which parts of the foodplant are available (Southwood, 1986) as may sclerenchymatous tissue around vascular bundles (e.g. Hagen and Chabot, 1986). Larger, tougher mouthparts give access to tougher foliage which might be unavailable to smaller larvae (Bernays and Chapman, 1970; Bernays, 1986). Yet they could also reduce the fine scale selectivity of species that feed by scraping the leaf surface, some of which prefer to feed on the upper side of the leaf and others which prefer the under side; the two surfaces may differ in as yet poorly understood ways in their food value for larvae. Similarly, leaf-mining larvae may find it more difficult to exclusively exploit tissues of the highest quality within leaves as they get larger. With a few notable exceptions (e.g. Kimmerer and Potter, 1987), very little is known about food selection by individual leaf-mining larvae.

We also expect free-living caterpillars to be confronted with very different trophic problems as they grow, even on nutritionally and toxicologically identical diets. Not only must the ratio of gut volume to gut surface area change as larvae grow, but so too must weight-specific feeding rates and weight-specific concentrations of enzymes important in the detoxification of secondary plant compounds (Gould, 1984). Hence the optimal behaviour for one species of caterpillar on one hostplant may vary markedly between the first and last instar, irrespective of all the other things that can change as larvae grow.

A change of diet on the coarsest possible scale—from one foodplant species to another—could be forced if a larva falls from a tree to the undergrowth or if its first foodplants senesce before growth is completed or if an autumn herb feeder overwinters to find different species predominating the following spring. Forcing larvae to switch foodplants can adversely affect growth (e.g. Schoonhoven and Meerman, 1978; Grabstein and Scriber, 1982; Karowe, 1989). However, there are times when sequential feeding on more than one species can increase fecundity (Barbosa *et al.*, 1986; see also Thompson, 1988). Amateur breeders have known this for a long time; the grey chi *Antitype chi* (Noctuidae) does best in captivity if given a different foodplant in each instar, finishing with lilac (Allan, 1949). It is very unclear how often polyphagous caterpillars in the field change hostplant species as they grow, thereby increasing overall fitness by exploiting a mixed diet.

Body size affects mobility, and mobility affects the area of the foodplant to which the larva has access. The fractal nature of plant surfaces (Lawton, 1986c) magnifies the relative problems confronting small and large larvae of the same species, making it much more difficult for small larvae to walk to new resting sites or feeding areas (Weiss and Murphy, 1988). Once again, the optimal solution to a particular problem will undoubtedly be different for small and large larvae of the same species, but this remains virtually unstudied. There are no simple rules. Larger larvae might well be physically capable of travelling faster or further but might avoid this if it increases vulnerability to enemies or risk of being parted from the foodplant. Larger larvae may also be capable of using food of lower quality and/or higher toxicity than smaller larvae for the reason outlined above. This could mean for some species that larger larvae need move shorter distances than smaller larvae. The amount they move affects what opportunity there is for assessing the food resources that are available and for selecting the best from among them. But whether mobility is a limiting factor at all during foraging by any instar is largely unknown.

While some larvae are capable of moving great distances, they do not seem to make the most of their mobility in searching for the best food. We have already mentioned the case-bearer *Coleophora serratella*. This moves only short distances on the same leaf or among neighbouring leaves between feeding bouts. Yet the larva moves a much greater distance just before ecdysis, taking it well clear of its conspicuous feeding damage during the prolonged ecdysis period (Bergelson *et al.*, 1986). Larvae of a small ermine moth *Yponomeuta vigintipunctatus* (Yponomeutidae) regularly move away from their web to deposit their faeces a few centimetres away; but they rarely move any distance from the web to feed (Kooi, 1988). Other species move considerable distances away from the leaves on which they have been feeding before resting; the tussock moth *Dasychira dorsipennatata* (Lymantriidae), for example, passes very many leaves during a movement over a metre or more (Heinrich, 1979). This behaviour could well reduce predation to birds which use leaf damage as a visual cue; but does this mobility also allow the larva to sample and select its food? Gregarious behaviour allows efficient location of leaves; trail marking by eastern tent caterpillars *Malacosoma americanum* (Lasiocampidae) means that all the tent can benefit from the most suitable leaves over a wide area, even though the searching efficiency of the small individual larvae is limited (Peterson, 1987).

Few studies have looked at the movements and feeding patterns of individual larvae over their whole larval period to see what opportunities they have to make foraging decisions (but see Feichtinger and Reavey, 1989). There are all sorts of questions to answer. We might predict from foraging theory that a more heterogeneous foodplant may be assessed more carefully and over more of its area before leaves are chosen. Leaf-tiers might be restricted in the distance they can move from their ties and so should be particularly careful in choosing where to site them. One advantage of being a hostplant specialist may be the evolution of fixed foraging rules that bring caterpillars more rapidly into contact with high quality foliage. Hostplants of different species differ markedly in architecture; for example, in the spatial arrangement of new and old leaves, fruiting bodies, etc. It may be much more difficult for an individual caterpillar of a polyphagous species to locate efficiently the best feeding sites on any one species of hostplant than it is for a hostplant specialist. But as far as we are aware, the problem has never been looked at.

Last but not least, these ideas on foraging are by no means restricted to feeding. They apply equally to whatever the larva is selecting, whether it is a safe site for ecdysis or pupation, or a leaf suitable to tie, or a beneficial microclimate.

11.5 THE CATERPILLAR'S FINAL WORD

Alice was right. Being so many different sizes *is* confusing, at least for a biologist! But Lewis Carroll's Caterpillar knew better. Our problem is to try to understand the world through the eyes of a caterpillar and grasp what it sees as important. In this chapter we have done no more than lay out the framework for an approach to the problem of fitness in leaf-eating insects. It most certainly is not the final word.

We have looked at some of the ways in which fitness is affected by what happens in the larval stage. In temperate regions, at least, the immature stages take up the bulk of the year for many species and, while more vulnerable stages might be passed more quickly, they can hardly be by-passed completely. Getting it right as a larva is critical. At the start of this chapter we considered how fitness could be maximized by an optimal trade-off between survival, fecundity and generation time over the whole life cycle. We suggest that, in practice, the contribution of the larval stage will often outweigh those of the egg, pupa and adult. After all, the larva is more often than not at a greater daily risk of mortality than the other stages; it is frequently the stage that assembles most, if not all, of the resources that translate into fecundity; and it is larval feeding that is one major constraint on timing. When the contribution to fitness is greater, the pressure to get it right is greater; the wonderful diversity of life styles among leaf-eating insects suggests that there are a multiplicity of ways of solving these problems. Whether simple rules underpin the avalanche of case histories is less clear.

One of the most surprising things to emerge from the chapter is how little biologists have thought about the determinants of fitness in so important a group of organisms. Insect herbivores dominate global diversity (Strong *et al.*, 1984). Yet most of the work relating to fitness in leaf-eating insects takes a narrow, microscopic view of the problem; studies on aspects of the feeding behaviour of individual caterpillars are legion. But is this really more important than, say, the selection of a safe pupation site—which hardly anybody studies? How do other events in the life history constrain what caterpillars do? And so on and so forth. Studying caterpillars is an adventure in wonderland but it also goes to the heart of fundamental problems in evolutionary biology.

ACKNOWLEDGEMENTS

We thank our colleagues in York and further afield—Phillip Ackery, Veronika Feichtinger, Alastair Fitter, Kevin Gaston, Dave Griffiths, Susan Hartley, Chris Rees, Mike Singer, Graciela Valladares and Dick Vane-Wright—for their ideas and suggestions while we were putting this chapter together. DR is supported by an SERC studentship.

References

Allan, P. B. M. (1949) *Larval Foodplants*, Watkins and Doncaster, London.

Atkinson, D. and Begon, M. (1988) Adult size variation in two co-occurring grasshopper species in a sand-dune habitat. *J. Anim. Ecol.*, **57**, 185–200.

Barbosa, P., Martinat, P. and Waldvogel, M. (1986) Development, fecundity and survival of the herbivore *Lymantria dispar* and the number of plant species in its diet. *Ecol. Entomol.*, **11**, 1–6.

Begon, M., Harper, J. L. and Townsend, C. R. (1986) *Ecology: Individuals, Populations and Communities*, Blackwell Scientific Publications, Oxford.

Bellows, T. S. Jr, Owens, J. C. and Huddleston, E. W. (1983) Plant species utilization by different life stages of the range caterpillar, *Hemileuca oliviae* (Lepidoptera: Saturniidae). *Environ. Entomol.*, **12**, 1315–1317.

Bergelson, J. M. and Lawton, J. H. (1988) Does foliage damage influence predation on the insect herbivores of birch? *Ecology*, **69**, 434–445.

Bergelson, J., Fowler, S. and Hartley, S. (1986) The effects of foliage damage on casebearing moth larvae, *Coleophora serratella*, feeding on birch. *Ecol. Entomol.*, **11**, 241–250.

Bernays, E. A. (1978) Tannins: an alternative viewpoint. *Entomol. Exp. Appl.*, **24**, 44–53.

Bernays, E. A. (1986) Diet-induced head allometry among foliage-chewing insects and its importance for graminivores. *Science*, **231**, 495–497.

Bernays, E. A. (1988) Host specificity in polyphagous insects: selection pressure from generalist predators. *Entomol. Exp. Appl.*, **49**, 131–140.

Bernays, E. A. and Barbehenn, R. (1987) Nutritional ecology of grass foliage-chewing insects. In *Nutritional Ecology of Insects, Mites, Spiders and related Invertebrates* (eds F. Slansky Jr and J. G. Rodriguez), Wiley, New York, 147–175.

Bernays, E. A. and Chapman, R. F. (1970) Food selection by *Chorthippus parallelus* (Zetterstedt) (Orthoptera: Acrididae) in the field. *J. Anim. Ecol.*, **39**, 383–394.

Blau, W. S. (1981) Latitudinal variation in the life histories of insects occupying disturbed habitats: a case study. In *Insect Life History Patterns. Habitat and Geographic Variation* (eds R. F. Denno and H. Dingle), Springer-Verlag, New York, 75–95.

Brower, L. P. (1958) Bird predation and foodplant specificity in closely related procryptic insects. *Amer. Nat.*, **92**, 183–187.

Brown, K. S. Jr (1985) Chemical ecology of dehydropyrrolizidine alkaloids in adult Ithomiinae (Lepidoptera: Nymphalidae). *Rev. Bras. Biol.*, **44**, 435–460.

Carter, D. J. and Hargreaves, B. (1986) *A Field Guide to Caterpillars of Butterflies and Moths in Britain and Europe*, Collins, London.

Carter, D. J., Kitching, I. J. and Scoble, M. J. (1988) The adaptable caterpillar. *The Entomol.*, **107**, 68–78.

Clancy, K. M. and Price, P. W. (1987) Rapid herbivore growth enhances enemy attack: sublethal plant defenses remain a paradox. *Ecology*, **68**, 733–737.

Cohen, R. W., Waldbauer, G. P., Friedman, S. and Schiff, N. M. (1987) Nutrient self-selection by *Heliothis zea* larvae: a time-lapse film study. *Entomol. Exp. Appl.*, **44**, 65–73.

Cohen, R. W., Waldbauer, G. P. and Friedman, S. (1988) Natural diets and self-selection: *Heliothis zea* larvae and maize. *Entomol. Exp. Appl.*, **46**, 161–171.

Coleman, J. S. and Jones, C. G. (1988) Plant stress and insect performance: cottonwood, ozone and a leaf beetle. *Oecologia*, **76**, 57–61.

Courtney, S. P. (1981) Coevolution of pierid butterflies and their cruciferous foodplants. III. *Anthocharis cardamines* (L.) survival, development and oviposition on different host plants. *Oecologia*, **51**, 91–96.

Courtney, S. P. (1982) Coevolution of pierid butterflies and their cruciferous foodplants. IV. Crucifer apparency and *Anthocharis cardamines* (L.) oviposition. *Oecologia*, **52**, 258–265.

Damman, H. (1987) Leaf quality and enemy avoidance by the larvae of a pyralid moth. *Ecology*, **68**, 88–97.

Danks, H. V. (1987) *Insect Dormancy: An Ecological Perspective*, Biological Survey of Canada (Terrestrial Arthropods), Ottawa.

Dement, W. A. and Mooney, H. A. (1974) Seasonal variation in the production of tannins and cyanogenic glucosides in the chaparral shrub, *Heteromeles arbutifolia*. *Oecologia*, **15**, 65–76.

Dempster, J. P. (1983) The natural control of populations of butterflies and moths. *Biol. Rev.*, **58**, 461–481.

Dennis, R. L. H. (1983) The 'edge effect' in butterfly oviposition: a simple calculus for an insect in a hurry. *Entomol. Gaz.*, **34**, 5–8.

Denno, R. F. and McClure, M. S. (eds) (1983) *Variable Plants and Herbivores in Natural and Managed Systems*, Academic Press, New York.

Dixon, A. F. G. and McKay, S. (1970) Aggregation in the sycamore aphid, *Drepanosiphum platanoides* (Schr.) (Hemiptera: Aphididae) and its relevance to the regulation of population growth. *J. Anim. Ecol.*, **39**, 439–454.

Du Merle, P. (1988) Phenological resistance of oaks to the green oak leafroller, *Tortrix viridana* (Lepidoptera: Tortricidae). In *Mechanisms of Woody Plant Defenses against Insects. Search for Pattern* (eds W. J. Mattson, J. Levieux and C. Bernard-Dagan), Springer-Verlag, New York. pp. 215–26.

Edley, M. T. and Law, R. (1988) Evolution of life histories and yields in experimental populations of *Daphnia magna*. *Biol. J. Linn. Soc.*, **34**, 309–327.

Edwards, P. J., Wratten, S. D. and Greenwood, S. (1986) Palatability of British trees to insects: constitutive and induced defences. *Oecologia*, **69**, 316–319.

Emmet, A. M. (1988) *A Field Guide to the Smaller British Lepidoptera*, British Entomological and Natural History Society, London.

Faeth, S. H. (1986) Indirect interactions between temporally separated herbivores, mediated by the host plant. *Ecology*, **67**, 474–494.

Fatzinger, C. W. and Merkel, E. P. (1985) Oviposition and feeding preferences of the southern pine coneworm (Lepidoptera: Pyralidae) for different host-plant materials and observations on monoterpenes as an oviposition stimulant. *J. Chem. Ecol.*, **11**, 689–699.

Feeny, P. (1970) Seasonal changes in oak leaf tannins and nutrients as a cause of spring feeding by winter moth caterpillars. *Ecology*, **51**, 565–581.

Feichtinger, V. E. and Reavey, D. (1989) Changes in movement, tying and feeding patterns as caterpillars grow: the case of the yellow horned moth. *Ecol. Entomol.*, **14**, 471–474.

Fields, P. G. and McNeill, J. N. (1988) Characteristics of the larval diapause in *Ctenucha virginica* (Lepidoptera: Arctiidae) *J. Insect. Physiol.*, **34**, 111–115.

Fitzgerald, T. D., Casey, T. and Joos, B. (1988) Daily foraging schedule of field colonies of the eastern tent caterpillar *Malacosoma americanum*. *Oecologia*, **76**, 574–578.

Forsberg, J. (1987) Size discrimination among conspecific hostplants in two pierid butterflies; *Pieris napi* L. and *Pontia daplidice* L. *Oecologia*, **72**, 52–57.

Fowler, S. V. and Lawton, J. H. (1984) Foliage preferences of birch herbivores: a field manipulation experiment. *Oikos*, **42**, 239–248.

Fowler, S. V. and Lawton, J. H. (1985) Rapidly induced defenses and talking trees: the devil's advocate position. *Amer. Nat.*, **126**, 181–195.

Fowler, S. V. and MacGarvin, M. (1986) The effects of leaf damage on the performance of insect herbivores on birch, *Betula pubescens*. *J. Anim. Ecol.*, **55**, 565–574.

García-Barros, E. (1988) Delayed ovarian maturation in the butterfly *Hipparchia semele* as a possible response to summer drought. *Ecol. Entomol.*, **13**, 391–398.

Gaston, K. J. and Reavey, D. (1989) Patterns in the life histories and feeding strategies of British macrolepidoptera. *Biol. J. Linn. Soc.*, **37**, 367–381.

Geyspitz, K. F. and Zarankina, A. I. (1963) Some features of the photoperiodic reaction of *Dasychira pudibunda* L. (Lepidoptera, Orgyidae). *Entomol. Rev.*, **42**, 14–19.

Gibberd, R., Edwards, P. J. and Wratten, S. D. (1988) Wound-induced changes in the acceptability of tree-foliage to Lepidoptera: within-leaf effects. *Oikos*, **51**, 43–47.

Goettel, M. S. and Philogène, B. J. R. (1978) Effects of photoperiod and temperature on the development of a univoltine population of the banded woolly-bear, *Pyrrharctia (Isia) isabella*. *J. Insect Physiol.*, **24**, 523–527.

Goldschmidt, R. (1940) *The Material Basis of Evolution*, Yale University Press, New Haven, CT.

Gould, F. (1984) Mixed function oxidases and herbivore polyphagy: the devil's advocate position. *Ecol. Entomol.*, **9**, 29–34.

Grabstein, E. M. and Scriber, J. M. (1982) Host-plant utilization by *Hyalophora cecropia* as affected by prior feeding experience. *Entomol. Exp. Appl.*, **32**, 262–268.

Greene, E. (1989) A diet-induced developmental polymorphism in a caterpillar. *Science*, **243**, 643–646.

Grison, P. A. (1952) Relation entre l'état physiologique de la planthôte, *Solanum tuberosum*, et la fécondité du doryphore, *Leptinotarsa decemlineata* Say. *Trans. 'Int. Congr. Entomol, 9th, 1951*, **1**, 331–337.

Griswold, M. J. and Trumble, J. T. (1985) Consumption and utilization of celery, *Apium graveolens*, by the beet armyworm *Spodoptera exigua*. *Entomol. Exp. Appl.*, **38**, 73–79.

Grossmueller, D. W. and Lederhouse, R. C. (1987) The role of nectar source distribution in habitat use and oviposition by the tiger swallowtail butterfly. *J. Lepid. Soc.*, **41**, 159–165.

Hagen, R. H. and Chabot, J. F. (1986) Leaf anatomy of maples (*Acer*) and host use by Lepidoptera larvae. *Oikos*, **47**, 335–345.

Hansen, R. M. and Ueckert, D. N. (1970) Dietary similarity of some primary consumers. *Ecology*, **51**, 640–648.

Hanson, F. E. (1983) The behavioral and neurophysiological basis of food-plant selection by lepidopterous larvae. In *Herbivorous Insects: Host-seeking Behavior and Mechanisms* (ed. S. Ahmad), Academic Press, New York, 3–23.

Hartley, S. E. (1987) Rapidly induced chemical changes in birth foliage: their biochemical nature and impact on insect herbivores. D.Phil. Thesis, University of York.

Hartley, S. E. (1988) The inhibition of phenolic biosynthesis in damaged and undamaged birch foliage and its effect on insect herbivores. *Oecologia*, **76**, 65–70.

Hartley, S. E. and Firn, R. D. (1989) Phenolic biosynthesis, leaf damage and insect herbivory in birch (*Betula pendula*). *J. Chem. Ecol.*, **15**, 275–283.

Hartley, S. E. and Lawton, J. H. (1987) Effects of different types of damage on the chemistry of birch foliage, and the responses of birch feeding insects. *Oecologia*, **74**, 432–437.

Haukioja, E. (1979) Antiherbivore strategies in mountain birch at the tree line. *Holarct. Ecol.*, **2**, 272–274.

Haukioja, E. and Hanhimäki, S. (1985) Rapid wound-induced resistance in white birch (*Betula pubescens*) foliage to the geometrid *Epirrita autumnata*: a comparison of trees and moths within and outside the outbreak range of the moth. *Oecologia*, **65**, 223–228.

Haukioja, E. and Neuvonen, S. (1985) The relationship between size and reproductive potential in male and female *Epirrita autumnata* (Lep., Geometridae). *Ecol. Entomol.*, **10**, 267–270.

Haukioja, E. and Niemelä, P. (1977) Retarded growth of a geometrid larva after mechanical damage to leaves of its host tree. *Ann. Zool. Fenn.*, **14**, 48–52.

Haukioja, E. and Niemelä, P. (1979) Birch leaves as a resource for herbivores: seasonal occurrence of increased resistance in the foliage after mechanical damage of adjacent leaves. *Oecologia*, **39**, 151–159.

Haukioja, E., Niemelä, P. and Iso-Iivari, L. (1978) Birch leaves as a resource for herbivores. II. Diurnal variation in the usability of leaves for *Oporinia autumnata* and *Dineura virididorsata*. *Rep. Kevo Subarct. Res. Stn.*, **14**, 21–24.

Haukioja, E., Suomela, J. and Neuvonen, S. (1985) Long-term inducible resistance in birth foliage: triggering cues and efficacy on a defoliator. *Oecologia*, **65**, 363–369.

Hawkins, B. A. and Lawton, J. H. (1987) Species richness for parasitoids of British phytophagous insects. *Nature*, **326**, 788–790.

Hawkins, B. A. (1988) Do galls protect endophytic herbivores from parasitoids? A comparison of galling and non-galling Diptera. *Ecol. Entomol.*, **13**, 473–477.

Hayes, J. L. (1982) A study of the relationships of diapause phenomena and other life history characteristics in temperate butterflies. *Amer. Nat.*, **120**, 160–170.

Heinrich, B. (1979) Foraging strategies of caterpillars. Leaf damage and possible predator avoidance strategies. *Oecologia*, **42**, 325–337.

Heinrich, B. and Collins, S. L. (1983) Caterpillar leaf damage and the game of hide and seek with birds. *Ecology*, **64**, 592–602.

Herrebout, W. M., Kuyten, P. J. and DeRuiter, L. (1963) Observations on colour patterns and behaviour of caterpillars feeding on Scots pine. *Arch. Neerl. Zool.*, **15**, 315–357.

Hinton, H. E. (1981) *Biology of Insect Eggs*, Pergamon Press, Oxford.

Holdren, C. E. and Ehrlich, P. R. (1982) Ecological determinants of food plant choice in the checkerspot butterfly *Euphydryas editha* in Colorado. *Oecologia*, **52**, 417–423.

Jaenike, J. (1978) On optimal oviposition behavior in phytophagous insects. *Theor. Popul. Biol.*, **14**, 350–356.

Janzen, D. H. (1984) Two ways to be a tropical big moth: Santa Rosa saturniids and sphingids. *Oxf. Stud. in Evol. Biol.*, **1**, 85–140.

Jones, C. G. and Coleman, J. S. (1988) Plant stress and insect behaviour: cotton-wood, ozone and the feeding and oviposition preference of a beetle. *Oecologia*, **76**, 51–56.

Jones, R. E. (1977) Search behaviour: a study of three caterpillar species. *Behaviour*, **60**, 237–259.

Jones, R. E. and Rienks, J. (1987) Reproductive seasonality in the tropical genus *Eurema* (Lepidoptera: Pieridae). *Biotropica*, **19**, 7–16.

Karowe, D. N. (1989) Facultative monophagy as a consequence of prior feeding experience: behavioral and physiological specialization in *Colias philodice* larvae. *Oecologia*, **78**, 106–111.

Kennedy, G. G., Farrar, R. R. and Riskallah, M. R. (1987) Induced tolerance of neonate *Heliothis zea* to host plant allelochemicals and carbaryl following incubation of eggs on foliage of *Lycopersicon hirsutum* f. *glabratum*. *Oecologia*, **73**, 615–620.

Kimmerer, T. W. and Potter, D. A. (1987) Nutritional quality of specific leaf tissues and selective feeding by a specialist leafminer. *Oecologia*, **71**, 548–551.

Kooi, R. E. (1988) Keeping the web tidy: hygienic behaviour of a small ermine moth (Lepidoptera: Yponomeutidae). *Entomol. Ber. Amst.*, **48**, 145–146.

Kukal, O. and Kevan, P. G. (1987) The influence of parasitism on the life history of a high arctic insect, *Gynaephora groenlandica* (Wöcke) (Lepidoptera: Lymantriidae). *Can. J. Zool.*, **65**, 156–163.

Lance, D. R. (1983) Host-seeking behaviour of the gypsy moth: the influence of polyphagy and highly apparent host plants. In *Herbivorous Insects: Host-seeking Behaviour and Mechanisms* (ed. S. Ahmad), Academic Press, New York, 201–224.

Lance, D. R., Elkinton, J. S. and Schwalbe, C. P. (1987) Behaviour of late-instar gypsy moth larvae in high and low density populations. *Ecol. Entomol.*, **12**, 267–273.

Lawton, J. H. (1986a) The effects of parasitoids on phytophagous insect communities. In *Insect Parasitoids* (eds J. Waage and D. Greathead), Symposia. of the Royal Entomological Society, London, Vol. 13, Academic Press, London, 265–287.

Lawton, J. H. (1986b) Food-shortage in the midst of apparent plenty?: the case for birch-feeding insects. *Proc. 3rd Eur. Congr. Entomol.*, 219–228.

Lawton, J. H. (1986c) Surface availability and insect community structure: the effects of architecture and fractal dimension of plants. In *Insects and the Plant Surface* (eds B. Juniper and T. R. E. Southwood), Edward Arnold, London, 317–331.

Lawton, J. H., Rashbrook, V. K. and Compton, S. G. (1988) Biocontrol of British bracken: the potential of two moths from Southern Africa. *Ann. Appl. Biol.*, **112**, 479–490.

Leather, S. R. (1985) Oviposition preferences in relation to larval growth rates and survival in the pine beauty moth, *Panolis flammea*. *Ecol. Entomol.*, **10**, 213–217.

Leather, S. R. (1988) Size, reproductive potential and fecundity in insects: things aren't as simple as they seem. *Oikos*, **51**, 386–389.

Lechowicz, M. J. and Mauffette, Y. (1986) Host preferences of the gypsy moth in eastern North American versus European forests. *Rev. Entomol. Qué.*, **31**, 43–51.

Lederhouse, R. C., Finke, M. D. and Scriber, J. M. (1982) The contributions of larval growth and pupal duration to protandry in the black swallowtail butterfly, *Papilio polyxenes. Oecologia*, **53**, 296–301.

Lewis, A. C. and Bernays, E. A. (1985) Feeding behaviour: selection of both wet and dry food for increased growth in *Schistocerca gregaria* nymphs. *Entomol. Exp. Appl.*, **37**, 105–112.

Lindroth, R. L. and Peterson, S. S. (1988) Effects of plant phenols on performance of southern armyworm larvae. *Oecologia*, **75**, 185–189.

Lloyd, M. (1984) Periodical cicadas. *Antenna*, **8**, 79–91.

Marquis, R. J. and Braker, H. E. (1987) Influence of method of presentation on results of plant-host preference tests with two species of grasshopper. *Entomol. Exp. Appl.*, **44**, 59–63.

Mattson, W. J. and Scriber, J. M. (1987) Nutritional ecology of insect folivores of woody plants: nitrogen, water, fiber, and mineral considerations. In *Nutritional Ecology of Insects, Mites, Spiders, and Related Invertebrates* (eds F. Slansky, Jr and J. G. Rodriguez), Wiley, New York, 105–146.

Mikkola, K. (1987) Protogyny during an outbreak in *Diaphora mendica* (Lepidoptera, Arctiidae). *Ann. Entomol. Fenn.*, **53**, 30–32.

Miller, F. D. Jr and Hart, E. R. (1987) Overwintering survivorship of pupae of the mimosa webworm, *Homadaula anisocentra* (Lepidoptera: Plutellidae), in an urban landscape. *Ecol. Entomol.*, **12**, 41–50.

Mitter, C., Futuyma, D. J., Schneider, J. C. and Hare, J. D. (1979) Genetic variation and host plant relations in a parthenogenetic moth. *Evolution*, **33**, 777–790.

Murphy, D. D., Menninger, M. S. and Ehrlich, P. R. (1984) Nectar source distribution as a determinant of oviposition host species in *Euphydryas chalcedona. Oecologia*, **62**, 269–271.

Myers, J. (1985) Effect of physiological condition of the host plant on the ovipositional choice of the cabbage white butterfly, *Pieris rapae. J. Anim. Ecol.*, **54**, 193–204.

Niemelä, P., Aro, E. M. and Haukioja, E. (1979) Birch leaves as a resource for herbivores. Damage induced increase in leaf phenols with trypsin inhibiting effects. *Rep. Kevo Subarct. Res. Stn*, **15**, 37–40.

Owen, D. (1980) *Camouflage and Mimicry*, Oxford University Press, Oxford.

Peterson, S. C. (1987) Communication of leaf suitability by gregarious eastern tent caterpillars (*Malacosoma americanum*). *Ecol. Entomol.*, **12**, 283–289.

Porter, K. (1982) Basking behaviour in larvae of the butterfly *Euphydryas aurinia*. *Oikos*, **38**, 308–312.

Porter, K. (1984) Sunshine, sex-ratio and behaviour of *Euphydryas aurinia* larvae. In *The Biology of Butterflies* (eds R. I. Vane-Wright and P. R. Ackery), Symposia of the Royal Entomological Society, London, Vol. 11, Academic Press, Orlando, FL, 309–311.

Potter, D. A. and Kimmerer, T. W. (1986) Seasonal allocation of defense investment in *Ilex opaca* Aiton and constraints on a specialist leafminer. *Oecologia*, **69**, 217–224.

Price, P. W., Bouton, C. E., Gross, P. *et al.* (1980) Interactions among three trophic levels: influence of plants on interactions between insect herbivores and natural enemies. *Ann. Rev. Ecol. Syst.*, **11**, 41–65.

Price, P. W., Waring, G. L. and Fernandes, G. W. (1986) Hypotheses on the adaptive nature of galls. *Proc. Entomol. Soc. Wash.*, **88**, 361–363.

Ramachandran, R. (1987) Influence of host-plants on the wind dispersal and the survival of an Australian geometrid caterpillar. *Entomol. Exp. Appl.*, **44**, 289–294.

Raupp, M. J. and Denno, R. F. (1983) Leaf age as a predictor of herbivore distribution and abundance. In *Variable Plants and Herbivores in Natural and Managed Systems* (eds R. F. Denno and M. S. McClure), Academic Press, New York, 91–124.

Rausher, M. D. (1979) Larval habitat suitability and oviposition preference in three related butterflies. *Ecology*, **60**, 503–511.

Rausher, M. D. and Papaj, D. R. (1983) Demographic consequences of discrimination among conspecific host plants by *Battus philenor* butterflies. *Ecology*, **64**, 1402–1410.

Risch, S. J. (1985) Effects of induced chemical changes on interpretation of feeding preference tests. *Entomol. Exp. Appl.*, **39**, 81–84.

Robertson, H. G. (1987) Oviposition site selection in *Cactoblastis cactorum* (Lepidoptera): constraints and compromises. *Oecologia*, **73**, 601–608.

Roff, D. A. (1981) On being the right size. *Amer. Nat.*, **118**, 405–422.

Roff, D. A. (1986) Predicting body size with life history models. *BioScience*, **36**, 316–323.

Root, R. B. (1973) Organization of a plant-arthropod association in simple and diverse habitats: the fauna of collards (*Brassica oleracea*). *Ecol. Monogr.*, **43**, 95–124.

Rossiter, M. C., Schultz, J. C. and Baldwin, I. T. (1988) Relationships among defoliation, red oak phenolics, and gypsy moth growth and reproduction. *Ecology*, **69**, 267–277.

Rothschild, M. (1963) Is the buff ermine (*Spilosoma lutea* (Hu.)) a mimic of the white ermine (*S. lubricipeda* (L.))? *Proc. R. Entomol. Soc. Lond. (A)*, **38**, 159–164.

Sargent, T. D. (1978) On the maintenance of stability in hindwing diversity among moths of the genus *Catocala* (Lepidoptera: Noctuidae). *Evolution*, **32**, 424–434.

Schneider, J. C. (1980) The role of parthenogenesis and female aptery in microgeographic, ecological adaptation in the fall cankerworm, *Alsophila pometaria* Harris (Lepidoptera: Geometridae). *Ecology*, **61**, 1082–1090.

Schoonhoven, L. M. (1987) What makes a caterpillar eat? The sensory code underlying feeding behavior. In *Perspectives of Chemoreception and Behavior* (eds R. F. Chapman, E. A. Bernays and J. G. Stoffolano, Jr), 69–97.

Schoonhoven, L. M. and Meerman, J. (1978) Metabolic cost of changes in diet and neutralization of allelochemics. *Entomol. Exp. Appl.*, **24**, 489–493.

Schultz, J. C. (1983a) Habitat selection and foraging tactics of caterpillars in heterogeneous trees. In *Variable Plants and Herbivores in Natural and Managed Systems* (eds R. F. Denno and M. S. McClure), Academic Press, New York, 61–90.

Schultz, J. C. (1983b) Impact of variable plant defensive chemistry on susceptibility of insects to natural enemies. In *Plant Resistance to Insects* (ed. P. A. Hedin), American Chemical Society, Washington, DC, 37–54.

Schultz, J. C. (1988) Plant responses induced by herbivores. *Trends Ecol. Evol.*, **3**, 45–49.

Schultz, J. C., Nothnagle, P. J. and Baldwin, I. T. (1982) Seasonal and individual variation in leaf quality of two northern hardwood tree species. *Amer. J. Bot.*, **69**, 753–759.

Schweitzer, D. F. (1979) Effects of foliage age on body weight and survival of larvae of the tribe Lithophanini (Lepidoptera: Noctuidae). *Oikos*, **32**, 403–408.

Scriber, J. M. and Hainze, J. H. (1987) Geographic invasion and abundance as facilitated by differential host plant utilization abilities. In *Insect Outbreaks*. (eds P. Barbosa and J. C. Schultz), Academic Press, San Diego, 433–468.

Scriber, J. M. and Slansky, F., Jr (1981) The nutritional ecology of immature insects. *Ann. Rev. Entomol.*, **26**, 183–211.

Sherman, P. W. and Watt, W. B. (1973) The thermal ecology of some *Colias* butterfly larvae. *J. Comp. Physiol.*, **83**, 25–40.

Shreeve, T. G. (1986) The effect of weather on the life cycle of the speckled wood butterfly. *Ecol. Entomol.*, **11**, 325–332.

Silkstone, B. E. (1987) The consequences of leaf damage for subsequent insect grazing on birch (*Betula* spp.): a field experiment. *Oecologia*, **74**, 149–152.

Simberloff, D. and Stiling, P. (1987) Larval dispersion and survivorship in a leaf-mining moth. *Ecology*, **68**, 1647–1657.

Sims, S. R. (1980) Diapause dynamics and host plant suitability of *Papilio zelicaon* (Lepidoptera: Papilionidae). *Amer. Midl. Nat.*, **103**, 375–384.

Sims, S. R. and Shapiro, A. M. (1983) Pupal color dimorphism in California *Battus philenor* (L.) (Papilionidae): mortality factors and selective advantage. *J. Lepid. Soc.*, **37**, 236–243.

Singer, M. C. (1982) Sexual selection for small size in male butterflies. *Amer. Nat.*, **119**, 440–443.

Singer, M. C. (1984) Butterfly-hostplant relationships: host quality, adult choice and larval success. In *The Biology of Butterflies* (eds R. I. Vane-Wright and P. R. Ackery), Symposia of the Royal Entomological Society, London, Vol. 11, Academic Press, Orlando, FL, 81–88.

Skinner, B. (1984) *Colour Identification Guide to Moths of the British Isles*, Penguin Books, Harmondsworth, Middlesex.

Slansky, F. Jr (1974) Relationship of larval food-plants and voltinism patterns in temperate butterflies. *Psyche*, **81**, 243–253.

Smiley, J. T. (1985) Are chemical barriers necessary for butterfly–host plant coevolution? *Oecologia*, **65**, 580–583.

Smith, D. A. S. (1984) Mate selection in butterflies: competition, coyness, choice and chauvinism. In *The Biology of Butterflies* (eds R. I. Vane-Wright and P. R. Ackery), Symposia of the Royal Entomological Society, London, Vol. 11, Academic Press, Orlando, FL, 225–244.

Soberon, J. M., Cordero, C. M., Benrey, B. B. *et al.* (1988) Patterns of oviposition by *Sandia xami* (Lepidoptera, Lycaenidae) in relation to food plant apparency. *Ecol. Entomol.*, **13**, 71–79.

South, R. (1961) *The Moths of the British Isles*, Frederick Warne, London.

Southwood, T. R. E. (1986) Plant surfaces and insects—an overview. In *Insects and the Plant Surface* (eds B. Juniper and T. R. E. Southwood), Edward Arnold, London, 1–22.

Stamp, N. E. and Bowers, M. D. (1988) Direct and indirect effects of predatory wasps (*Polistes* sp: Vespidae) on gregarious caterpillars (*Hemileuca lucina*: Saturniidae). *Oecologia*, **75**, 619–624.

Stiling, P. (1988) Density-dependent processes and key factors in insect populations. *J. Anim. Ecol.*, **57**, 581–593.

Strong, D. R., Lawton, J. H. and Southwood, T. R. E. (1984) *Insects on Plants. Community Patterns and Mechanisms*, Blackwell Scientific Publications, Oxford.

Tabashnik, B. E. and Slansky, F. Jr (1987) Nutritional ecology of forb foliage-chewing insects. In *Nutritional Ecology of Insects, Mites, Spiders, and Related Invertebrates* (eds F. Slansky, Jr and J. G. Rodriguez), Wiley, New York, 71–103.

Tauber, M. J., Tauber, C. A. and Masaki, S. (1986) *Seasonal Adaptations of Insects*, Oxford University Press, New York.

Taylor, F. (1980) Timing in the life histories of insects. *Theor. Popul. Biol.*, **18**, 112–124.

Thompson, J. N. (1988) Evolutionary ecology of the relationship between oviposition preference and performance of offspring in phytophagous insects. *Entomol. Exp. Appl.*, **47**, 3–14.

Tostowaryk, W. (1972) The effect of prey defense on the functional response of *Podisus modestus* (Hemiptera: Pentatomidae) to the densities of the sawflies *Neodiprion swainei* and *N. pratti banksianae* (Hymenoptera: Neodiprionidae). *Can. Entomol.*, **104**, 61–69.

Varley, G. C., Gradwell, G. R. and Hassell, M. P. (1973) *Insect Population Ecology. An Analytical Approach*, Blackwell Scientific Publications, Oxford.

Via, S. (1986) Genetic covariance between oviposition preference and larval performance in an insect herbivore. *Evolution*, **40**, 778–785.

Visscher, S. N. (1987) Plant growth hormones: their physiological effects on a rangeland grasshopper (*Aulocara elliotti*). In *Insects–Plants* (eds V. Labeyrie, G. Fabres and D. Lachaise), Proceedings of the 6th International Symposium on Insect–Plant Relations, Dr W. Junk, Dordrecht, 37–41.

Waldbauer, G. P. and Sheldon, J. K. (1971) Phenological relationships of some aculeate Hymenoptera, their dipteran mimics, and insectivorous birds. *Evolution*, **25**, 371–382.

Waldbauer, G. P., Cohen, R. W. and Friedman, S. (1984). Self-selection of an optimal nutrient mix from defined diets by larvae of the corn earworm, *Heliothis zea* (Boddie). *Physiol. Zool.*, **57**, 590–597.

Warrington, S. (1985) Consumption rates and utilization efficiencies of four species of polyphagous Lepidoptera feeding on sycamore leaves. *Oecologia*, **67**, 460–463.

Weiss, S. B. and Murphy, D. D. (1988) Fractal geometry and caterpillar dispersal: or how many inches can an inchworm inch? *Funct. Ecol.*, **2**, 116–118.

Weiss, S. B., White, R. R., Murphy, D. D. and Ehrlich, P. R. (1987) Growth and dispersal of larvae of the checkerspot butterfly *Euphydryas editha*. *Oikos*, **50**, 161–166.

West, C. (1985) Factors underlying the late seasonal appearance of the lepidopterous leaf-mining guild on oak. *Ecol. Entomol.*, **10**, 111–120.

White, R. R. (1986) Pupal mortality in the bay checkerspot butterfly (Lepidoptera: Nymphalidae). *J. Res. Lepid.*, **25**, 52–62.

Wiklund, C. (1975a) The evolutionary relationship between adult oviposition preferences and larval host plant range in *Papilio machaon* L. *Oecologia*, **18**, 185–197.

Wiklund, C. (1975b) Pupal colour polymorphism in *Papilio machaon* and the survival in the field of cryptic versus non-cryptic pupae. *Trans. R. Entomol. Soc. Lond.*, **127**, 73–84.

Wiklund, C. and Persson, A. (1983) Fecundity, and the relation between egg weight variation to offspring fitness in the speckled wood butterfly, *Parage aegeria*, or why don't butterfly females lay more eggs? *Oikos*, **40**, 53–63.

Williams, E. H. and Bowers, M. D. (1987) Factors affecting host-plant use by the montane butterfly *Euphydryas gillettii*. *Amer. Midl. Nat.*, **118**, 153–161.

Williams, G. C. (1966) *Adaptation and Natural Selection*, Princeton University Press, Princeton.

Williams, K. S. (1983) The coevolution of *Euphydryas chalcedona* butterflies and their larval host plants. III. Ovipositional behaviour and host plant quality. *Oecologia*, **56**, 336–340.

Winokur, L. (1988) Influence of a rearing protocol on the life cycle and survival in *Parage aegeria* (L.) (Lepidoptera: Satyridae). *Entomol. Gaz.*, **39**, 113–122.

Wint, W. (1983) The role of alternative host-plant species in the life of a polyphagous moth, *Operophtera brumata* (Lepidoptera: Geometridae). *J. Anim. Ecol.*, **52**, 439–450.

Wipking, W. and Neumann, D. (1986) Polymorphism in the larval hibernation strategy of the burnet moth, *Zygaena trifolii*. In *The Evolution of Insect Life Cycles* (eds F. Taylor and R. Karban), Springer-Verlag, New York, 125–134.

Wratten, S. D. Edwards, P. J. and Dunn, I. (1984) Wound-induced changes in the palatability of *Betula pubescens* and *Betula pendula*. *Oecologia*, **61**, 372–375.

Zucoloto, F. S. (1987) Feeding habits of *Ceratitis capitata* (Diptera: Tephritidae): can larvae recognise a nutritionally effective diet? *J. Insect. Physiol.*, **33**, 349–353.

Index